PEARSON EDEXCEL INTERNATIONAL A LEVEL

MECHANICS 2

Student Book

Series Editors: Joe Skrakowski and Harry Smith

Authors: Greg Attwood, Dave Berry, Ian Bettison, Alan Clegg, Gill Dyer, Jane Dyer, Keith Gallick, Susan Hooker, Michael Jennings, Mohammed Ladak, Jean Littlewood, Bronwen Moran, James Nicholson, Su Nicholson, Laurence Pateman, Keith Pledger, Joe Skrakowski, Harry Smith, Jack Williams

Published by Pearson Education Limited, 80 Strand, London, WC2R 0RL.

www.pearsonglobalschools.com

Copies of official specifications for all Pearson qualifications may be found on the website: https://qualifications.pearson.com

Text © Pearson Education Limited 2018
Edited by Lyn Imeson and Eric Pradel
Typeset by Tech-Set Ltd, Gateshead, UK
Original illustrations © Pearson Education Limited 2018
Illustrated by © Tech-Set Ltd, Gateshead, UK
Cover design by © Pearson Education Limited 2018

Cover images: *Front*: **Getty Images:** Werner Van Steen
Inside front cover: **Shutterstock.com:** Dmitry Lobanov

First published 2018

21 20
10 9 8 7 6 5 4 3

British Library Cataloguing in Publication Data
A catalogue record for this book is available from the British Library

ISBN 978 1 292244 76 1

Acknowledgments

(Key: b-bottom; c-centre; l-left; r-right; t-top)

Alamy: Alvey & Towers Picture Library 50, Cultura RM 85, Teo Moreno Moreno 108;
Shutterstock: joan_bautista 138, Fer Gregory 24, Mark Herreid 1

All other images © Pearson Education Limited 2018
All artwork © Pearson Education Limited 2018

Endorsement Statement
In order to ensure that this resource offers high-quality support for the associated Pearson qualification, it has been through a review process by the awarding body. This process confirms that this resource fully covers the teaching and learning content of the specification or part of a specification at which it is aimed. It also confirms that it demonstrates an appropriate balance between the development of subject skills, knowledge and understanding, in addition to preparation for assessment.

Endorsement does not cover any guidance on assessment activities or processes (e.g. practice questions or advice on how to answer assessment questions) included in the resource, nor does it prescribe any particular approach to the teaching or delivery of a related course.

While the publishers have made every attempt to ensure that advice on the qualification and its assessment is accurate, the official specification and associated assessment guidance materials are the only authoritative source of information and should always be referred to for definitive guidance.

Pearson examiners have not contributed to any sections in this resource relevant to examination papers for which they have responsibility.

Examiners will not use endorsed resources as a source of material for any assessment set by Pearson. Endorsement of a resource does not mean that the resource is required to achieve this Pearson qualification, nor does it mean that it is the only suitable material available to support the qualification, and any resource lists produced by the awarding body shall include this and other appropriate resources.

ABOUT THIS BOOK

The following three themes have been fully integrated throughout the Pearson Edexcel International Advanced Level in Mathematics series, so they can be applied alongside your learning.

1. Mathematical argument, language and proof

- Rigorous and consistent approach throughout
- Notation boxes explain key mathematical language and symbols

2. Mathematical problem-solving

- Hundreds of problem-solving questions, fully integrated into the main exercises
- Problem-solving boxes provide tips and strategies
- Challenge questions provide extra stretch

The Mathematical Problem-Solving Cycle

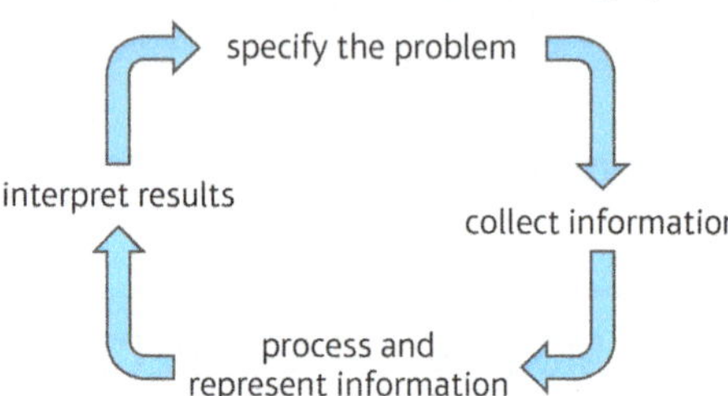

3. Transferable skills

- Transferable skills are embedded throughout this book, in the exercises and in some examples
- These skills are signposted to show students which skills they are using and developing

Finding your way around the book

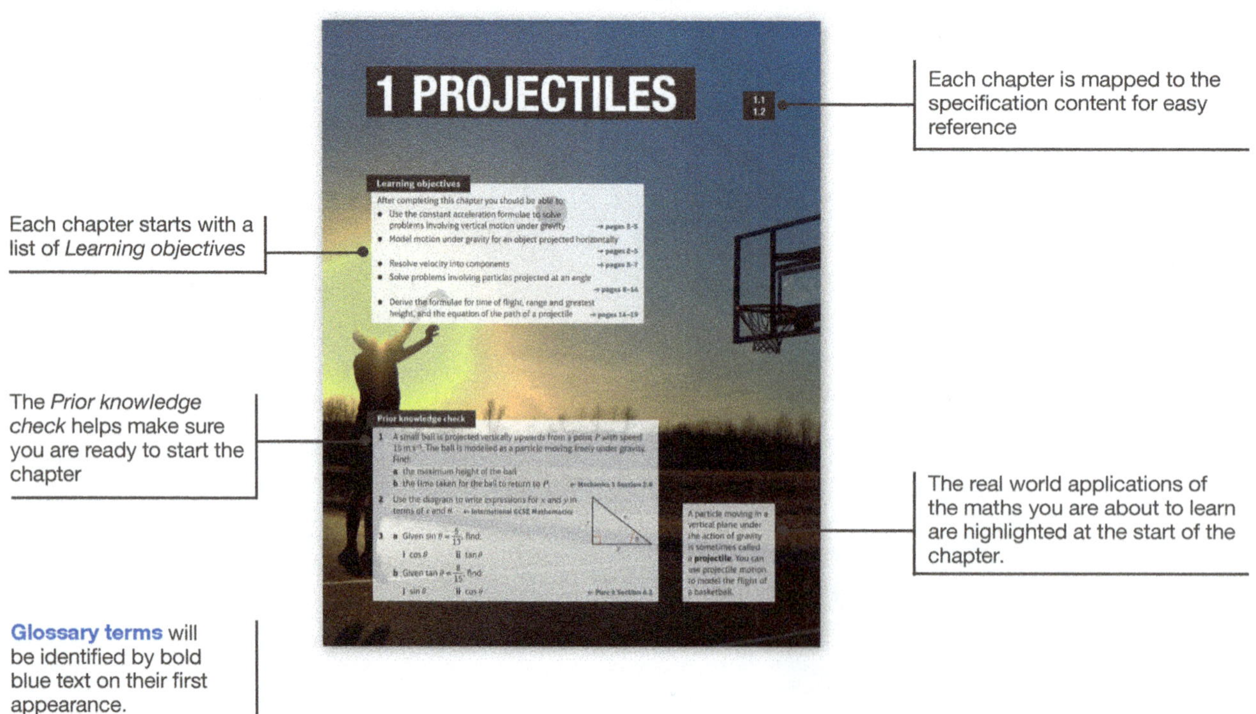

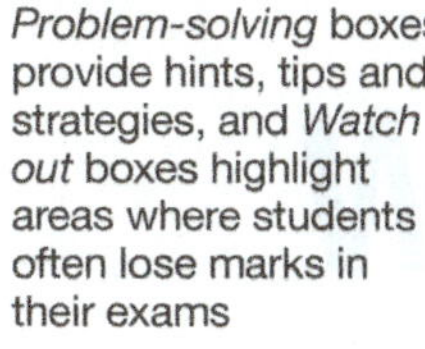

Problem-solving boxes provide hints, tips and strategies, and *Watch out* boxes highlight areas where students often lose marks in their exams

Step-by-step worked examples focus on the key types of questions you'll need to tackle

Exercise questions are carefully graded so they increase in difficulty and gradually bring you up to exam standard

Exercises are packed with exam-style questions to ensure you are ready for the exams

Each section begins with explanation and key learning points

Challenge boxes give you a chance to tackle some more difficult questions

Transferable skills are signposted where they naturally occur in the exercises and examples

Exam-style questions are flagged with Ⓔ

Problem-solving questions are flagged with Ⓟ

Each chapter ends with a *Chapter review* and a *Summary of key points*

After every few chapters, a *Review exercise* helps you consolidate your learning with lots of exam-style questions

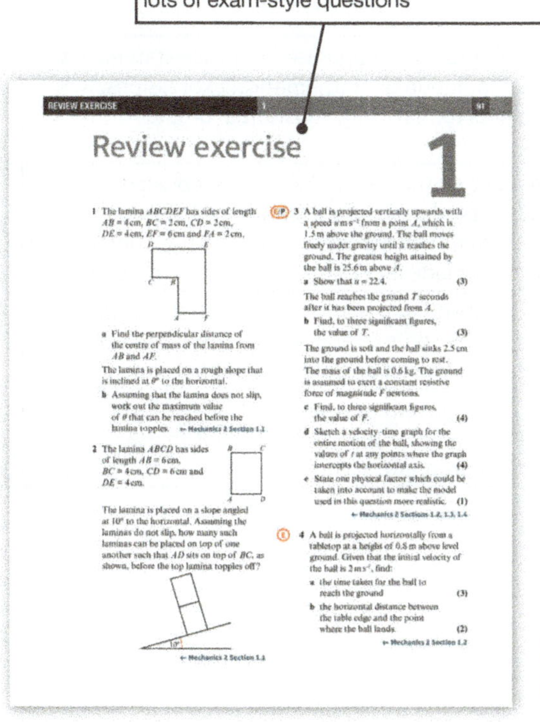

Review exercise 1

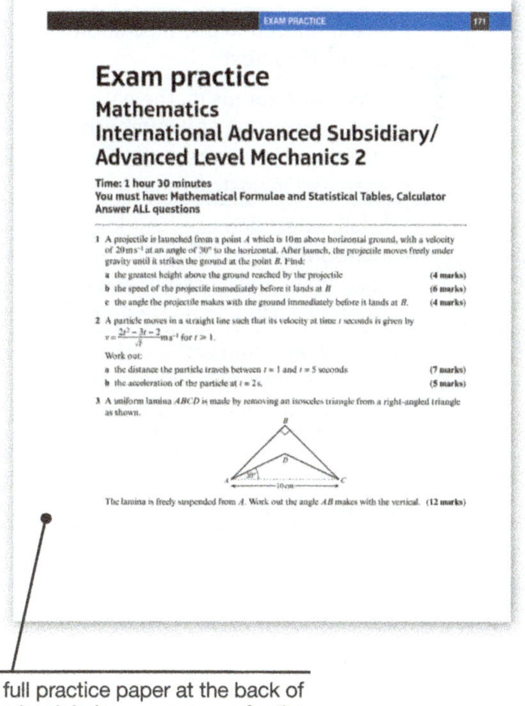

Exam practice

Mathematics

International Advanced Subsidiary/ Advanced Level Mechanics 2

Time: 1 hour 30 minutes

You must have: Mathematical Formulae and Statistical Tables, Calculator

Answer ALL questions

A full practice paper at the back of the book helps you prepare for the real thing

QUALIFICATION AND ASSESSMENT OVERVIEW

Qualification and content overview

Mechanics 2 (M2) is a **optional** unit in the following qualifications:

International Advanced Subsidiary in Further Mathematics

International Advanced Level in Mathematics

International Advanced Level in Further Mathematics

Assessment overview

The following table gives an overview of the assessment for this unit.

We recommend that you study this information closely to help ensure that you are fully prepared for this course and know exactly what to expect in the assessment.

Unit	Percentage	Mark	Time	Availability
M2: Mechanics 2 Paper code WME02/01	$33\frac{1}{3}$ % of IAS $16\frac{2}{3}$ % of IAL	75	1 hour 30 mins	January, June and October First assessment June 2019

IAS: International Advanced Subsidiary, IAL: International Advanced A Level.

Assessment objectives and weightings		Minimum weighting in IAS and IAL
AO1	Recall, select and use their knowledge of mathematical facts, concepts and techniques in a variety of contexts.	30%
AO2	Construct rigorous mathematical arguments and proofs through use of precise statements, logical deduction and inference and by the manipulation of mathematical expressions, including the construction of extended arguments for handling substantial problems presented in unstructured form.	30%
AO3	Recall, select and use their knowledge of standard mathematical models to represent situations in the real world; recognise and understand given representations involving standard models; present and interpret results from such models in terms of the original situation, including discussion of the assumptions made and refinement of such models.	10%
AO4	Comprehend translations of common realistic contexts into mathematics; use the results of calculations to make predictions, or comment on the context; and, where appropriate, read critically and comprehend longer mathematical arguments or examples of applications.	5%
AO5	Use contemporary calculator technology and other permitted resources (such as formulae booklets or statistical tables) accurately and efficiently; understand when not to use such technology, and its limitations. Give answers to appropriate accuracy.	5%

Relationship of assessment objectives to units

P2	Assessment objective				
	AO1	AO2	AO3	AO4	AO5
Marks out of 75	20–25	20–25	10–15	7–12	5–10
%	$26\frac{2}{3}$–$33\frac{1}{3}$	$26\frac{2}{3}$–$33\frac{1}{3}$	$13\frac{1}{3}$–20	$9\frac{1}{3}$–16	$6\frac{2}{3}$–$13\frac{1}{3}$

Calculators

Students may use a calculator in assessments for these qualifications. Centres are responsible for making sure that calculators used by their students meet the requirements given in the table below.

Students are expected to have available a calculator with at least the following keys: +, –, ×, ÷, π, x^2, $\sqrt{x}$, $\frac{1}{x}$, x^y, $\ln x$, e^x, $x!$, sine, cosine and tangent and their inverses in degrees and decimals of a degree, and in radians; memory.

Prohibitions

Calculators with any of the following facilities are prohibited in all examinations:

- databanks
- retrieval of text or formulae
- built-in symbolic algebra manipulations
- symbolic differentiation and/or integration
- language translators
- communication with other machines or the internet

Extra online content

Whenever you see an *Online* box, it means that there is extra online content available to support you.

SolutionBank

SolutionBank provides worked solutions for questions in the book. Download all the solutions as a PDF or quickly find the solution you need online.

Use of technology

Explore topics in more detail, visualise problems and consolidate your understanding. Use pre-made GeoGebra activities or Casio resources for a graphic calculator.

Online Find the point of intersection graphically using technology.

GeoGebra

GeoGebra-powered interactives

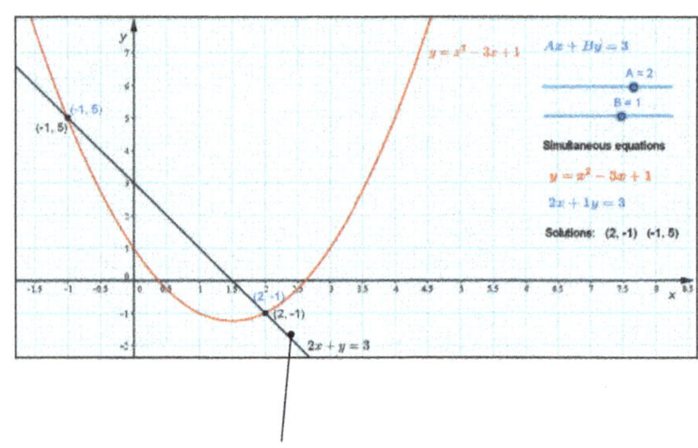

Interact with the mathematics you are learning using GeoGebra's easy-to-use tools

CASIO®

Graphic calculator interactives

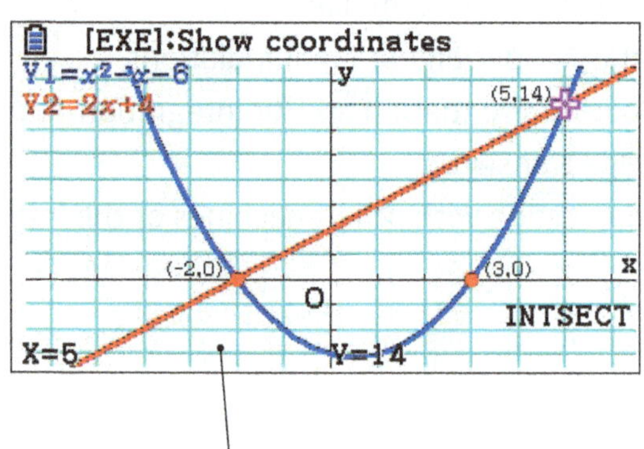

Explore the mathematics you are learning and gain confidence in using a graphic calculator

Calculator tutorials

Our helpful video tutorials will guide you through how to use your calculator in the exams. They cover both Casio's scientific and colour graphic calculators.

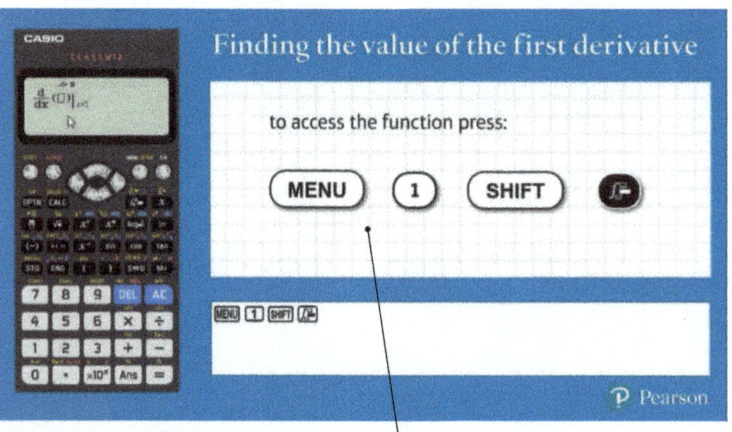

Step-by-step guide with audio instructions on exactly which buttons to press and what should appear on your calculator's screen

Online Work out each coefficient quickly using the ${}^nC_r$ and power functions on your calculator.

1 PROJECTILES

1.1
1.2

Learning objectives

After completing this chapter you should be able to:

- Use the constant acceleration formulae to solve problems involving vertical motion under gravity → pages 2–5
- Model motion under gravity for an object projected horizontally → pages 2–5
- Resolve velocity into components → pages 5–7
- Solve problems involving particles projected at an angle → pages 8–14
- Derive the formulae for time of flight, range and greatest height, and the equation of the path of a projectile → pages 14–19

Prior knowledge check

1 A small ball is projected vertically upwards from a point P with speed 15 m s^{-1}. The ball is modelled as a particle moving freely under gravity. Find:

a the maximum height of the ball

b the time taken for the ball to return to P. ← Mechanics 1 Section 2.6

2 Use the diagram to write expressions for x and y in terms of v and θ. ← International GCSE Mathematics

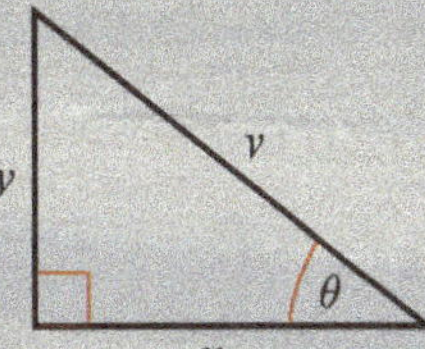

3 **a** Given $\sin\theta = \frac{5}{13}$, find:

i $\cos\theta$ **ii** $\tan\theta$

b Given $\tan\theta = \frac{8}{15}$, find:

i $\sin\theta$ **ii** $\cos\theta$ ← Pure 2 Section 6.2

A particle moving in a vertical plane under the action of gravity is sometimes called a **projectile**. You can use projectile motion to model the flight of a basketball.

1.1 Horizontal projection

You can **model** the **motion** of a **projectile** as a **particle** being acted on by a single **force**, **gravity**. In this model you ignore the effects of **air resistance** and any rotational movement (i.e. spinning) on the particle.

You can analyse the motion of a projectile by considering its **horizontal** motion and its **vertical** motion separately. Because gravity acts vertically downwards, there is **no force** acting on the particle in the horizontal direction.

- **The horizontal motion of a projectile is modelled as having constant velocity ($a = 0$). You can use the formula $s = vt$.**

The force due to gravity is modelled as being constant, so the vertical acceleration is constant.

- **The vertical motion of a projectile is modelled as having constant acceleration due to gravity ($a = g$).**

Use $g = 9.8\text{ m s}^{-2}$ unless the question specifies a different value.

Links You can use the **constant acceleration** formulae for the vertical motion of a projectile:

$v = u + at$

$s = \left(\frac{u+v}{2}\right)t$ $s = ut + \frac{1}{2}at^2$

$v^2 = u^2 + 2as$ $s = vt - \frac{1}{2}at^2$

← Mechanics 1 Section 2.5

Example 1 SKILLS PROBLEM-SOLVING

A particle is **projected** horizontally at 25 m s^{-1} from a point 78.4 metres above a horizontal surface. Find:

a the time taken by the particle to reach the surface

b the horizontal distance travelled in that time.

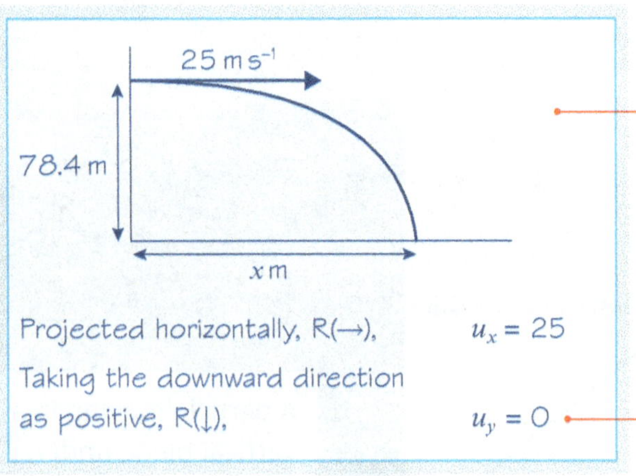

Projected horizontally, R(→), $u_x = 25$

Taking the downward direction as positive, R(↓), $u_y = 0$

First draw a diagram showing all the information given in the question.

Notation u_x is the initial horizontal velocity. u_y is the initial vertical velocity.

The particle is projected horizontally so $u_y = 0$.

a R(↓), $u = 0$, $s = 78.4$, $a = 9.8$, $t = ?$

$s = ut + \frac{1}{2}at^2$

$78.4 = 0 + \frac{1}{2} \times 9.8 \times t^2$

$78.4 = 4.9t^2$

$\frac{78.4}{4.9} = t^2$

$t^2 = 16$ so $t = 4\text{ s}$

b R(→), $u = 25$, $s = x$, $t = 4$

$s = vt$

$x = 25 \times 4$ so $x = 100\text{ m}$

Watch out The sign of g (**positive** or **negative**) depends on which direction is chosen as positive.
Positive direction downwards: $g = 9.8\text{ m s}^{-2}$
Positive direction upwards: $g = -9.8\text{ m s}^{-2}$

The time taken must be positive so choose the positive square root.

Your answer to part **a** tells you the time taken for the particle to hit the surface. The horizontal motion has constant velocity so you can use: distance = **speed** × time.

Example 2

A particle is projected horizontally with a velocity of 15 m s^{-1}. Find:

a the horizontal and vertical **components** of the **displacement** of the particle from the point of projection after 3 seconds

b the distance of the particle from the point of projection after 3 seconds.

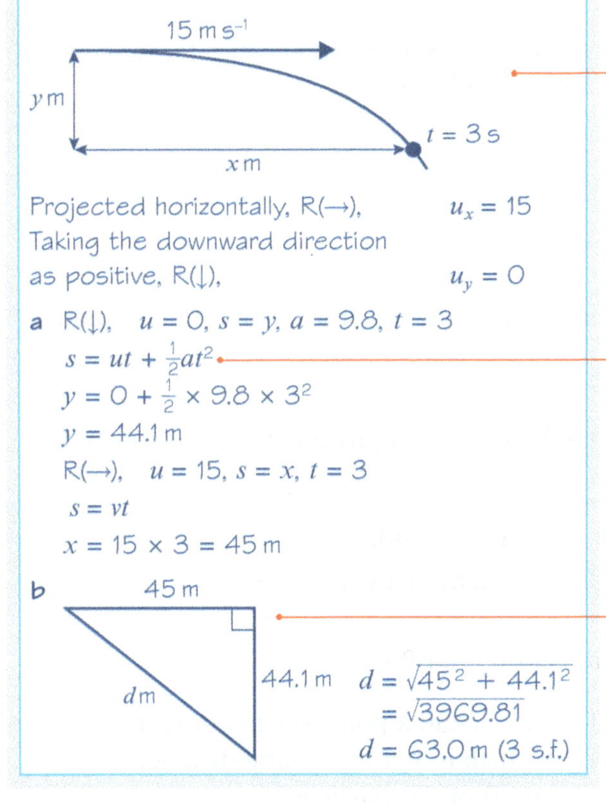

Projected horizontally, R(→), $u_x = 15$

Taking the downward direction as positive, R(↓), $u_y = 0$

a R(↓), $u = 0$, $s = y$, $a = 9.8$, $t = 3$

$s = ut + \frac{1}{2}at^2$

$y = 0 + \frac{1}{2} \times 9.8 \times 3^2$

$y = 44.1\text{ m}$

R(→), $u = 15$, $s = x$, $t = 3$

$s = vt$

$x = 15 \times 3 = 45\text{ m}$

b

$d = \sqrt{45^2 + 44.1^2}$

$= \sqrt{3969.81}$

$d = 63.0\text{ m}$ (3 s.f.)

Draw a diagram based on the information in the question.

Use $s = ut + \frac{1}{2}at^2$ to find the vertical distance. This is the same distance as the particle would travel in 3 seconds if it was dropped and fell under the action of gravity.

The distance travelled is the **magnitude** of the displacement **vector**. Sketch a right-angled triangle showing the components and use Pythagoras' Theorem.

Example 3

A particle is projected horizontally with a speed of $U\,\text{m s}^{-1}$ from a point 122.5 m above a horizontal **plane**. The particle hits the plane at a point which is at a horizontal distance of 90 m from the starting point. Find the initial speed of the particle.

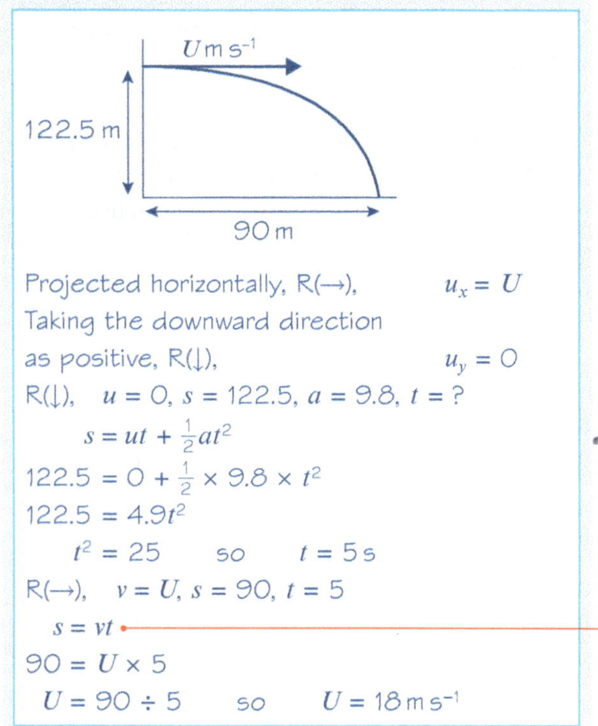

Projected horizontally, R(→), $u_x = U$

Taking the downward direction as positive, R(↓), $u_y = 0$

R(↓), $u = 0$, $s = 122.5$, $a = 9.8$, $t = ?$

$s = ut + \frac{1}{2}at^2$

$122.5 = 0 + \frac{1}{2} \times 9.8 \times t^2$

$122.5 = 4.9t^2$

$t^2 = 25$ so $t = 5\,\text{s}$

R(→), $v = U$, $s = 90$, $t = 5$

$s = vt$

$90 = U \times 5$

$U = 90 \div 5$ so $U = 18\,\text{m s}^{-1}$

Substitute $t = 5$ into the equation for horizontal motion to find U.

Problem-solving

Many projectile problems can be solved by first using the **vertical motion** to find the total time taken.

Exercise 1A

SKILLS PROBLEM-SOLVING

1 A particle is projected horizontally at $20\,\text{m s}^{-1}$ from a point h metres above horizontal ground. It lands on the ground 5 seconds later. Find:

a the value of h

b the horizontal distance travelled between the time the particle is projected and the time it hits the ground.

2 A particle is projected horizontally with a velocity of $18\,\text{m s}^{-1}$. Find:

a the horizontal and vertical components of the displacement of the particle from the point of projection after 2 seconds

b the distance of the particle from the point of projection after 2 seconds.

3 A particle is projected horizontally with a speed of $U\,\text{m s}^{-1}$ from a point 160 m above a horizontal plane. The particle hits the plane at a point which is at a horizontal distance of 95 m from the point of projection. Find the initial speed of the particle.

4 A particle is projected horizontally from a point A which is 16 m above horizontal ground. The particle strikes the ground at a point B which is at a horizontal distance of 140 m from A. Find the **speed of projection** of the particle.

(P) **5** A particle is projected horizontally with velocity $20\,\text{m}\,\text{s}^{-1}$ along a flat **smooth** table-top from a point 2 m from the table edge. The particle then leaves the table-top which is at a height of 1.2 m from the floor. Work out the total time taken for the particle to travel from the point of projection until it lands on the floor.

(E) **6** A darts player throws darts at a dartboard which hangs vertically. The motion of a dart is modelled as that of a particle moving freely under gravity. The darts move in a vertical plane which is **perpendicular** to the plane of the dartboard. A dart is thrown horizontally with an initial velocity of $14\,\text{m}\,\text{s}^{-1}$. It hits the dartboard at a point which is 9 cm below the level from which it was thrown.

Find the horizontal distance from the point where the dart was thrown to the dartboard. **(4 marks)**

(E/P) **7** A particle of **mass** 2.5 kg is projected along a horizontal **rough surface** with a velocity of $5\,\text{m}\,\text{s}^{-1}$. After travelling a distance of 2 m the ball leaves the rough surface as a projectile and lands on the ground which is 1.2 m vertically below. Given that the total time taken for the ball to travel from the initial point of projection to the point when it lands is 1.0 seconds, find:

a the time for which the particle is in contact with the surface **(4 marks)**

b the **coefficient of friction** between the particle and the surface **(6 marks)**

c the horizontal distance travelled from the point of projection to the point where the particle hits the ground. **(3 marks)**

1.2 Horizontal and vertical components

Suppose a particle is projected with initial velocity U, at an angle α above the horizontal. The angle α is called the **angle of projection**.

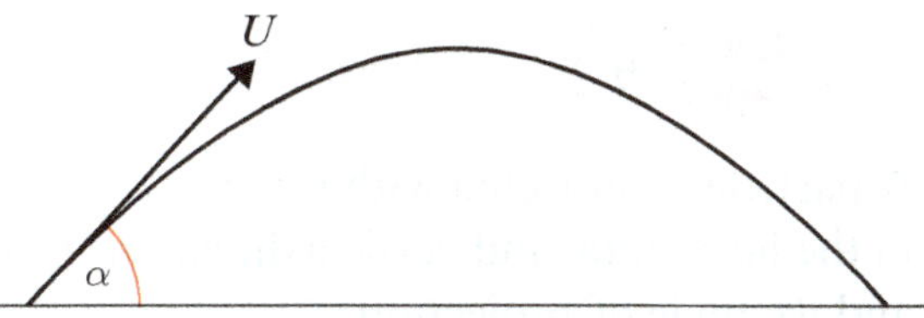

You can **resolve** the velocity into **components** that act horizontally and vertically:

$\cos\alpha = \frac{u_x}{U}$ so $u_x = U\cos\alpha$

$\sin\alpha = \frac{u_y}{U}$ so $u_y = U\sin\alpha$

Links This is the same technique as you use to resolve forces into components.
← **Mechanics 1 Section 5.1**

- **When a particle is projected with initial velocity U, at an angle α above the horizontal:**
  - **The horizontal component of the initial velocity is $U\cos\alpha$.**
  - **The vertical component of the initial velocity is $U\sin\alpha$.**

Example 4 SKILLS PROBLEM-SOLVING

A particle is projected from a point on a horizontal plane with an initial velocity of $40\,\mathrm{m\,s^{-1}}$ at an angle α above the horizontal, where $\tan\alpha = \frac{3}{4}$.

a Find the horizontal and vertical components of the initial velocity.

Given that the vectors **i** and **j** are unit vectors acting in a plane, horizontally and vertically respectively,

b express the initial velocity as a vector in terms of **i** and **j**.

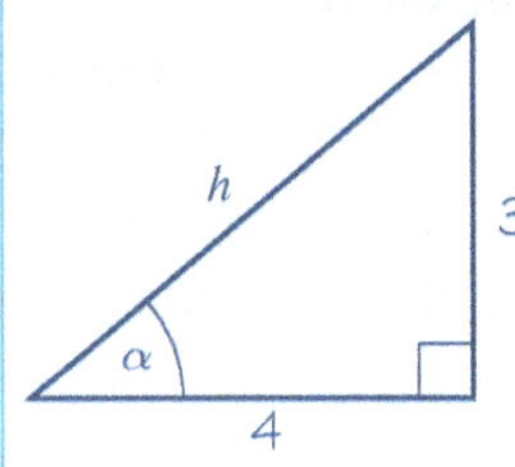

a $\tan\alpha = \frac{3}{4}$ so $h = \sqrt{3^2 + 4^2} = 5$

$\sin\alpha = \frac{3}{5}$ $\cos\alpha = \frac{4}{5}$

R(→), $u_x = u\cos\alpha = 40 \times \frac{4}{5} = 32\,\mathrm{m\,s^{-1}}$

R(↑), $u_y = u\sin\alpha = 40 \times \frac{3}{5} = 24\,\mathrm{m\,s^{-1}}$

b $U = (32\mathbf{i} + 24\mathbf{j})\,\mathrm{m\,s^{-1}}$

Problem-solving

When you are given a value for $\tan\alpha$, you can find the values of $\cos\alpha$ and $\sin\alpha$ without working out the value of α. Here $\tan\alpha = \frac{3}{4} = \frac{\text{opp}}{\text{adj}}$, so sketch a right-angled triangle with **opposite** side = 3 and **adjacent** side = 4.

Online Find $\cos\alpha$ and $\sin\alpha$ using your calculator.

You can write velocity as a vector using **i**–**j** notation. Remember to include units.

Example 5

A particle is projected with velocity $\mathbf{U} = (3\mathbf{i} + 5\mathbf{j})\,\mathrm{m\,s^{-1}}$, where **i** and **j** are the unit vectors in the horizontal and vertical directions respectively. Find the initial speed of the particle and its angle of projection.

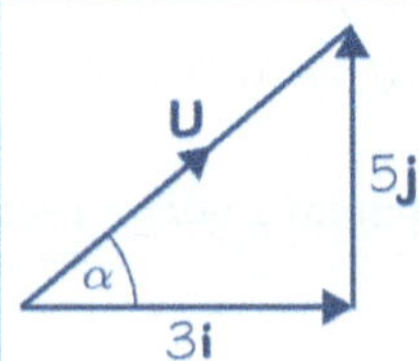

Speed, $|\mathbf{U}| = \sqrt{3^2 + 5^2} = \sqrt{34}\,\mathrm{m\,s^{-1}}$

$\tan\alpha = \frac{5}{3}$

$\alpha = 59.0°$ (3 s.f.)

Initial speed is $\sqrt{34}\,\mathrm{m\,s^{-1}}$ and the particle is projected at an angle of 59.0° above the horizontal.

Speed is the magnitude of the velocity vector. If the initial velocity is $(p\mathbf{i} + q\mathbf{j})\,\mathrm{m\,s^{-1}}$, the initial speed is $\sqrt{p^2 + q^2}$.

When an initial velocity is given in the form $(p\mathbf{i} + q\mathbf{j})\,\mathrm{m\,s^{-1}}$, the values of p and q are the horizontal and vertical components of the velocity respectively.

Exercise 1B SKILLS PROBLEM-SOLVING

In this exercise, **i** and **j** are unit vectors acting in a vertical plane, horizontally and vertically respectively.

1 A particle is projected from a point on a horizontal plane with an initial velocity of $25\,\text{m s}^{-1}$ at an angle of 40° above the horizontal.

a Find the horizontal and vertical components of the initial velocity.

b Express the initial velocity as a vector in the form $(p\mathbf{i} + q\mathbf{j})\,\text{m s}^{-1}$.

2 A particle is projected from a cliff top with an initial velocity of $18\,\text{m s}^{-1}$ at an angle of 20° below the horizontal.

a Find the horizontal and vertical components of the initial velocity.

b Express the initial velocity as a vector in the form $(p\mathbf{i} + q\mathbf{j})\,\text{m s}^{-1}$.

3 A particle is projected from a point on level ground with an initial velocity of $35\,\text{m s}^{-1}$ at an angle α above the horizontal, where $\tan\alpha = \frac{5}{12}$.

a Find the horizontal and vertical components of the initial velocity.

b Express the initial velocity as a vector in terms of **i** and **j**.

4 A particle is projected from the top of a building with an initial velocity of $28\,\text{m s}^{-1}$ at an angle θ below the horizontal, where $\tan\theta = \frac{7}{24}$.

a Find the horizontal and vertical components of the initial velocity.

b Express the initial velocity as a vector in terms of **i** and **j**.

5 A particle is projected with initial velocity $\mathbf{U} = (6\mathbf{i} + 9\mathbf{j})\,\text{m s}^{-1}$.
Find the initial speed of the particle and its angle of projection.

6 A particle is projected with initial velocity $\mathbf{U} = (4\mathbf{i} - 5\mathbf{j})\,\text{m s}^{-1}$.
Find the initial speed of the particle and its angle of projection.

(P) **7** A particle is projected with initial velocity $\mathbf{U} = (3k\mathbf{i} + 2k\mathbf{j})\,\text{m s}^{-1}$.

a Find the angle of projection.

Given the initial speed is $3\sqrt{13}\,\text{m s}^{-1}$,

b find the value of k.

1.3 Projection at any angle

You can solve problems involving particles projected at any angle by resolving the initial velocity into horizontal and vertical components.

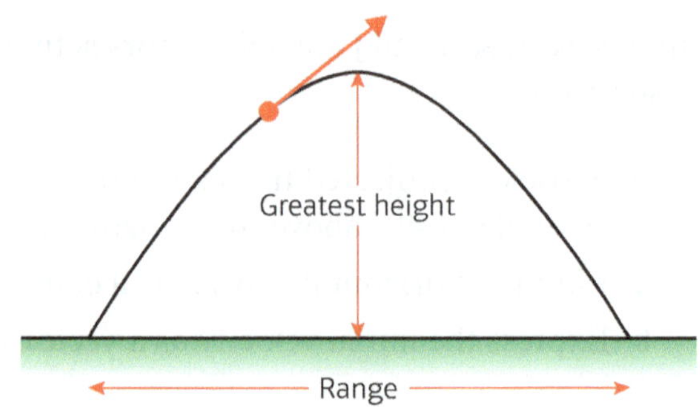

The distance from the point where the particle was projected to the point where it strikes the horizontal plane is called the **range**.

The time the particle takes to move from its point of projection to the point where it strikes the horizontal plane is called the **time of flight** of the particle.

- **A projectile reaches its point of greatest height when the vertical component of its velocity is equal to 0.**

Example 6 SKILLS PROBLEM-SOLVING

A particle P is projected from a point O on a horizontal plane with speed 28 m s^{-1} and with angle of **elevation** 30°. After projection, the particle moves freely under gravity until it strikes the plane at a point A. Find:

a the greatest height above the plane reached by P **b** the time of flight of P **c** the distance OA.

Resolving the velocity of projection horizontally and vertically:

R(→), $u_x = 28\cos 30° = 24.248...$

R(↑), $u_y = 28\sin 30° = 14$

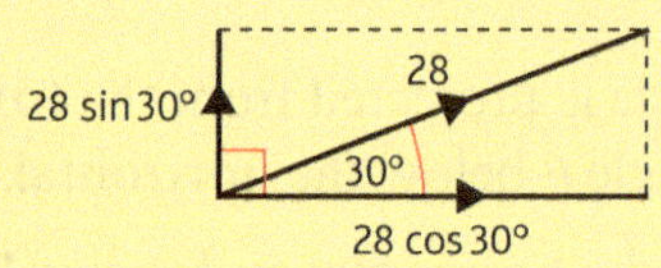

a Taking the upward direction as positive:

R(↑), $u = 14,\ v = 0,\ a = -9.8,\ s = ?$

At the highest point the vertical component of the velocity is zero.

$v^2 = u^2 + 2as$

$0^2 = 14^2 - 2 \times 9.8 \times s$

The vertical motion is motion with constant acceleration.

$s = \frac{14^2}{2 \times 9.8} = 10$

The greatest height above the plane reached by P is 10 m.

b R(↑), $s = 0,\ u = 14,\ a = -9.8,\ t = ?$

When the particle strikes the plane, it is at the same height (zero) as when it started.

$s = ut + \frac{1}{2}at^2$

$0 = 14t - 4.9t^2$

$= t(14 - 4.9t)$

$t = 0$ or $t = \frac{14}{4.9} = 2.857...$

$t = 0$ corresponds to the point where P was projected and can be ignored.

The time of flight is 2.9 s (2 s.f.)

Watch out In this example the value for g was given to 2 significant figures so your answers should be given to 2 significant figures.

c R(→), distance = speed × time

There is no horizontal acceleration.

$= 28\cos 30° \times 2.857...$

Use the unrounded value for the time of flight.

$= 69.282...$

$OA = 69\text{ m}$ (2 s.f.)

Example 7

A particle is projected from a point O with speed V m s^{-1} and at an angle of elevation of θ, where $\tan\theta = \frac{4}{3}$. The point O is 42.5 m above a horizontal plane. The particle strikes the plane at a point A, 5 s after it is projected.

a Show that $V = 20$.

b Find the distance between O and A.

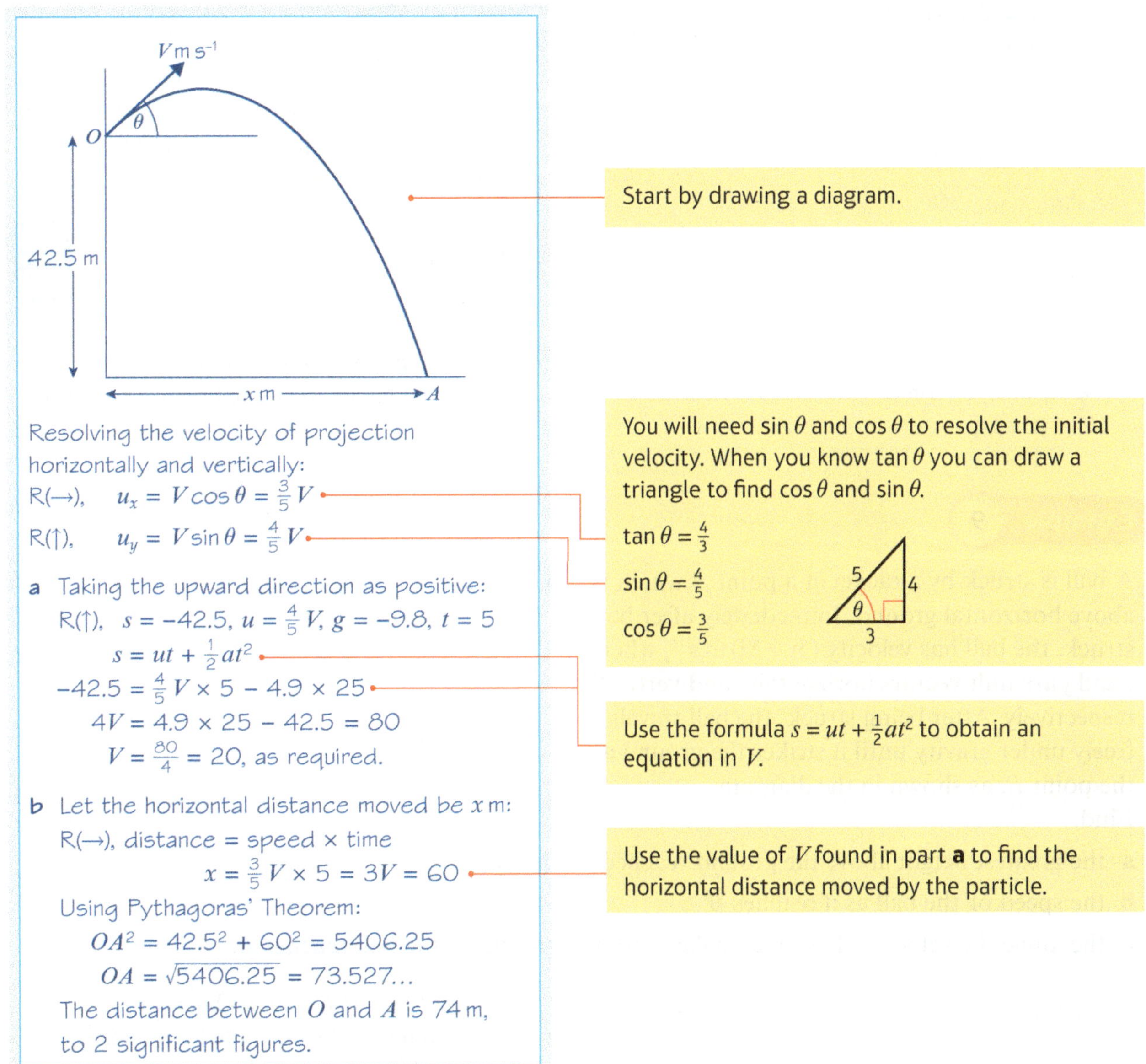

Example 8

A particle is projected from a point O with speed 35 m s^{-1} at an angle of elevation of 30°. The particle moves freely under gravity.

Find the length of time for which the particle is 15 m or more above O.

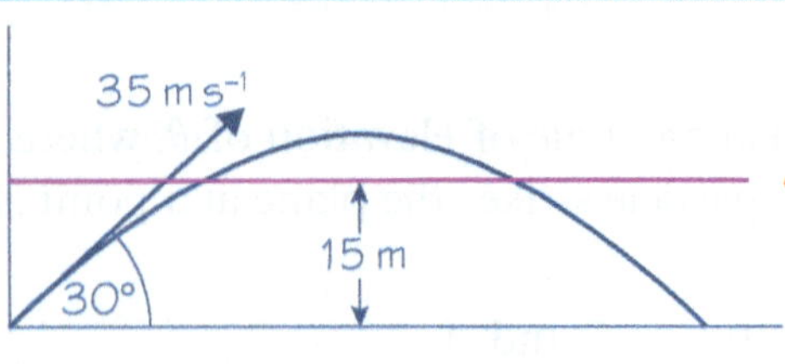

The particle is 15 m above O twice. First on the way up and then on the way down.

Resolving the initial velocity vertically:

In this example the horizontal component of the initial velocity is not used.

R(↑), $u_y = 35 \sin 30° = 17.5$

$s = 15$, $u = 17.5$, $a = -9.8$, $t = ?$

$$s = ut + \frac{1}{2}at^2$$

$$15 = 17.5t - 4.9t^2$$

$4.9t^2 - 17.5t + 15 = 0$

Form a quadratic equation in t to find the two times when the particle is 15 m above O. Between these two times, the particle will be more than 15 m above O.

Multiplying by 10:

$49t^2 - 175t + 150 = 0$

$(7t - 10)(7t - 15) = 0$

$$t = \frac{10}{7}, \frac{15}{7}$$

Online Use your calculator to solve a quadratic equation.

$$\frac{15}{7} - \frac{10}{7} = \frac{5}{7} = 0.71 \text{ s (2 s.f.)}$$

The particle is 15 m or more above O for 0.71 s (2 s.f.).

You should give this answer as a decimal to 2 significant figures, 0.71 s, following previous use of $g = 9.8 \text{ m s}^{-2}$.

Example 9

A ball is struck by a racket at a point A which is 2 m above horizontal ground. Immediately after being struck, the ball has velocity $(5\mathbf{i} + 8\mathbf{j}) \text{ m s}^{-1}$, where $\mathbf{i}$ and $\mathbf{j}$ are unit vectors horizontally and vertically respectively. After being struck, the ball travels freely under gravity until it strikes the ground at the point B, as shown in the diagram.

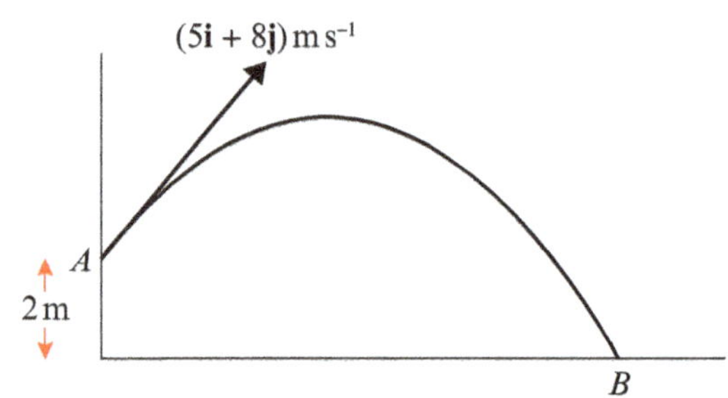

Find:

a the greatest height above the ground reached by the ball

b the speed of the ball as it reaches B

c the angle the velocity of the ball makes with the ground as the ball reaches B.

a Taking the upward direction as positive:

The velocity of projection has been given as a vector in terms of $\mathbf{i}$ and $\mathbf{j}$. The horizontal component is 5 and the vertical component is 8.

R(↑), $u = 8$, $v = 0$, $a = -9.8$, $s = ?$

$v^2 = u^2 + 2as$

$0^2 = 8^2 - 2 \times 9.8 \times s$

$$s = \frac{64}{19.6} = 3.265\ldots$$

This is the greatest height above the point of projection. You need to add 2 m to find the height above the ground.

The greatest height above the ground reached by the ball is $2 + 3.265\ldots = 5.3$ m, to 2 significant figures.

b The horizontal component of the velocity of the ball at B is 5 m s^{-1}.
The vertical component of the velocity of the ball at B is given by:
R(↑), $s = -2$, $u = 8$, $a = -9.8$, $v = ?$
$v^2 = u^2 + 2as$
$= 8^2 + 2 \times (-9.8) \times (-2) = 103.2$
The speed at B is given by:
$v^2 = 5^2 + 103.2 = 128.2$
$v = \sqrt{128.2}$
The speed of the ball as it reaches B is 11 m s^{-1}, to 2 significant figures.

c The angle is given by:
$\tan\theta = \frac{\sqrt{103.2}}{5} \Rightarrow \theta = 64°$ (2 s.f.)
The angle the velocity of the ball makes with the ground as the ball reaches B is 64°, to the nearest degree.

The horizontal motion is motion with constant speed, so the horizontal component of the velocity never changes.

There is no need to find the square root of 103.2 at this point, as you need v^2 in the next stage of the calculation.

As the ball reaches B, its velocity has two components as shown below.

The magnitude (speed) and direction of the velocity are found using trigonometry and Pythagoras' Theorem.

Exercise 1C

SKILLS PROBLEM-SOLVING

In this exercise, **i** and **j** are unit vectors acting in a vertical plane, horizontally and vertically respectively.

Whenever a numerical value of g is required, take $g = 9.8\text{ m s}^{-2}$ unless otherwise stated.

1 A particle is projected with speed 35 m s^{-1} at an angle of elevation of 60°. Find the time the particle takes to reach its greatest height.

2 A ball is projected from a point 5 m above horizontal ground with speed 18 m s^{-1} at an angle of elevation of 40°. Find the height of the ball above the ground 2 s after projection.

3 A stone is projected from a point above horizontal ground with speed 32 m s^{-1}, at an angle of 10° below the horizontal. The stone takes 2.5 s to reach the ground. Find:

a the height of the point of projection above the ground

b the distance from the point on the ground vertically below the point of projection to the point where the stone reaches the ground.

4 A projectile is launched from a point on horizontal ground with speed 150 m s^{-1} at an angle of 10° above the horizontal. Find:

a the time the projectile takes to reach its highest point above the ground

b the range of the projectile.

5 A particle is projected from a point O on a horizontal plane with speed 20 m s^{-1} at an angle of elevation of 45°. The particle moves freely under gravity until it strikes the ground at a point X. Find:

a the greatest height above the plane reached by the particle

b the distance OX.

(P) 6 A ball is projected from a point A on level ground with speed 24 m s^{-1}. The ball is projected at an angle θ to the horizontal where $\sin\theta = \frac{4}{5}$. The ball moves freely under gravity until it strikes the ground at a point B. Find:

a the time of flight of the ball

b the distance from A to B.

(P) 7 A particle is projected with speed 21 m s^{-1} at an angle of elevation α. Given that the greatest height reached above the point of projection is 15 m, find the value of α, giving your answer to the nearest degree.

8 A particle P is projected from the origin with velocity $(12\mathbf{i} + 24\mathbf{j})\text{ m s}^{-1}$, where $\mathbf{i}$ and $\mathbf{j}$ are horizontal and vertical unit vectors respectively. The particle moves freely under gravity. Find:

a the position vector of P after 3 s

b the speed of P after 3 s.

(P) 9 A stone is thrown with speed 30 m s^{-1} from a window which is 20 m above horizontal ground. The stone hits the ground 3.5 s later. Find:

a the angle of projection of the stone

b the horizontal distance from the window to the point where the stone hits the ground.

(E/P) 10 A ball is thrown from a point O on horizontal ground with speed $U\text{ m s}^{-1}$ at an angle of elevation of θ, where $\tan\theta = \frac{3}{4}$. The ball strikes a vertical wall which is 20 m from O at a point which is 3 m above the ground. Find:

a the value of U **(6 marks)**

b the time from the instant the ball is thrown to the instant that it strikes the wall. **(2 marks)**

(E/P) 11 A particle P is projected from a point A with position vector $20\mathbf{j}$ m with respect to a fixed origin O. The velocity of projection is $(5u\mathbf{i} + 4u\mathbf{j})\text{ m s}^{-1}$. The particle moves freely under gravity, passing through a point B, which has position vector $(k\mathbf{i} + 12\mathbf{j})$ m, where k is a constant, before reaching the point C on the x-axis, as shown in the diagram.

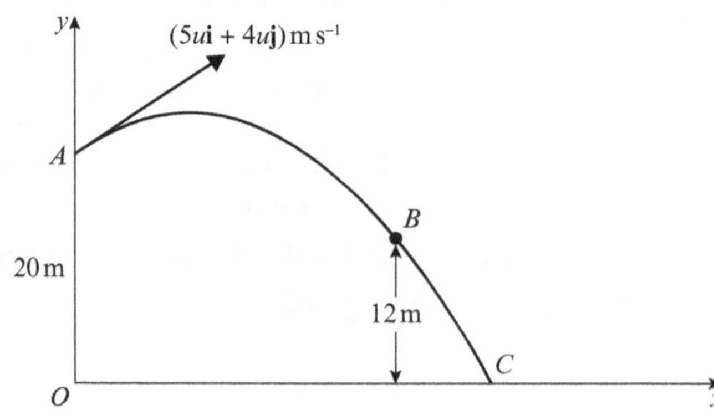

The particle takes 4 s to move from A to B. Find:

a the value of u **(4 marks)**

b the value of k **(2 marks)**

c the angle the velocity of P makes with the x-axis as it reaches C. **(6 marks)**

Watch out When finding a square root involving use of $g = 9.8\,\text{m s}^{-2}$ to work out an answer, an exact surd (irrational number) answer is **not** acceptable.

(E) **12** A stone is thrown from a point A with speed $30\,\text{m s}^{-1}$ at an angle of 15° below the horizontal. The point A is 14 m above horizontal ground. The stone strikes the ground at the point B, as shown in the diagram. Find:

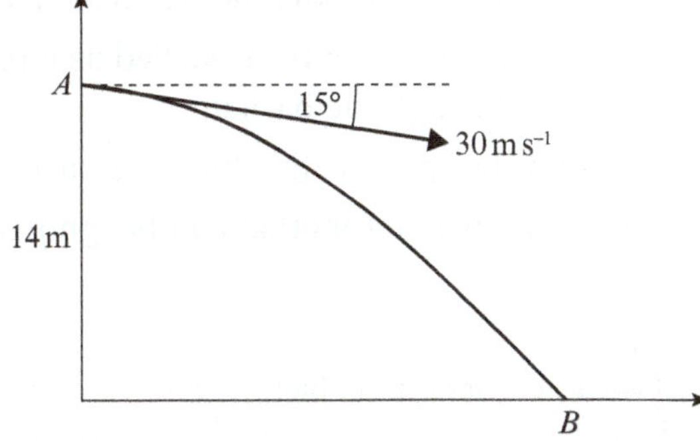

a the time the stone takes to travel from A to B **(6 marks)**

b the distance AB. **(2 marks)**

(E/P) **13** A particle is projected from a point on level ground with speed $U\,\text{m s}^{-1}$ and angle of elevation α. The maximum height reached by the particle is 42 m above the ground and the particle hits the ground 196 m from its point of projection.

Find the value of α and the value of U. **(9 marks)**

(E/P) **14** In this question use $g = 10\,\text{m s}^{-2}$.

An object is projected with speed $U\,\text{m s}^{-1}$ from a point A at the top of a vertical building. The point A is 25 m above the ground. The object is projected at an angle α above the horizontal, where $\tan\alpha = \frac{5}{12}$. The object hits the ground at the point B, which is at a horizontal distance of 42 m from the foot of the building, as shown in the diagram. The object is modelled as a particle moving freely under gravity.

Find:

a the value of U **(6 marks)**

b the time taken by the object to travel from A to B **(2 marks)**

c the speed of the object when it is 12.4 m above the ground, giving your answer to 2 significant figures. **(5 marks)**

15 An object is projected from a fixed origin O with velocity $(4\mathbf{i} + 5\mathbf{j})\text{ m s}^{-1}$. The particle moves freely under gravity and passes through the point P with position vector $k(\mathbf{i} - \mathbf{j})$ m, where k is a positive constant.

a Find the value of k. **(6 marks)**

b Find:

i the speed of the object at the instant when it passes through P

ii the direction of motion of the object at the instant when it passes through P. **(7 marks)**

16 A basketball player is standing on the floor 10 m from the basket. The height of the basket is 3.05 m, and he shoots the ball from a height of 2 m, at an angle of 40° above the horizontal.

The basketball can be modelled as a particle moving in a vertical plane. Given that the ball passes through the basket,

a find the speed with which the basketball is thrown. **(6 marks)**

b State two factors that can be ignored by modelling the basketball as a particle. **(2 marks)**

Challenge

A vertical tower is 85 m high. A stone is projected at a speed of 20 m s^{-1} from the top of a tower at an angle of α below the horizontal. At the same time, a second stone is projected horizontally at a speed of 12 m s^{-1} from a window in the tower 45 m above the ground.

Given that the two stones move freely under gravity in the same vertical plane, and that they collide in mid-air, show that the time that elapses between the moment they are projected and the moment they collide is 2.5 s.

1.4 Projectile motion formulae

You need to be able to derive general formulae related to the motion of a particle which is projected from a point on a horizontal plane and moves freely under gravity.

Example 10  REASONING/ARGUMENTATION

A particle is projected from a point on a horizontal plane with an initial velocity U, at an angle α above the horizontal, and moves freely under gravity until it hits the plane at point B. Given that the acceleration due to gravity is g, find expressions for:

a the time of flight, T

b the range, R, on the horizontal plane.

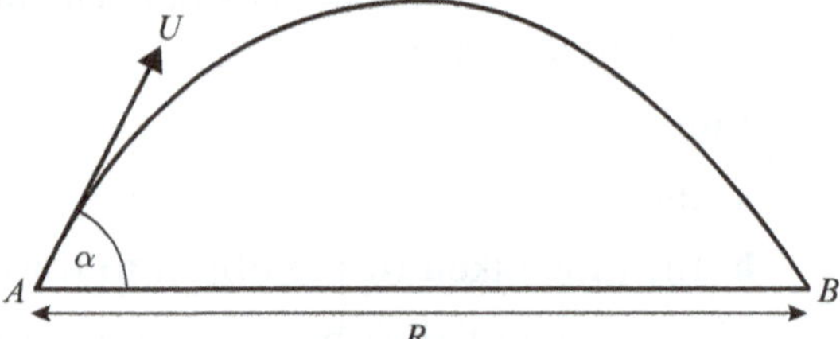

Taking the upward direction as positive and resolving the velocity of projection:

R(↑), $u_y = U\sin\alpha$

R(→), $u_x = U\cos\alpha$

a Considering vertical motion:

R(↑), $u = U\sin\alpha,\ s = 0,\ a = -g,\ t = T$

$s = ut + \frac{1}{2}at^2$

$0 = (U\sin\alpha)T - \frac{1}{2} \times g \times T^2$

$0 = T\left(U\sin\alpha - \frac{gT}{2}\right)$

either $T = 0$ (at A) or $U\sin\alpha - \frac{gT}{2} = 0$

so $T = \frac{2U\sin\alpha}{g}$

b Considering horizontal motion:

R(→), $v_x = U\cos\alpha,\ s = R,\ t = T$

$s = v_x t$

$R = U\cos\alpha \times T$ using $T = \frac{2U\sin\alpha}{g}$

$R = U\cos\alpha \times \frac{2U\sin\alpha}{g} = \frac{2U^2\sin\alpha\cos\alpha}{g}$

Using $2\sin\alpha\cos\alpha \equiv \sin 2\alpha$:

$R = \frac{U^2\sin 2\alpha}{g}$

Online Explore the parametric equations for the path of a particle and their Cartesian form, both algebraically and graphically using technology.

When the particle reaches the horizontal plane, the vertical displacement is 0.

Taking out the factor T, one solution is $T = 0$ which is at the start of the motion.

Problem-solving

Follow the same steps as you would if you were given values of U and α and asked to find the time of flight and the range. The answer will be an algebraic expression in terms of U and α instead of a numerical value.

Substitute for T in the equation $R = U\cos\alpha \times T$

$U\cos\alpha \times \frac{2U\sin\alpha}{g} = \frac{U\cos\alpha}{1} \times \frac{2U\sin\alpha}{g}$

Use the double-angle formula for $\sin 2\alpha$.

Notation g is usually left as a letter in the formulae for projectile motion.

Example 11

A particle is projected from a point with speed U at an angle of elevation α and moves freely under gravity. When the particle has moved a horizontal distance x, its height above the point of projection is y.

a Show that $y = x\tan\alpha - \frac{gx^2}{2u^2}(1 + \tan^2\alpha)$.

A particle is projected from a point O on a horizontal plane, with speed $28\,\text{m s}^{-1}$ at an angle of elevation α. The particle passes through a point B, which is at a horizontal distance of 32 m from O and at a height of 8 m above the plane.

b Find the two possible values of α, giving your answers to the nearest degree.

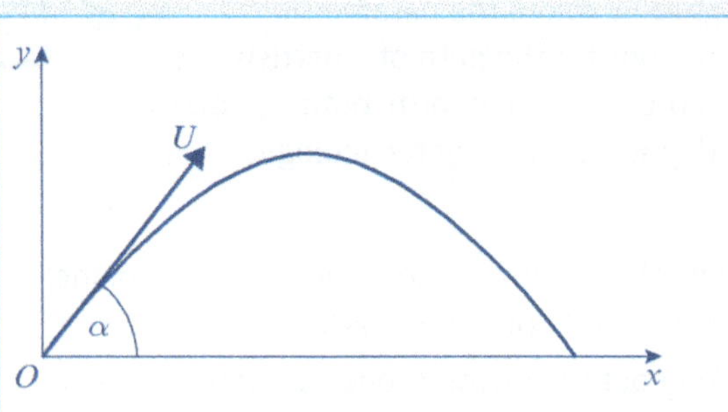

a R(→), $u_x = U\cos\alpha$
R(↑), $u_y = U\sin\alpha$

For the horizontal motion:
R(→), $s = vt$
$x = u\cos\alpha \times t$ **(1)**

For the vertical motion, taking upward as positive:
R(↑), $s = ut + \frac{1}{2}at^2$
$y = U\sin\alpha \times t - \frac{1}{2}gt^2$ **(2)**

Rearranging **(1)** to make t the subject of the formula:
$t = \dfrac{x}{U\cos\alpha}$ **(3)**

Substituting **(3)** into **(2)**:
$$y = U\sin\alpha \times \frac{x}{U\cos\alpha} - \frac{1}{2}g\left(\frac{x}{U\cos\alpha}\right)^2$$

Using $\tan\alpha \equiv \dfrac{\sin\alpha}{\cos\alpha}$ and $\dfrac{1}{\cos\alpha} \equiv \sec\alpha$,
$$y = x\tan\alpha - \frac{gx^2}{2U^2}\sec^2\alpha$$

Using $\sec^2\alpha \equiv 1 + \tan^2\alpha$,
$$y = x\tan\alpha - \frac{gx^2}{2u^2}(1 + \tan^2\alpha)\text{, as required.}$$

b Using the result in **a** with $U = 28$, $x = 32$, $y = 8$ and $g = 9.8$
$8 = 32\tan\alpha - 6.4(1 + \tan^2\alpha)$

Rearranging as a quadratic in $\tan\alpha$:
$6.4\tan^2\alpha - 32\tan\alpha + 14.4 = 0$
$4\tan^2\alpha - 20\tan\alpha + 9 = 0$
$(2\tan\alpha - 1)(2\tan\alpha - 9) = 0$
$\tan\alpha = \frac{1}{2}, \frac{9}{2}$

$\alpha = 27°$ and $77°$, to the nearest degree

Resolve the velocity of projection horizontally and vertically.

You have obtained two equations, labelled **(1)** and **(2)**. Both equations contain t and the result you have been asked to show has no t in it. You must eliminate (i.e. remove) t using substitution.

If the upward direction is taken as positive, the vertical acceleration is $-g$.

(1) and **(2)** are parametric equations describing the path of the particle. You can eliminate the parameter, t, to find the Cartesian form of the path.

To obtain a quadratic expression in $\tan\alpha$, you need to use the identity $\sec^2\alpha \equiv 1 + \tan^2\alpha$.

You should use your calculator to check the solutions to this equation.

There are two possible angles of elevation for which the particle will pass through B. This sketch illustrates the two paths.

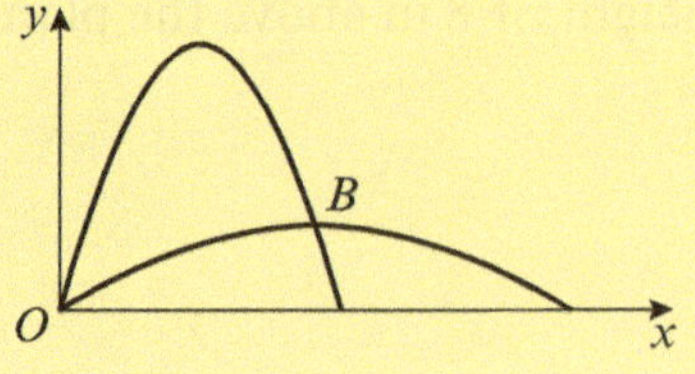

- For a particle which is projected from a point on a horizontal plane with an initial velocity U at an angle α above the horizontal, and that moves freely under gravity:
  - Time of flight $= \frac{2U\sin\alpha}{g}$
  - Time to reach greatest height $= \frac{U\sin\alpha}{g}$
  - Range on horizontal plane $= \frac{U^2\sin 2\alpha}{g}$
  - Equation of trajectory: $y = x\tan\alpha - gx^2\frac{(1+\tan^2\alpha)}{2U^2}$

  where y is the vertical height of the particle, x is the horizontal distance from the point of projection, and g is the acceleration due to gravity.

Watch out You need to know how to derive the equations. But be careful of using them in projectile problems. They are hard to memorise, and it is usually safer to answer projectile problems using the techniques covered in Section 1.3.

Hint The equation for the trajectory of the particle is a **quadratic** equation for y in x. This proves that the path of a projectile moving freely under gravity is a quadratic curve, or **parabola**.

Exercise 1D

SKILLS REASONING/ARGUMENTATION

Whenever a numerical value of g is required, take $g = 9.8\text{ m s}^{-2}$ unless otherwise stated.

(P) **1** A particle is launched from a point on a horizontal plane with initial velocity $U\text{ m s}^{-1}$ at an angle of elevation α. The particle moves freely under gravity until it strikes the plane. The greatest height of the particle is h m.

Show that $h = \frac{U^2\sin^2\alpha}{2g}$.

(P) **2** A particle is projected from a point with speed 21 m s^{-1} at an angle of elevation α and moves freely under gravity. When the particle has moved a horizontal distance x m, its height above the point of projection is y m.

a Show that $y = x\tan\alpha - \frac{x^2}{90\cos^2\alpha}$.

b Given that $y = 8.1$ when $x = 36$, find the value of $\tan\alpha$.

(P) **3** A particle is launched from a point on a horizontal plane with initial speed $U\text{ m s}^{-1}$ at an angle of elevation α. The particle moves freely under gravity until it strikes the plane. The range of the particle is R m.

a Show that the time of flight of the particle is $\frac{2U\sin\alpha}{g}$ seconds.

b Show that $R = \frac{U^2\sin 2\alpha}{g}$.

c Deduce that, for a fixed u, the greatest possible range is when $\alpha = 45°$.

d Given that $R = \frac{2U^2}{5g}$, find the two possible values of the angle of elevation at which the particle could have been launched.

(P) **4** A firework is launched vertically with a speed of $v\,\text{m s}^{-1}$. When it reaches its maximum height, the firework explodes into two parts, which are projected horizontally in opposite directions, each with speed $2v\,\text{m s}^{-1}$.
Show that the two parts of the firework land a distance $\frac{4v^2}{g}$ m apart.

(E/P) **5** In this question use $g = 10\,\text{m s}^{-2}$.
A particle is projected from a point O with speed U at an angle of elevation α above the horizontal and moves freely under gravity. When the particle has moved a horizontal distance x, its height above O is y.

a Show that $y = x\tan\alpha - \frac{gx^2}{2U^2\cos^2\alpha}$. **(4 marks)**

A boy throws a stone from a point P at the end of a pier. The point P is 15 m above sea level. The stone is projected with a speed of $8\,\text{m s}^{-1}$ at an angle of elevation of 40°. By modelling the stone as a particle moving freely under gravity,

b find the horizontal distance of the stone from P when the stone is 2 m above sea level. **(5 marks)**

(E/P) **6** A particle is projected from a point with speed U at an angle of elevation α above the horizontal and moves freely under gravity. When it has moved a horizontal distance x, its height above the point of projection is y.

a Show that $y = x\tan\alpha - \frac{gx^2}{2U^2}(1 + \tan^2\alpha)$. **(5 marks)**

An athlete throws a javelin from a point P at a height of 2 m above horizontal ground. The javelin is projected at an angle of elevation of 45° with a speed of $30\,\text{m s}^{-1}$.
By modelling the javelin as a particle moving freely under gravity,

b find, to 3 significant figures, the horizontal distance of the javelin from P when it hits the ground **(5 marks)**

c find, to 2 significant figures, the time elapsed from the point the javelin is thrown to the point when it hits the ground. **(2 marks)**

(E/P) **7** A girl playing volleyball on horizontal ground hits the ball towards the net 9 m away from a point 1.5 m above the ground. The ball moves in a vertical plane which is perpendicular to the net. The ball just passes over the top of the net, which is 2.4 m above the ground, as shown in the diagram.
The ball is modelled as a particle projected with initial speed $U\,\text{m s}^{-1}$ from point O, 1.5 m above the ground at an angle α to the horizontal.

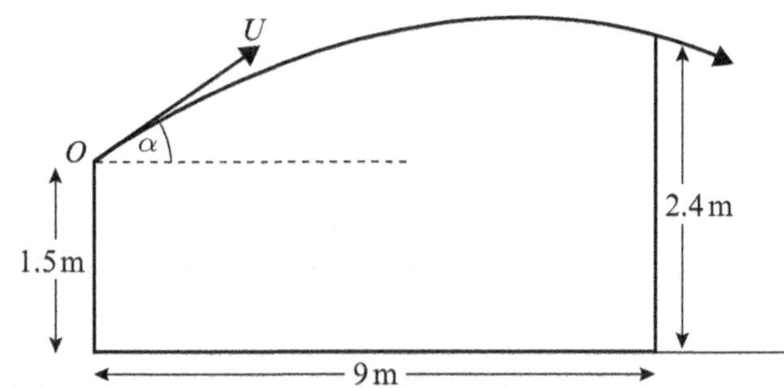

a By writing down expressions for the horizontal and vertical distances from O to the ball, t seconds after it was hit, show that when the ball passes over the net:

$$0.9 = 9\tan\alpha - \frac{81g}{2U^2\cos^2\alpha}$$ **(6 marks)**

Given that $\alpha = 30°$,

b find the speed of the ball as it passes over the net. **(6 marks)**

E/P **8** In this question, **i** and **j** are unit vectors in a horizontal and upward vertical direction respectively. An object is projected from a fixed point A on horizontal ground with velocity $(k\mathbf{i} + 2k\mathbf{j})\text{ m s}^{-1}$, where k is a positive constant. The object moves freely under gravity until it strikes the ground at B, where it immediately comes to rest. Relative to O, the position vector of a point on the path of the object is $(x\mathbf{i} + y\mathbf{j})$ m.

a Show that $y = 2x - \dfrac{gx^2}{2k^2}$. **(5 marks)**

Given that $AB = R$ m and the maximum vertical height of the object above the ground is H m,

b using the result in part **a**, or otherwise, find, in terms of k and g,

i R **ii** H **(6 marks)**

Challenge

A stone is projected from a point on a straight sloping hill. Given that the hill slopes downwards at an angle of 45°, and that the stone is projected at an angle of 45° above the horizontal with speed $U\text{ m s}^{-1}$, show that the stone lands a distance $\dfrac{2\sqrt{2}\,U^2}{g}$ m down the hill.

Chapter review 1

Whenever a numerical value of g is required, take $g = 9.8\text{ m s}^{-2}$ unless otherwise stated.

1 A ball is thrown vertically downwards from the top of a tower with speed 6 m s^{-1}. The ball strikes the ground with speed 25 m s^{-1}. Find the time the ball takes to move from the top of the tower to the ground.

2 A child drops a ball from a point at the top of a cliff which is 82 m above the sea. The ball is initially at rest. Find:

a the time taken for the ball to reach the sea **b** the speed with which the ball hits the sea.

c State one physical factor which has been ignored in making your calculation.

P **3** The velocity–time graph represents the motion of a particle moving in a straight line, accelerating from velocity u at time 0 to velocity v at time t.

a Use the graph to show that:

i $v = u + at$ **ii** $s = \left(\dfrac{u+v}{2}\right)t$

b Hence show that:

i $v^2 = u^2 + 2as$ **ii** $s = ut + \frac{1}{2}at^2$ **iii** $s = vt - \frac{1}{2}at^2$

E/P **4** A particle is projected vertically upwards with a speed of 30 m s^{-1} from a point A. The point B is h metres above A. The particle moves freely under gravity and is above B for 2.4 s. Calculate the value of h. **(5 marks)**

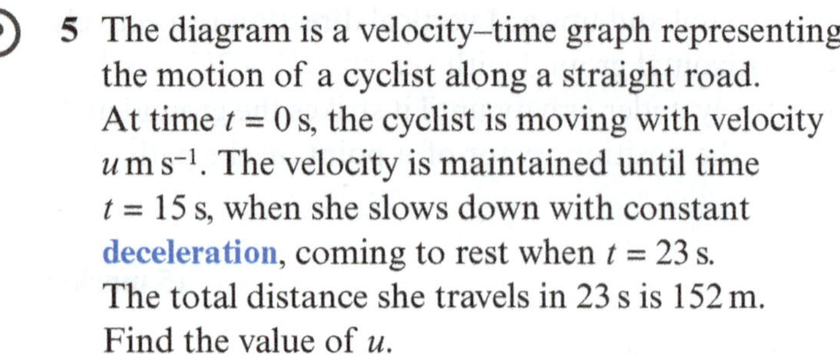

(P) **5** The diagram is a velocity–time graph representing the motion of a cyclist along a straight road. At time $t = 0$ s, the cyclist is moving with velocity $u\,\text{m s}^{-1}$. The velocity is maintained until time $t = 15$ s, when she slows down with constant **deceleration**, coming to rest when $t = 23$ s. The total distance she travels in 23 s is 152 m. Find the value of u.

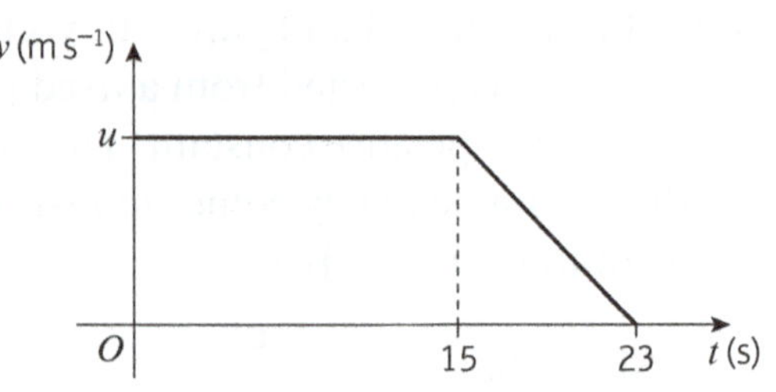

6 A particle P is projected from a point O on a horizontal plane with speed $42\,\text{m s}^{-1}$ and with angle of elevation 45°. After projection, the particle moves freely under gravity until it strikes the plane. Find:

a the greatest height above the plane reached by P

b the time of flight of P.

7 A stone is thrown horizontally with speed $21\,\text{m s}^{-1}$ from a point P on the edge of a cliff h metres above sea level. The stone lands in the sea at a point Q, where the horizontal distance of Q from the cliff is 56 m.

Calculate the value of h.

(E) **8** A ball is thrown from a window above a horizontal lawn. The velocity of projection is $15\,\text{m s}^{-1}$ and the angle of elevation is α, where $\tan\alpha = \frac{4}{3}$. The ball takes 4 s to reach the lawn. Find:

a the horizontal distance between the point of projection and the point where the ball hits the lawn **(3 marks)**

b the vertical height above the lawn from which the ball was thrown. **(3 marks)**

(E) **9** A projectile is fired with velocity $40\,\text{m s}^{-1}$ at an angle of elevation of 30° from a point A on horizontal ground. The projectile moves freely under gravity until it reaches the ground at the point B. Find:

a the distance AB **(5 marks)**

b the speed of the projectile at the first instant when it is 15 m above the ground. **(5 marks)**

(E/P) **10** A projectile P is projected from a point on a horizontal plane with speed U at an angle of elevation θ.

a Show that the range of the projectile is $\dfrac{U^2 \sin 2\theta}{g}$. **(6 marks)**

b Hence find, as θ varies, the maximum range of the projectile. **(2 marks)**

c Given that the range of the projectile is $\dfrac{2U^2}{3g}$, find the two possible value of θ. Give your answers to the nearest 0.1°. **(3 marks)**

E **11**

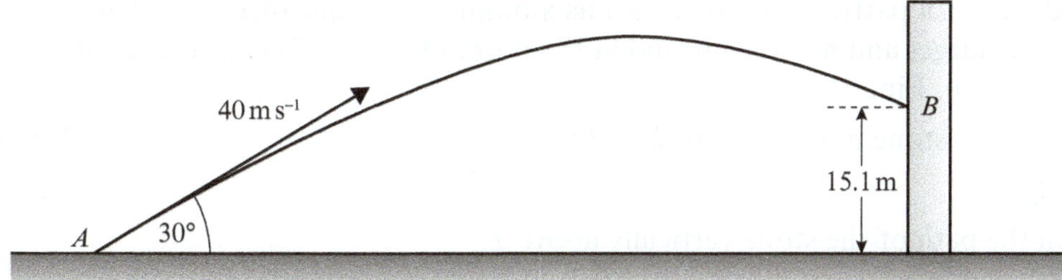

A golf ball is driven from a point A with a speed of $40\,\text{m}\,\text{s}^{-1}$ at an angle of elevation of 30°. On its downward flight, the ball hits a tree at a height 15.1 m above the level of A, as shown in the diagram above. Find:

a the time taken by the ball to reach its greatest height above A **(3 marks)**

b the time taken by the ball to travel from A to B **(6 marks)**

c the speed with which the ball hits the tree. **(5 marks)**

E/P **12** A particle P is projected from a fixed origin O with velocity $(12\mathbf{i} + 5\mathbf{j})\,\text{m}\,\text{s}^{-1}$. The particle moves freely under gravity and passes through the point A with position vector $\lambda(2\mathbf{i} - \mathbf{j})$ m, where λ is a positive constant.

a Find the value of λ. **(6 marks)**

b Find:

i the speed of P at the instant when it passes through A

ii the direction of motion of P at the instant when it passes through A. **(7 marks)**

E/P **13** In this question use $g = 10\,\text{m}\,\text{s}^{-2}$.

A boy plays a game at a fairground. He needs to throw a ball through a hole in a vertical target to win a prize. The motion of the ball is modelled as that of a particle moving freely under gravity. The ball moves in a vertical plane which is perpendicular to the plane of the target. The boy throws the ball horizontally at the same height as the hole with a speed of $10\,\text{m}\,\text{s}^{-1}$. It hits the target at a point 20 cm below the hole.

a Find the horizontal distance from the point where the ball was thrown to the target. **(4 marks)**

The boy throws the ball again with the same speed and at the same distance from the target.

b Work out the possible angles above the horizontal the boy could throw the ball so that it passes through the hole. **(6 marks)**

E/P **14** In this question use $g = 10\,\text{m}\,\text{s}^{-2}$.

A stone is thrown from a point P at a target, which is on horizontal ground. The point P is 10 m above the point O on the ground. The stone is thrown from P with speed $20\,\text{m}\,\text{s}^{-1}$ at an angle of α below the horizontal, where $\tan\alpha = \frac{3}{4}$.

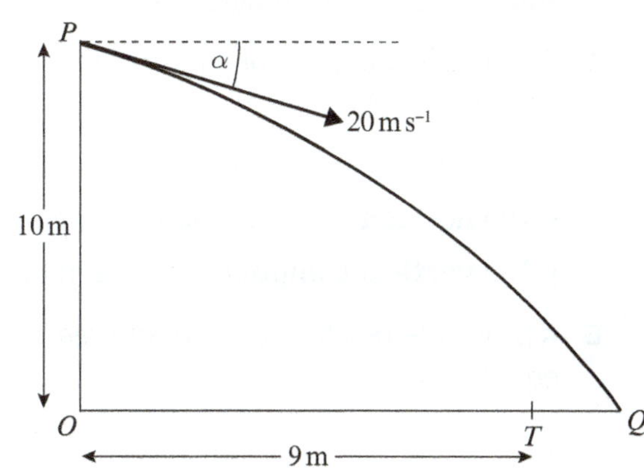

The stone is modelled as a particle and the target as a point T. The distance OT is 9 m. The stone misses the target and hits the ground at the point Q, where OTQ is a straight line, as shown in the diagram. Find:

a the time taken by the stone to travel from P to Q **(5 marks)**

b the distance TQ. **(4 marks)**

The point A is on the path of the stone vertically above T.

c Find the speed of the stone at A. **(5 marks)**

15 A vertical mast is 32 m high. Two balls P and Q are projected at the same time. Ball P is projected horizontally from the top of the mast with speed 18 m s^{-1}. Ball Q is projected from the bottom of the mast with speed 30 m s^{-1} at an angle α above the horizontal. The balls move freely under gravity in the same vertical plane and collide in mid-air. By considering the horizontal motion of each ball,

a prove that $\cos\alpha = \frac{3}{5}$. **(4 marks)**

b Find the time which elapses between the instant when the balls are projected and the instant when they collide. **(4 marks)**

Challenge

A cruise ship is 250 m long, and is accelerating forwards in a straight line at a constant rate of 1.5 m s^{-2}. A golfer stands at the stern (back) of the cruise ship and hits a golf ball towards the bow (front). Given that the golfer hits the golf ball at an angle of elevation of 60°, and that the ball lands directly on the bow of the cruise ship, find the speed, v, with which the golfer hits the ball.

Problem-solving

You need to calculate the initial speed of the ball relative to the golfer. This is the speed the ball would appear to be travelling at if you were standing on the ship.

Summary of key points

1. The force of **gravity** causes all objects to accelerate towards the Earth. If you ignore the effects of air resistance, this acceleration is constant. It does not depend on the mass of the object.
2. An object moving vertically freely under gravity can be modelled as a particle with a constant downward acceleration of $g = 9.8\text{ m s}^{-2}$.
3. The **horizontal** motion of a projectile is modelled as having **constant velocity** ($a = 0$). You can use the formula $s = vt$.
4. The **vertical** motion of a projectile is modelled as having **constant acceleration** due to gravity ($a = g$).
5. When a particle is projected with initial velocity U, at an angle α above the horizontal:
   - The **horizontal component** of the initial velocity is $U\cos\alpha$.
   - The **vertical component** of the initial velocity is $U\sin\alpha$.
6. A projectile reaches its point of greatest height when the vertical component of its velocity is equal to 0.

7 For a particle which is projected from a point on a horizontal plane with an initial velocity U at an angle α above the horizontal, and that moves freely under gravity:

- Time of flight $= \dfrac{2U\sin\alpha}{g}$
- Time to reach greatest height $= \dfrac{U\sin\alpha}{g}$
- Range on horizontal plane $= \dfrac{U^2\sin 2\alpha}{g}$
- Equation of trajectory: $y = x\tan\alpha - gx^2\dfrac{(1+\tan^2\alpha)}{2U^2}$

  where y is the vertical height of the particle, x is the horizontal distance from the point of projection, and g is the acceleration due to gravity.

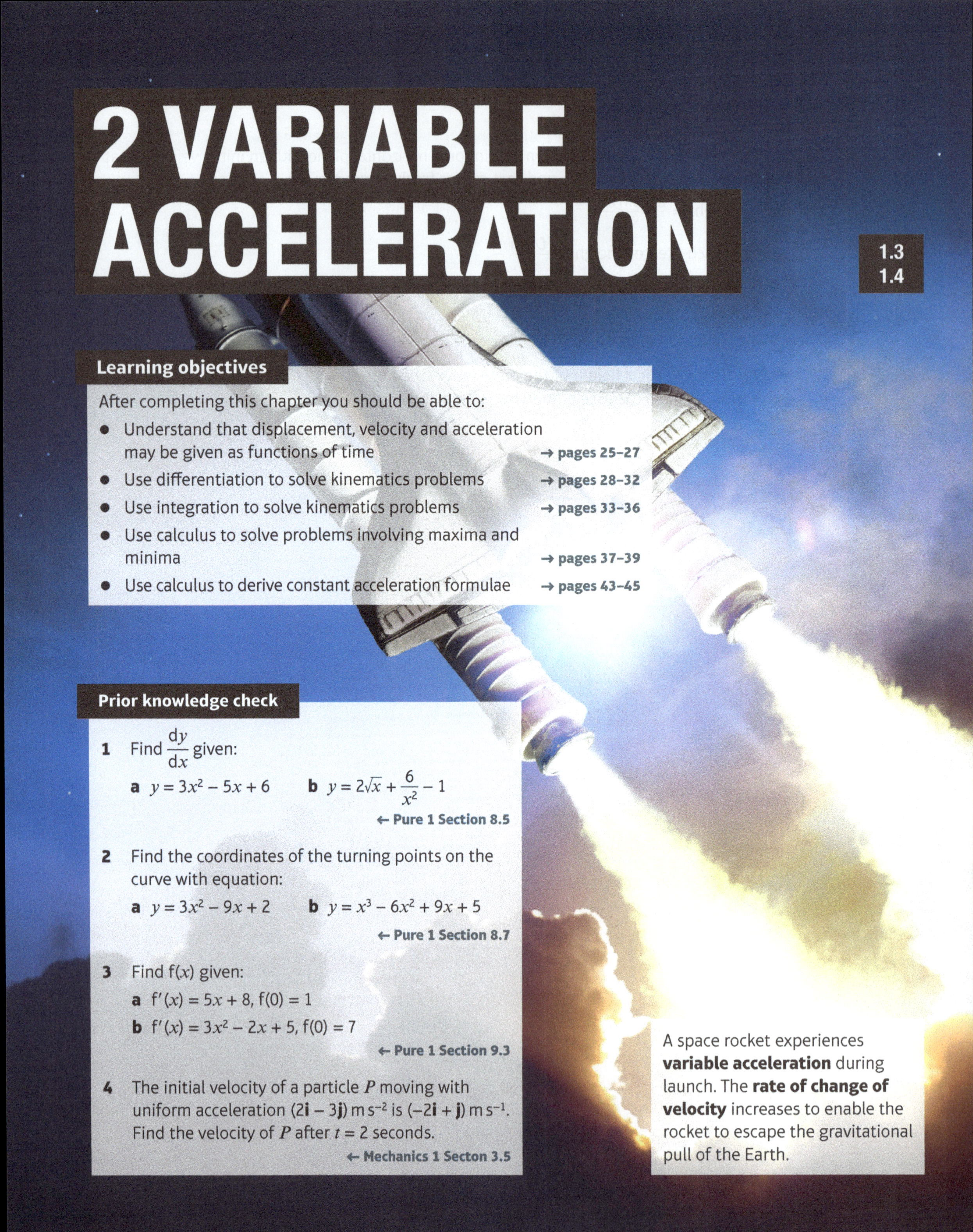

2 VARIABLE ACCELERATION

1.3
1.4

Learning objectives

After completing this chapter you should be able to:

- Understand that displacement, velocity and acceleration may be given as functions of time → pages 25–27
- Use differentiation to solve kinematics problems → pages 28–32
- Use integration to solve kinematics problems → pages 33–36
- Use calculus to solve problems involving maxima and minima → pages 37–39
- Use calculus to derive constant acceleration formulae → pages 43–45

Prior knowledge check

1 Find $\frac{\mathrm{d}y}{\mathrm{d}x}$ given:

a $y = 3x^2 - 5x + 6$ **b** $y = 2\sqrt{x} + \frac{6}{x^2} - 1$

← Pure 1 Section 8.5

2 Find the coordinates of the turning points on the curve with equation:

a $y = 3x^2 - 9x + 2$ **b** $y = x^3 - 6x^2 + 9x + 5$

← Pure 1 Section 8.7

3 Find f(x) given:

a $\mathrm{f}'(x) = 5x + 8$, $\mathrm{f}(0) = 1$

b $\mathrm{f}'(x) = 3x^2 - 2x + 5$, $\mathrm{f}(0) = 7$

← Pure 1 Section 9.3

4 The initial velocity of a particle P moving with uniform acceleration $(2\mathbf{i} - 3\mathbf{j})\,\mathrm{m\,s^{-2}}$ is $(-2\mathbf{i} + \mathbf{j})\,\mathrm{m\,s^{-1}}$. Find the velocity of P after $t = 2$ seconds.

← Mechanics 1 Secton 3.5

A space rocket experiences **variable acceleration** during launch. The **rate of change of velocity** increases to enable the rocket to escape the gravitational pull of the Earth.

2.1 Functions of time

If the acceleration of a moving particle is **variable**, it changes with time and can be expressed as a **function** of time.

In the same way, velocity and displacement can also be expressed as functions of time.

Links Acceleration is the gradient of a velocity–time graph.
← **Mechanics 1 Section 3.2**

These velocity–time graphs represent the motion of a particle travelling in a straight line. They show examples of increasing and decreasing acceleration.

Increasing acceleration

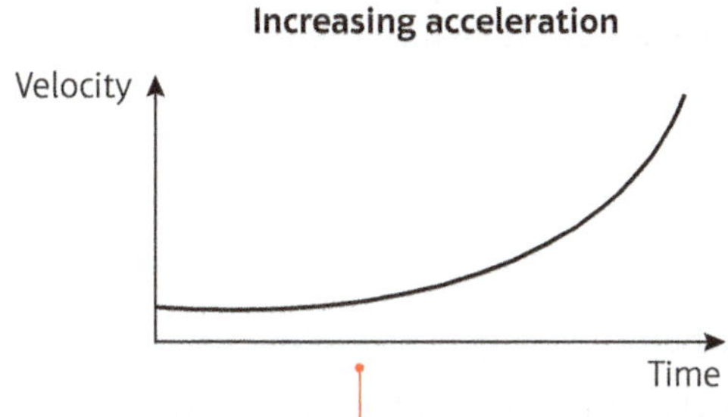

The rate of increase of velocity is increasing with time and the gradient of the curve is increasing.

Decreasing acceleration

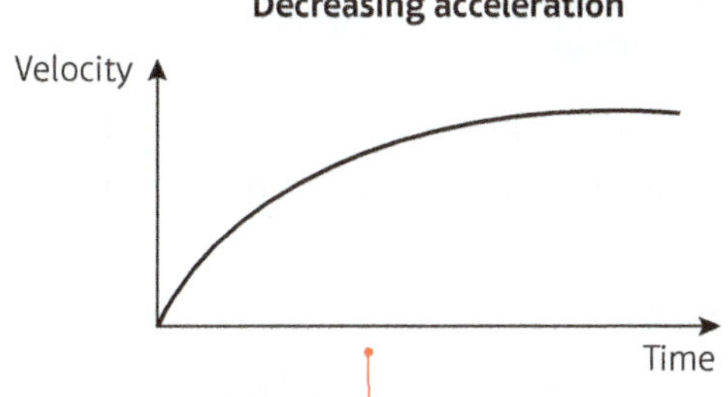

The rate of increase of velocity is decreasing with time and the gradient of the curve is decreasing.

Example 1

A **body** moves in a straight line, such that its displacement, s metres, from a point O at time t seconds is given by $s = 2t^3 - 3t$ for $t > 0$. Find:

a s when $t = 2$ **b** the time taken for the particle to return to O.

a $s = 2 \times 2^3 - 3 \times 2$
$= 16 - 6 = 10$ metres

Substitute $t = 2$ into the equation for s.

b $2t^3 - 3t = 0$

When the particle returns to the starting point, the displacement is equal to zero.

$t(2t^2 - 3) = 0$ either $t = 0$ or $2t^2 = 3$

$\Rightarrow t^2 = \frac{3}{2}$ so $t = \pm\sqrt{\frac{3}{2}}$ seconds

Time taken to return to $O = \sqrt{\frac{3}{2}}$ seconds

Answer is $+\sqrt{\frac{3}{2}}$ as equation is valid only for $t > 0$.

Example 2

A toy train travels along a straight track, leaving the start of the track at time $t = 0$. It then returns to the start of the track. The distance, s metres, from the start of the track at time t seconds is modelled by $s = 4t^2 - t^3$, where $0 \leqslant t \leqslant 4$.

Explain the restriction $0 \leqslant t \leqslant 4$.

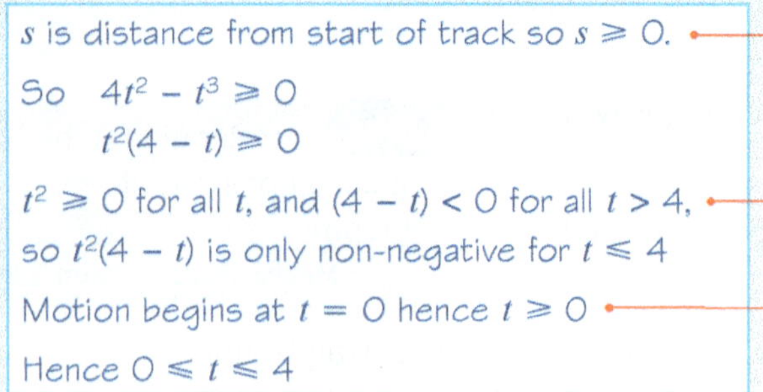

s is distance from start of track so $s \geqslant 0$.

So $4t^2 - t^3 \geqslant 0$

$t^2(4 - t) \geqslant 0$

$t^2 \geqslant 0$ for all t, and $(4 - t) < 0$ for all $t > 4$, so $t^2(4 - t)$ is only non-negative for $t \leqslant 4$

Motion begins at $t = 0$ hence $t \geqslant 0$

Hence $0 \leqslant t \leqslant 4$

Use the initial conditions given.

Distance is a **scalar** quantity and must be $\geqslant 0$.

The restriction $t \geqslant 0$ is due to the motion beginning at $t = 0$, not due to the function.

Problem-solving

You could also sketch the graph of $s = 4t^2 - t^3$ to show the values of t for which the model is valid.

Example 3 SKILLS PROBLEM-SOLVING

A body moves in a straight line such that its velocity, $v\,\text{m s}^{-1}$, at time t seconds is given by $v = 2t^2 - 16t + 24$, for $t \geqslant 0$.

Find:

a the initial velocity

b the values of t when the body is **instantaneously** at rest

c the value of t when the velocity is $64\,\text{m s}^{-1}$

d the greatest speed of the body in the **interval** $0 \leqslant t \leqslant 5$.

a $v = 0 - 0 + 24$

$v = 24\,\text{m s}^{-1}$

The initial velocity means the velocity at $t = 0$.

b $2t^2 - 16t + 24 = 0$

$t^2 - 8t + 12 = 0$

$(t - 6)(t - 2) = 0$

The body is at rest when $t = 2$ seconds and $t = 6$ seconds.

The body is at rest when $v = 0$, so solve the quadratic equation when $v = 0$.

c $2t^2 - 16t + 24 = 64$

$2t^2 - 16t - 40 = 0$

$t^2 - 8t - 20 = 0$

$(t - 10)(t + 2) = 0$

Either $t = 10$ or $t = -2$

Velocity $= 64\,\text{m s}^{-1}$ when $t = 10$ seconds

Rearrange the quadratic equation to make it equal to zero and factorise.

The equation for velocity is valid for $t \geqslant 0$, so $t = -2$ is not a valid solution.

d

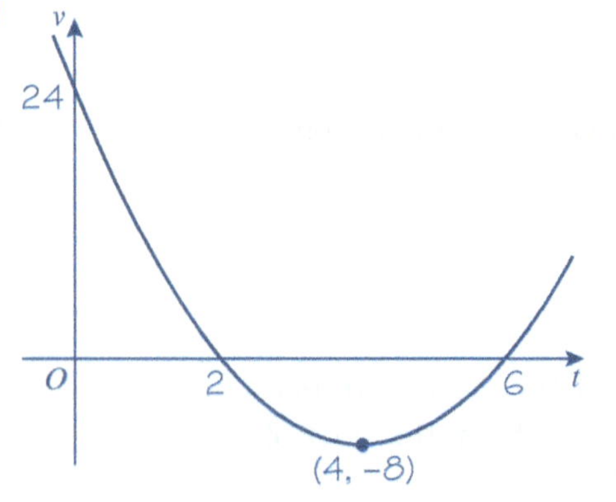

When $t = 4$, $v = 2(4)^2 - 16(4) + 24 = -8$

So in $0 \leqslant t \leqslant 5$ range of v is $-8 \leqslant v \leqslant 24$

Greatest speed is $24\,\text{m s}^{-1}$

Sketch a velocity–time graph for the motion of the body. You can use the **symmetry** of the quadratic curve to determine the position of the turning point. ← Mechanics 1 Section 2.2

Watch out You need to find the greatest **speed**. This could occur when the velocity is positive or negative, so find the range of values taken by v in the interval $0 \leqslant t \leqslant 5$.

Online Explore the solution using technology.

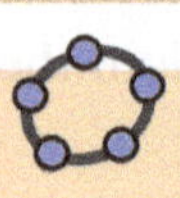

Exercise 2A SKILLS PROBLEM-SOLVING

1 A body moves in a straight line such that its displacement, s metres, at time t seconds is given by $s = 9t - t^3$. Find:

a s when $t = 1$

b the values of t when $s = 0$.

2 A particle P moves on the x-axis. At time t seconds the displacement, s metres, is given by $s = 5t^2 - t^3$. Find:

a the change in displacement between $t = 2$ and $t = 4$

b the change in displacement in the third second.

Hint The third second is the time between $t = 2$ and $t = 3$.

3 A particle moves in a straight line such that its velocity, $v\,\text{m s}^{-1}$, at time t seconds is given by $v = 3 + 5t - t^2$ for $t \geqslant 0$. Find:

a the velocity of the particle when $t = 1$

b the greatest speed of the particle in the interval $0 \leqslant t \leqslant 4$

c the velocity of the particle when $t = 7$ and describe the direction of motion of the particle at this time.

4 At time $t = 0$, a toy car is at point P. It moves in a straight line from point P and then returns to P. Its distance from P, s m, at time t seconds can be modelled by $s = \frac{1}{5}(4t - t^2)$. Find:

a the maximum displacement

b the time taken for the toy car to return to P

c the total distance travelled

d the values of t for which the model is valid.

5 A body moves in a straight line such that its velocity, $v\,\text{m s}^{-1}$, at time t seconds is given by $v = 3t^2 - 10t + 8$, for $t \geqslant 0$. Find:

a the initial velocity

b the values of t when the body is instantaneously at rest

c the values of t when the velocity is $5\,\text{m s}^{-1}$

d the greatest speed of the body in the interval $0 \leqslant t \leqslant 2$.

E 6 A particle P moves on the x-axis. At time t seconds the velocity of P is $v\,\text{m s}^{-1}$ in the direction of x increasing, where $v = 8t - 2t^2$. When $t = 0$, P is at the origin O. Find:

a the time taken for the particle to come to instantaneous rest **(2 marks)**

b the greatest speed of the particle in the interval $0 \leqslant t \leqslant 4$. **(3 marks)**

E 7 At time $t = 0$, a particle moves in a straight horizontal line from a point O, then returns to the starting point. The distance, s metres, from the point O at time t seconds is given by:

$$s = 3t^2 - t^3, \quad 0 \leqslant t \leqslant T$$

Given that the model is valid when $s \geqslant 0$, find the value of T. Explain your answer. **(3 marks)**

E 8 A particle P moves on the x-axis. At time t seconds the velocity of P is $v\,\text{m s}^{-1}$ in the direction of x increasing, where:

$$v = \frac{1}{5}(3t^2 - 10t + 3), \qquad x \geqslant 0$$

a Find the values of t when P is instantaneously at rest. **(3 marks)**

b Determine the greatest speed of P in the interval $0 \leqslant t \leqslant 3$. **(4 marks)**

2.2 Using differentiation

Velocity is the **rate of change** of displacement.

- If the displacement, s, is expressed as a function of t, then the velocity, v, can be expressed as $v = \frac{ds}{dt}$

Links The gradient of a displacement–time graph represents the velocity.
← Mechanics 1 Section 2.1

In the same way, acceleration is the rate of change of velocity.

- If the velocity, v, is expressed as a function of t, then the acceleration, a, can be expressed as $a = \frac{dv}{dt} = \frac{d^2s}{dt^2}$

Links The gradient of a velocity–time graph represents the acceleration.
← Mechanics 1 Section 2.3

$\frac{d^2s}{dt^2}$ is the second **derivative** (or second-order derivative) of s with respect to t.
← Pure 1 Section 8.7

Example 4 SKILLS PROBLEM-SOLVING

A particle P is moving on the x-axis. At time t seconds, the displacement x metres from O is given by $x = t^4 - 32t + 12$. Find:

a the velocity of P when $t = 3$

b the value of t for which P is instantaneously at rest

c the acceleration of P when $t = 1.5$.

a $x = t^4 - 32t + 12$

$v = \frac{dx}{dt} = 4t^3 - 32$ — You find the velocity by differentiating the displacement.

When $t = 3$,

$v = 4 \times 3^3 - 32 = 76$ — To find the velocity when $t = 3$, you substitute $t = 3$ into the expression.

The velocity of P when $t = 3$ is $76\,\text{m s}^{-1}$ in the direction of x increasing.

b $v = 4t^3 - 32 = 0$ — The particle is at rest when $v = 0$. You substitute $v = 0$ into your expression for v and solve the resulting equation to find t.

$t^3 = \frac{32}{4} = 8$

$t = 2$

c $v = 4t^3 - 32$

$a = \frac{dv}{dt} = 12t^2$ — You find the acceleration by differentiating the velocity.

When $t = 1.5$,

$a = 12 \times 1.5^2 = 27$

The acceleration of P when $t = 1.5$ is $27\,\text{m s}^{-2}$.

You can use calculus to determine **maximum** and **minimum** values of displacement, velocity and acceleration.

Example 5 SKILLS PROBLEM-SOLVING

A particle of mass 6 kg is moving on the positive x-axis. At time t seconds, the displacement, s metres, of the particle from the origin is given by $s = 2t^{\frac{3}{2}} + \frac{e^{-2t}}{3}$, where $t \geqslant 0$.

a Find the velocity of the particle when $t = 1.5$.

Given that the particle is acted on by a single force of variable magnitude F N which acts in the direction of the positive x-axis,

b find the value of F when $t = 2$.

a $v = \frac{ds}{dt} = 3t^{\frac{1}{2}} - \frac{2e^{-2t}}{3}$ m s^{-1}

When $t = 1.5$ seconds:

$v = 3 \times 1.5^{0.5} - \frac{2e^{-3}}{3}$

$= 3.64$ m s^{-1} (3 s.f.)

b $a = \frac{dv}{dt} = 1.5t^{-0.5} + \frac{4e^{-2t}}{3}$ m s^{-2}

When $t = 2$ seconds:

$a = 1.5 \times 2^{-0.5} + \frac{4e^{-4}}{3}$

$= 1.0850\ldots$ m s^{-2}

$F = ma = 6 \times 1.0850\ldots = 6.51$ N (3 s.f.)

If $y = ae^{kt}$ then $\frac{dy}{dt} = kae^{kt}$

Differentiate to find an expression for v and substitute $t = 1.5$.

Problem-solving

You know the mass of the particle, so if you find the acceleration you can use $F = ma$ to find the magnitude of the force acting on it. Differentiate the velocity, then substitute $t = 2$ to find the acceleration when $t = 2$ seconds.

Example 6 SKILLS PROBLEM-SOLVING

A child is playing with a yo-yo. The yo-yo leaves the child's hand at time $t = 0$ and travels vertically in a straight line before returning to the child's hand. The distance, s m, of the yo-yo from the child's hand after time t seconds is given by $s = 0.6t + 0.4t^2 - 0.2t^3$, $0 \leqslant t \leqslant 3$.

a Justify the restriction $0 \leqslant t \leqslant 3$.

b Find the maximum distance of the yo-yo from the child's hand, correct to 3 s.f.

a $s = 0.6t + 0.4t^2 - 0.2t^3$

$= 0.2t(3 + 2t - t^2)$

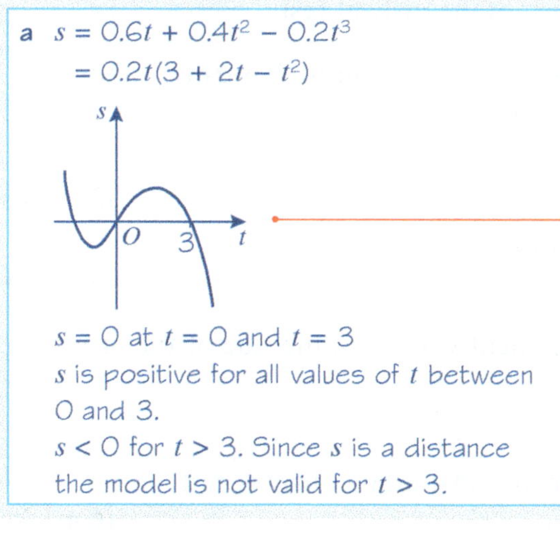

$s = 0$ at $t = 0$ and $t = 3$

s is positive for all values of t between 0 and 3.

$s < 0$ for $t > 3$. Since s is a distance the model is not valid for $t > 3$.

s is a cubic function with a negative coefficient of t^3, and roots at $t = 0$, $t = 3$ and $t = -1$.

← Pure 1 Section 4.1

Comment on the value of s at the limits of the range **and** the behaviour of s within the range.

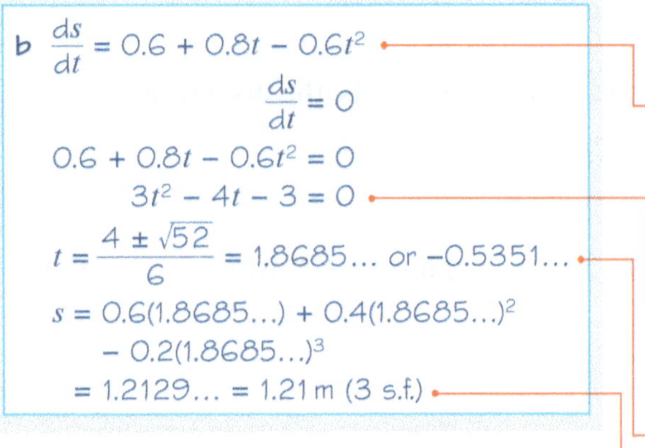

You know from your answer to part **a** that the maximum value of s must occur at the turning point. To find the turning point, differentiate and set $\frac{ds}{dt} = 0$. ← **Pure 2 Section 7.2**

Multiply each term by −5 to obtain an equation with a positive t^2 term and integer coefficients. This makes your working easier.

Use the quadratic formula to solve the equation and take the positive value of t.

Substitute this value of t back into the original equation to find the corresponding value of s. Remember to use unrounded values in your calculation, and check that your answer makes sense in the context of the question.

Exercise 2B SKILLS PROBLEM-SOLVING

1 Find an expression for **i** the velocity and **ii** the acceleration of a particle given that the displacement is given by:

a $s = 4t^4 - \frac{1}{t}$ **b** $x = \frac{2}{3}t^3 + \frac{1}{t^2}$ **c** $s = (3t^2 - 1)(2t + 5)$ **d** $x = \frac{3t^4 - 2t^3 + 5}{2t}$

2 A particle is moving in a straight line. At time t seconds, its displacement, x m, from a fixed point O on the line is given by $x = 2t^3 - 8t$. Find:

a the velocity of the particle when $t = 3$ **b** the acceleration of the particle when $t = 2$.

(P) **3** A particle P is moving on the x-axis. At time t seconds (where $t \geqslant 0$), the velocity of P is $v\,\text{m}\,\text{s}^{-1}$ in the direction of x increasing, where $v = 12 - t - t^2$.

Find the acceleration of P when P is instantaneously at rest.

(P) **4** A particle is moving in a straight line. At time t seconds, its displacement, x m, from a fixed point O on the line is given by $x = 4t^3 - 39t^2 + 120t$.

Find the distance between the two points where P is instantaneously at rest.

(E/P) **5** A particle P moves in a straight line. At time t seconds the acceleration of P is $a\,\text{m}\,\text{s}^{-2}$ and the velocity $v\,\text{m}\,\text{s}^{-1}$ is given by $v = kt - 3t^2$, where k is a constant.

The initial acceleration of P is $4\,\text{m}\,\text{s}^{-2}$.

a Find the value of k. **(3 marks)**

b Using the value of k found in part **a**, find the acceleration when P is instantaneously at rest. **(3 marks)**

(E/P) **6** The print head on a printer moves such that its displacement s cm from the side of the printer at time t seconds is given by:

$\frac{1}{4}(4t^3 - 15t^2 + 12t + 30), 0 \leqslant t \leqslant 3$

Find the distance between the points when the print head is instantaneously at rest, in cm to 1 decimal place. **(6 marks)**

7 A particle P moves in a straight line such that its distance, s m, from a fixed point O at time t is given by:

$s = 0.4t^3 - 0.3t^2 - 1.8t + 5, 0 \leqslant t \leqslant 3$

The diagram shows the displacement–time graph of the motion of P.

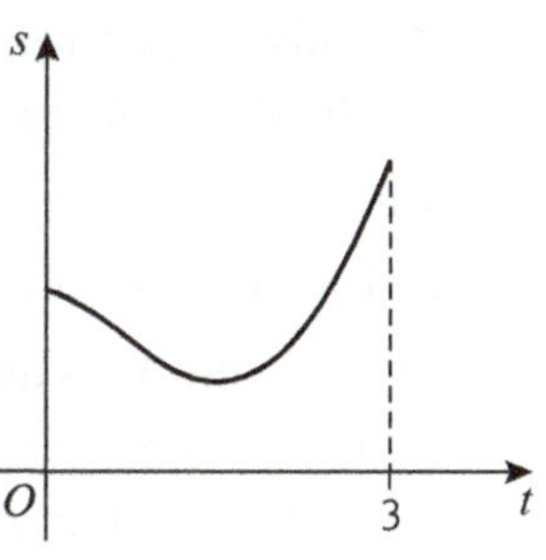

a Determine the time at which P is moving with minimum velocity.

b Find the displacement of P from O at this time.

c Find the velocity of P at this time.

(P) 8 A body starts at rest and moves in a straight line. At time t seconds the displacement of the body from its starting point, s m, is given by:

$s = 4t^3 - t^4, 0 \leqslant t \leqslant 4.$

a Show that the body returns to its starting position at $t = 4$.

b Explain why s is always non-negative.

c Find the maximum displacement of the body from its starting point.

Hint Write $s = t^3(4 - t)$ and consider the sign of each factor in the range $0 \leqslant t \leqslant 4$.

9 At time $t = 0$ a particle P leaves the origin O and moves along the x-axis. At time t seconds the velocity of P is v m s^{-1}, where:

$v = t^2(6 - t)^2, t \geqslant 0$

a Sketch a velocity–time graph for the motion of P.

b Find the maximum value of v and the time at which it occurs.

(P) 10 A particle P moves along the x-axis. Its velocity, v m s^{-1} in the positive x-direction at time t seconds, is given by:

$v = 2t^2 - 3t + 5, t \geqslant 0$

a Show that P never comes to rest.

b Find the minimum velocity of P.

(E/P) 11 A particle P starts at the origin O at time $t = 0$ and moves along the x-axis. At time t seconds the distance of the particle, s m, from the origin is given by:

$s = \frac{9t^2}{2} - t^3, 0 \leqslant t \leqslant 4.5$

a Sketch a displacement–time graph for the motion of P. **(2 marks)**

b Hence justify the restriction $0 \leqslant t \leqslant 4.5$. **(2 marks)**

c Find the maximum distance of the particle from O. **(5 marks)**

d Find the magnitude of the acceleration of the particle at this point. **(3 marks)**

(E/P) 12 A train moves in a straight line along a 4 km test track. The motion of the train is modelled as a particle travelling in a straight line, and the distance, s m, of the train from the start of the track after time t seconds is given by $s = 3.6t + 1.76t^2 - 0.02t^3, 0 \leqslant t \leqslant 90$. Show that the train never reaches the end of the track. **(7 marks)**

13 A body, M, of mass 5 kg moves along the positive x-axis. The displacement, s metres, of the body at time t seconds is given by $s = 3t^{\frac{2}{3}} + 2e^{-3t}$, where $t \geqslant 0$.

Find:

a the velocity of M when $t = 0.5$ **b** the acceleration of M when $t = 3$.

Given that M is acted on by a single force of variable magnitude F N which acts in the direction of the x-axis,

c find the value of F when $t = 3$ seconds.

(P) **14** A particle P moves in a straight line so that, at time t seconds, its displacement, s m, from a fixed point O on the line is given by

$$s = \begin{cases} \frac{1}{2}t, & 0 \leqslant t \leqslant 6 \\ \sqrt{t+3}, & t > 6 \end{cases}$$

Find:

a the velocity of P when $t = 4$ **b** the velocity of P when $t = 22$.

(P) **15** A particle P moves in a straight line so that, at time t seconds, its displacement from a fixed point O on the line is given by

$$s = \begin{cases} 3^t + 3t, & 0 \leqslant t \leqslant 3 \\ 24t - 36, & 3 < t \leqslant 6 \\ -252 + 96t - 6t^2, & t > 6 \end{cases}$$

Find:

a the velocity of P when $t = 2$ **b** the velocity of P when $t = 10$

c the greatest positive displacement of P from O

d the values of s when the speed of P is 18 m s^{-1}.

(E/P) **16** A runner takes part in a race in which competitors have to sprint 200 m in a straight line. At time t seconds after starting, her displacement, s, from the starting position is modelled as:

$$s = k\sqrt{t}, 0 \leqslant t \leqslant T$$

Given that the runner completes the race in 25 seconds,

a find the value of k and the value of T **(2 marks)**

b find the speed of the runner when she crosses the finish line **(3 marks)**

c criticise this model for small values of t. **(2 marks)**

(E/P) **17** A particle P is moving in a straight line. At time t seconds, where $t \geqslant 0$, the acceleration of P is a m s^{-2} and the velocity v m s^{-1} of P is given by

$$v = 2 + 8 \sin kt$$

where k is a constant.

The initial acceleration of P is 4 m s^{-2}.

a Find the value of k. **(3 marks)**

Using the value of k found in part **a**,

b find, in terms of π, the values of t in the interval $0 \leqslant t \leqslant 4\pi$ for which $a = 0$ **(2 marks)**

c show that $4a^2 = 64 - (v - 2)^2$ **(5 marks)**

d find the maximum velocity and the maximum acceleration. **(2 marks)**

2.3 Using integration

Integration is the opposite process to **differentiation**. You can integrate acceleration with respect to time to find velocity, and you can integrate velocity with respect to time to find displacement.

Links The area under a velocity–time graph represents the displacement.
← Mechanics 1 Section 2.2

- Differentiate ↓ / Integrate ↑

displacement $= s = \int v\,\mathrm{d}t$

$\frac{\mathrm{d}s}{\mathrm{d}t} =$ velocity $= v = \int a\,\mathrm{d}t$

$\frac{\mathrm{d}v}{\mathrm{d}t} = \frac{\mathrm{d}^2 s}{\mathrm{d}t^2} =$ acceleration $= a$

Example 7 SKILLS PROBLEM-SOLVING

A particle is moving on the x-axis. At time $t = 0$, the particle is at the point where $x = 5$. The velocity of the particle at time t seconds (where $t \geqslant 0$) is $(6t - t^2)\,\mathrm{m\,s^{-1}}$. Find:

a an expression for the displacement of the particle from O at time t seconds

b the distance of the particle from its starting point when $t = 6$.

a $x = \int v\,\mathrm{d}t$

$= 3t^2 - \frac{t^3}{3} + c$, where c is a constant of integration.

When $t = 0$, $x = 5$

$5 = 3 \times 0^2 - \frac{0^3}{3} + c = c \Rightarrow c = 5$

The displacement of the particle from O after t seconds is $\left(3t^2 - \frac{t^3}{3} + 5\right)$ m.

b Using the result in **a**, when $t = 6$

$x = 3 \times 6^2 - \frac{6^3}{3} + 5 = 41$

The distance from the starting point is $(41 - 5)$ m $= 36$ m.

You integrate the velocity to find the displacement. You must remember to add the constant of integration.
← Pure 1 Section 9.1

This information enables you to find the value of the constant of integration.
← Pure 1 Section 9.3

Example 8

A particle is moving in a straight line with acceleration at time t seconds given by

$a = \cos 2\pi t\ \mathrm{m\,s^{-2}}$, where $t \geqslant 0$

The velocity of the particle at time $t = 0$ is $\frac{1}{2\pi}\,\mathrm{m\,s^{-1}}$. Find:

a an expression for the velocity at time t seconds

b the maximum speed

c the distance travelled in the first 3 seconds.

a $v = \int \cos 2\pi t \, dt$

$= \frac{1}{2\pi}\sin 2\pi t + c$

When $t = 0$, $v = \frac{1}{2\pi}$ so $c = \frac{1}{2\pi}$

$v = \left(\frac{1}{2\pi}\sin 2\pi t + \frac{1}{2\pi}\right) \text{m s}^{-1}$

b Maximum speed $= \frac{1}{2\pi} \times 1 + \frac{1}{2\pi} = \frac{2}{2\pi} = \frac{1}{\pi} \text{m s}^{-1}$

c $s = \frac{1}{2\pi}\int_0^3 (\sin 2\pi t + 1)\, dt$

$= \frac{1}{2\pi}\left[-\frac{1}{2\pi}\cos 2\pi t + t\right]_0^3$

$= \frac{1}{2\pi}\left[\left(-\frac{1}{2\pi} + 3\right) - \left(-\frac{1}{2\pi}\right)\right]$

$= \frac{3}{2\pi}$ m or 0.477 m (3 s.f.)

$\int \cos at \, dt = \frac{1}{a}\sin at + c$

Substitute $t = 0$, $v = \frac{1}{2\pi}$ into the equation to find c.

The maximum value of $\sin x$ is 1.

To find the distance travelled in the first 3 seconds, integrate v between $t = 0$ and $t = 3$:

$\int_0^3 \left(\frac{1}{2\pi}\sin 2\pi t + \frac{1}{2\pi}\right) dt = \frac{1}{2\pi}\int_0^3 (\sin 2\pi t + 1)\, dt$

Example 9

A particle travels in a straight line. After t seconds its velocity, v m s^{-1}, is given by $v = 5 - 3t^2$, $t \geq 0$. Find the distance travelled by the particle in the third second of its motion.

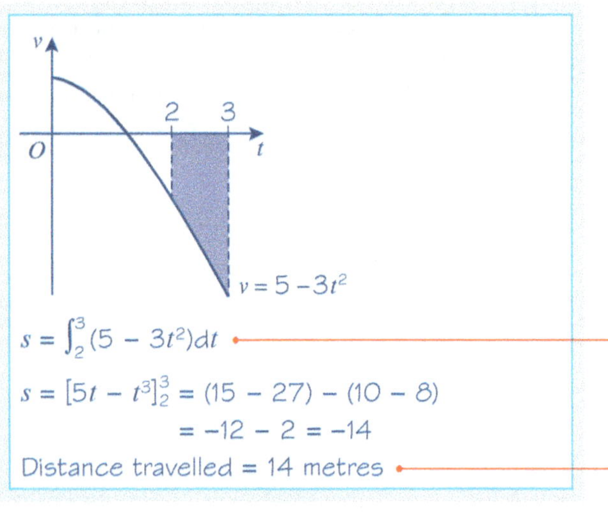

$s = \int_2^3 (5 - 3t^2)\, dt$

$s = [5t - t^3]_2^3 = (15 - 27) - (10 - 8)$

$= -12 - 2 = -14$

Distance travelled = 14 metres

Watch out Before using definite integration to find the distance travelled, check that v doesn't **change sign** in the interval you are considering. A sketch of the velocity–time graph can help.

The distance travelled is the area under the velocity–time graph. Use definite integration to find it. ← Pure 2 Section 8.2

The velocity is negative between $t = 2$ and $t = 3$ so the **displacement** will be negative. You are asked to find the distance travelled so give the positive numerical value of the displacement.

Exercise 2C

SKILLS PROBLEM-SOLVING

1 A particle is moving in a straight line. Given that $s = 0$ when $t = 0$, find an expression for the displacement of the particle if the velocity is given by:

a $v = 3t^2 - 1$ **b** $v = 2t^3 - \frac{3t^2}{2}$ **c** $v = 2\sqrt{t} + 4t^2$

2 A particle is moving in a straight line. Given that $v = 0$ when $t = 0$, find an expression for the velocity of the particle if the acceleration is given by:

a $a = 8t - 2t^2$ **b** $a = 6 + \frac{t^2}{3}$

3 A particle P is moving on the x-axis. At time t seconds, the velocity of P is $(8 + 2t - 3t^2)\text{ m s}^{-1}$ in the direction of x increasing. At time $t = 0$, P is at the point where $x = 4$. Find the distance of P from O when $t = 1$.

4 A particle P is moving on the x-axis. At time t seconds, the acceleration of P is $(16 - 2t)\text{ m s}^{-2}$ in the direction of x increasing. The velocity of P at time t seconds is $v\text{ m s}^{-1}$. When $t = 0$, $v = 6$ and when $t = 3$, $x = 75$. Find:

a v in terms of t **b** the value of x when $t = 0$.

(P) 5 A particle P is moving in a straight line. At time t seconds, its velocity, $v\text{ m s}^{-1}$, is given by $v = 6t^2 - 51t + 90$. When $t = 0$ the displacement is 0. Find the distance between the two points where P is instantaneously at rest.

(P) 6 At time t seconds, where $t \geqslant 0$, the velocity $v\text{ m s}^{-1}$ of a particle P moving in a straight line is given by $v = 12 + t - 6t^2$. When $t = 0$, P is at a point O on the line. Find the distance of P from O when $v = 0$.

(P) 7 A particle P is moving on the x-axis. At time t seconds, the velocity of P is $(4t - t^2)\text{ m s}^{-1}$ in the direction of x increasing. At time $t = 0$, P is at the origin O. Find:

a the value of x at the instant when $t > 0$ and P is at rest

b the total distance moved by P in the interval $0 \leqslant t \leqslant 5$.

Problem-solving

You will need to consider the motion when v is positive and negative separately.

(P) 8 A particle P is moving on the x-axis. At time t seconds, the velocity of P is $(6t^2 - 26t + 15)\text{ m s}^{-1}$ in the direction of x increasing. At time $t = 0$, P is at the origin O. In the subsequent motion P passes through O twice. Find the two non-zero values of t when P passes through O.

(P) 9 A particle P moves along the x-axis. At time t seconds (where $t \geqslant 0$) the velocity of P is $(3t^2 - 12t + 5)\text{ m s}^{-1}$ in the direction of x increasing. When $t = 0$, P is at the origin O. Find:

a the values of t when P is again at O

b the distance travelled by P in the interval $2 \leqslant t \leqslant 3$.

(P) 10 A particle P moves on the x-axis. The acceleration of P at time t seconds, $t \geqslant 0$, is $(4t - 3)\text{ m s}^{-2}$ in the positive x-direction. When $t = 0$, the velocity of P is 4 m s^{-1} in the positive x-direction. When $t = T$ $(T \neq 0)$, the velocity of P is 4 m s^{-1} in the positive x-direction. Find the value of T. **(6 marks)**

(E) 11 A particle P travels in a straight line such that its acceleration at time t seconds is $(t - 3)\text{ m s}^{-2}$. The velocity of P at time t seconds is $v\text{ m s}^{-1}$. When $t = 0$, $v = 4$. Find:

a v in terms of t **(4 marks)**

b the values of t when P is instantaneously at rest **(3 marks)**

c the distance between the two points at which P is instantaneously at rest. **(4 marks)**

(E/P) 12 A particle travels in a straight line such that its acceleration, $a\text{ m s}^{-2}$, at time t seconds is given by $a = 6t + 2$. When $t = 2$ seconds, the displacement, s, is 10 metres and when $t = 3$ seconds the displacement is 38 metres. Find:

a the displacement when $t = 4$ seconds **(6 marks)**

b the velocity when $t = 4$ seconds. **(2 marks)**

Problem-solving

You need to use integration to find expressions for the velocity and displacement, then substitute in the given values. Use simultaneous equations to find the values of the constants of integration.

13 A particle P moves in a straight line. The acceleration, a, of P at time t seconds is given by $a = (1 - \sin \pi t)\,\text{m s}^{-2}$, where $t \geqslant 0$.

When $t = 0$, the velocity of P is $0\,\text{m s}^{-1}$ and its displacement is $0\,\text{m}$. Find expressions for:

a the velocity at time t seconds

b the displacement at time t seconds.

14 A particle moving in a straight line has acceleration a, given by

$$a = \sin 3\pi t\,\text{m s}^{-2},\ t \geqslant 0$$

At time t seconds the particle has velocity $v\,\text{m s}^{-1}$ and displacement $s\,\text{m}$. Given that when $t = 0$, $v = \dfrac{1}{3\pi}$ and $s = 1$, find:

a an expression for v in terms of t

b the maximum speed of the particle

c an expression for s in terms of t.

(P) **15** An object moves in a straight line from a point O. At time t seconds the object has acceleration, a, where

$$a = -\cos 4\pi t\,\text{m s}^{-2},\ 0 \leqslant t \leqslant 4$$

When $t = 0$, the velocity of the object is $0\,\text{m s}^{-1}$ and its displacement is $0\,\text{m}$. Find:

a an expression for the velocity at time t seconds

b the maximum speed of the object

c an expression for the displacement of the object at time t seconds

d the maximum distance of the object from O

e the number of times the object changes direction during its motion.

Problem-solving

In part **e**, consider the number of times the velocity changes sign.

(P) **16** A particle moves in a straight line. At time t seconds after it begins its motion, the acceleration of the particle is $3\sqrt{t}\,\text{m s}^{-2}$ where $t > 0$.

Given that after 1 second the particle is moving with velocity $2\,\text{m s}^{-1}$, find the time taken for the particle to travel 16 m.

(E/P) **17** A particle P moves on the x-axis. At time t seconds the velocity of P is $v\,\text{m s}^{-1}$ in the direction of x increasing, where v is given by

$$v = \begin{cases} 10t - 2t^{\frac{3}{2}}, & 0 \leqslant t \leqslant 4 \\ 24 - \left(\dfrac{t-4}{2}\right)^4, & t > 4 \end{cases}$$

When $t = 0$, P is at the origin O.

Find:

a the greatest speed of P in the interval $0 \leqslant t \leqslant 4$ **(4 marks)**

b the distance of P from O when $t = 4$ **(3 marks)**

c the time at which P is instantaneously at rest for $t > 4$ **(1 mark)**

d the total distance travelled by P in the first 10 seconds of its motion. **(7 marks)**

Challenge

The motion of a robotic arm moving along a straight track is modelled using the equations:

$$v = \frac{t^2}{2} + 2, 0 \leqslant t \leqslant k \text{ and } v = 10 + \frac{t}{3} - \frac{t^2}{12}, k \leqslant t \leqslant 10$$

The diagram shows a sketch of the velocity–time graph of the motion of the arm.

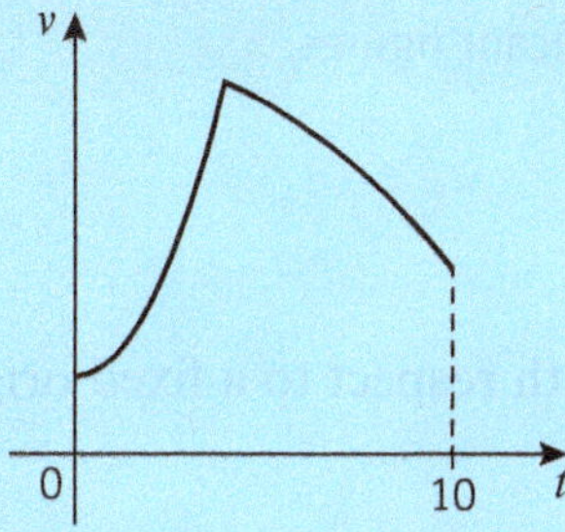

Work out the total distance travelled by the robotic arm.

2.4 Differentiating vectors

You can use calculus with vectors to solve problems involving motion in two dimensions with variable acceleration.

To differentiate a vector quantity in the form $\mathrm{f}(t)\mathbf{i} + \mathrm{g}(t)\mathbf{j}$ you differentiate each function of time separately.

- If $\mathbf{r} = x\mathbf{i} + y\mathbf{j}$, then $\mathbf{v} = \frac{\mathrm{d}\mathbf{r}}{\mathrm{d}t} = \dot{\mathbf{r}} = \dot{x}\mathbf{i} + \dot{y}\mathbf{j}$

  and $\mathbf{a} = \frac{\mathrm{d}\mathbf{v}}{\mathrm{d}t} = \frac{\mathrm{d}^2\mathbf{r}}{\mathrm{d}t^2} = \ddot{\mathbf{r}} = \ddot{x}\mathbf{i} + \ddot{y}\mathbf{j}$

Notation Dot notation is a short-hand for differentiation with respect to time:

$\dot{x} = \frac{\mathrm{d}x}{\mathrm{d}t}$ and $\dot{y} = \frac{\mathrm{d}y}{\mathrm{d}t}$

$\ddot{x} = \frac{\mathrm{d}^2x}{\mathrm{d}t^2}$ and $\ddot{y} = \frac{\mathrm{d}^2y}{\mathrm{d}t^2}$

Example 10 SKILLS PROBLEM-SOLVING

A particle P of mass 0.8 kg is acted on by a single force $\mathbf{F}$ N. Relative to a fixed origin O, the position vector of P at time t seconds is $\mathbf{r}$ metres, where

$$\mathbf{r} = 2t^3\mathbf{i} + 50t^{-\frac{1}{2}}\mathbf{j}, t \geqslant 0$$

Find:

a the speed of P when $t = 4$

b the acceleration of P as a vector when $t = 2$

c $\mathbf{F}$ when $t = 2$.

a $\mathbf{v} = \dot{\mathbf{r}} = (6t^2\mathbf{i} - 25t^{-\frac{3}{2}}\mathbf{j})\,\text{m s}^{-1}$
When $t = 4$: $\mathbf{v} = \left(96\mathbf{i} - \frac{25}{8}\mathbf{j}\right)\text{m s}^{-1}$
Speed $= \sqrt{96^2 + \left(\frac{25}{8}\right)^2} = 96.1\,\text{m s}^{-1}$ (3 s.f.)

Differentiate $2t^3$ and $50t^{-\frac{1}{2}}$ separately to find the **i**- and **j**-components of the velocity.

The speed is the magnitude of **v**.

b $\mathbf{a} = \ddot{\mathbf{r}} = \left(12t\mathbf{i} + \frac{75}{2}t^{-\frac{5}{2}}\mathbf{j}\right)\text{m s}^{-2}$
When $t = 2$: $\mathbf{a} = 24\mathbf{i} + 6.6291\ldots\mathbf{j}\,\text{m s}^{-2}$

$\mathbf{a} = \ddot{\mathbf{r}} = \frac{d^2\mathbf{r}}{dt^2}$ or alternatively, $\mathbf{a} = \dot{\mathbf{v}} = \frac{d\mathbf{v}}{dt}$

c $\mathbf{F} = m\mathbf{a} = 0.8(24\mathbf{i} + 6.6291\ldots\mathbf{j})$
$= (19.2\mathbf{i} + 5.30\mathbf{j})\,\text{N}$ (3 s.f.)

Use $\mathbf{F} = m\mathbf{a}$ and round each coefficient to 3 significant figures.

Exercise 2D SKILLS PROBLEM-SOLVING

1 At time t seconds, a particle P has position vector $\mathbf{r}$ m with respect to a fixed origin O, where

$\mathbf{r} = (3t - 4)\mathbf{i} + (t^3 - 4t)\mathbf{j}$, $t \geqslant 0$

Find:

a the velocity of P when $t = 3$

b the acceleration of P when $t = 3$.

2 A particle P of mass 3 grams moving in a plane is acted on by a force **F** N. Its velocity at time t seconds is given by $\mathbf{v} = (t^2\mathbf{i} + (2t - 3)\mathbf{j})\,\text{m s}^{-1}$, $t \geqslant 0$.

Find **F** when $t = 4$.

(P) 3 In this question, **i** and **j** are the unit vectors east and north respectively.

A particle P is moving in a plane. At time t seconds, the position vector of P, $\mathbf{r}$ m, relative to a fixed origin O is given by $\mathbf{r} = 5e^{-3t}\mathbf{i} + 2\mathbf{j}$, $t \geqslant 0$.

a Find the time at which the particle is directly north-east of O.

b Find the speed of the particle at this time.

c Explain why the particle is always moving directly west.

(E) 4 At time t seconds, a particle P has position vector $\mathbf{r}$ m with respect to a fixed origin O, where

$\mathbf{r} = 4t^2\mathbf{i} + (24t - 3t^2)\mathbf{j}$, $t \geqslant 0$

a Find the speed of P when $t = 2$. **(3 marks)**

b Show that the acceleration of P is a constant and find the magnitude of this acceleration. **(3 marks)**

(E) 5 A particle P is initially at a fixed origin O. At time $t = 0$, P is projected from O and moves so that, at time t seconds after projection, its position vector $\mathbf{r}$ m relative to O is given by

$\mathbf{r} = (t^3 - 12t)\mathbf{i} + (4t^2 - 6t)\mathbf{j}$, $t \geqslant 0$

Find:

a the speed of projection of P **(5 marks)**

b the value of t at the instant when P is moving parallel to **j** **(3 marks)**

c the position vector of P at the instant when P is moving parallel to **j**. **(3 marks)**

The motion of the particle is due to it being acted on by a single variable force, **F** N.

d Given that the mass of the particle is 0.5 kg, find the magnitude of **F** when $t = 5$ s. **(4 marks)**

E/P 6 A particle P is moving in a plane. At time t seconds, the position vector of P, $\mathbf{r}$ m, is given by $\mathbf{r} = (3t^2 - 6t + 4)\mathbf{i} + (t^3 + kt^2)\mathbf{j}$, where k is a constant.

When $t = 3$, the speed of P is $12\sqrt{5}$ m s^{-1}.

a Find the two possible values of k. **(6 marks)**

b For each of these values of k, find the magnitude of the acceleration of P when $t = 1.5$. **(4 marks)**

E 7 Relative to a fixed origin O, the position vector of a particle P at time t seconds is $\mathbf{r}$ metres, where

$$\mathbf{r} = 6t^2\mathbf{i} + t^{\frac{5}{2}}\mathbf{j},\ t \geqslant 0$$

At the instant when $t = 4$, find:

a the speed of P **(5 marks)**

b the acceleration of P, giving your answer as a vector. **(2 marks)**

E/P 8 A particle P moves in a horizontal plane. At time t seconds, the position vector of P is $\mathbf{r}$ metres relative to a fixed origin O where $\mathbf{r}$ is given by

$$\mathbf{r} = (18t - 4t^3)\mathbf{i} + ct^2\mathbf{j},\ t \geqslant 0,$$

where c is a positive constant. When $t = 1.5$, the speed of P is 15 m s^{-1}. Find:

a the value of c **(6 marks)**

b the acceleration of P when $t = 1.5$. **(3 marks)**

E 9 At time t seconds, a particle P has position vector $\mathbf{r}$ metres relative to a fixed origin O, where

$$\mathbf{r} = (2t^2 - 3t)\mathbf{i} + (5t + t^2)\mathbf{j},\ t \geqslant 0$$

Show that the acceleration of P is constant and find its magnitude. **(5 marks)**

E/P 10 A particle P moves in a horizontal plane. At time t seconds, the position vector of P is $\mathbf{r}$ metres relative to a fixed origin O, and $\mathbf{r}$ is given by $\mathbf{r} = (20t - 2t^3)\mathbf{i} + kt^2\mathbf{j}$, $t \geqslant 0$, where k is a positive constant. When $t = 2$, the speed of P is 16 m s^{-1}. Find:

a the value of k **(6 marks)**

b the acceleration of P at the instant when it is moving parallel to $\mathbf{j}$. **(4 marks)**

2.5 Integrating vectors

You can integrate vectors in the form $f(t)\mathbf{i} + g(t)\mathbf{j}$ by integrating each function of time separately.

- $\mathbf{v} = \int \mathbf{a}\,dt$ and $\mathbf{r} = \int \mathbf{v}\,dt$

Watch out When you integrate a vector, the constant of integration will also be a vector. Write it in the form $\mathbf{c} = p\mathbf{i} + q\mathbf{j}$.

Example 11 SKILLS PROBLEM-SOLVING

A particle P is moving in a plane. At time t seconds, its velocity $\mathbf{v}$ m s^{-1} is given by

$$\mathbf{v} = 3t\mathbf{i} + \tfrac{1}{2}t^2\mathbf{j},\ t \geqslant 0$$

When $t = 0$, the position vector of P with respect to a fixed origin O is $(2\mathbf{i} - 3\mathbf{j})$ m. Find the position vector of P at time t seconds.

$\mathbf{r} = \int \mathbf{v}\,dt = \int \left(3t\mathbf{i} + \frac{1}{2}t^2\mathbf{j}\right)dt$

$= \frac{3t^2}{2}\mathbf{i} + \frac{t^3}{6}\mathbf{j} + \mathbf{c}$

When $t = 0$, $\mathbf{r} = 2\mathbf{i} - 3\mathbf{j}$:

$2\mathbf{i} - 3\mathbf{j} = 0\mathbf{i} + 0\mathbf{j} + \mathbf{c}$

$\mathbf{c} = 2\mathbf{i} - 3\mathbf{j}$

Hence

$\mathbf{r} = \frac{3t^2}{2}\mathbf{i} + \frac{t^3}{6}\mathbf{j} + 2\mathbf{i} - 3\mathbf{j} = \left(\frac{3t^2}{2} + 2\right)\mathbf{i} + \left(\frac{t^3}{6} - 3\right)\mathbf{j}$

The position vector of P at time t seconds is $\left(\left(\frac{3t^2}{2} + 2\right)\mathbf{i} + \left(\frac{t^3}{6} - 3\right)\mathbf{j}\right)$ m.

You integrate $3t$ and $\frac{1}{2}t^2$ in the usual way, using $\int t^n\,dt = \frac{t^{n+1}}{n+1}$. You must include the constant of integration, which is a vector, **c**.

You are given an **initial condition** (or **boundary condition**) which allows you to find **c**. Substitute $t = 0$ and $\mathbf{r} = 2\mathbf{i} - 3\mathbf{j}$ into the integrated expression and solve to find **c**.

← **Pure 1 Section 9.3**

Collect together the terms in **i** and **j** to complete your answer.

Example 12

A particle P is moving in a plane so that, at time t seconds, its acceleration is $(4\mathbf{i} - 2t\mathbf{j})\,\text{m s}^{-2}$. When $t = 3$, the velocity of P is $6\mathbf{i}\,\text{m s}^{-1}$ and the position vector of P is $(20\mathbf{i} + 3\mathbf{j})$ m with respect to a fixed origin O. Find:

a the angle between the direction of motion of P and **i** when $t = 2$

b the distance of P from O when $t = 0$.

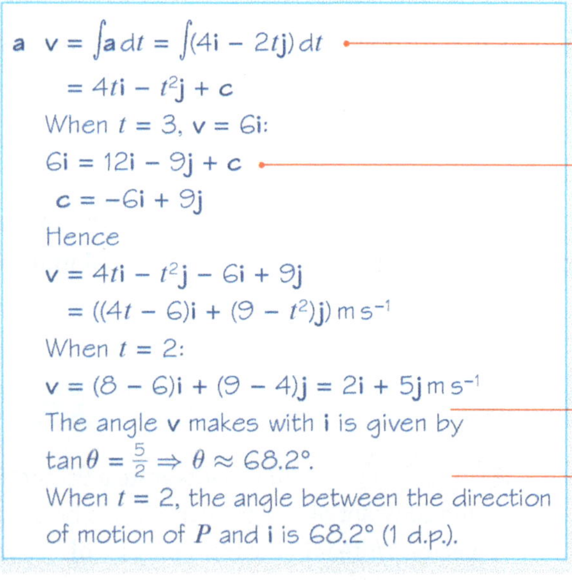

a $\mathbf{v} = \int \mathbf{a}\,dt = \int (4\mathbf{i} - 2t\mathbf{j})\,dt$

$= 4t\mathbf{i} - t^2\mathbf{j} + \mathbf{c}$

When $t = 3$, $\mathbf{v} = 6\mathbf{i}$:

$6\mathbf{i} = 12\mathbf{i} - 9\mathbf{j} + \mathbf{c}$

$\mathbf{c} = -6\mathbf{i} + 9\mathbf{j}$

Hence

$\mathbf{v} = 4t\mathbf{i} - t^2\mathbf{j} - 6\mathbf{i} + 9\mathbf{j}$

$= ((4t - 6)\mathbf{i} + (9 - t^2)\mathbf{j})\,\text{m s}^{-1}$

When $t = 2$:

$\mathbf{v} = (8 - 6)\mathbf{i} + (9 - 4)\mathbf{j} = 2\mathbf{i} + 5\mathbf{j}\,\text{m s}^{-1}$

The angle **v** makes with **i** is given by

$\tan\theta = \frac{5}{2} \Rightarrow \theta \approx 68.2°$.

When $t = 2$, the angle between the direction of motion of P and **i** is 68.2° (1 d.p.).

The direction of motion of P is the direction of the velocity vector of P. Your first step is to find the velocity by integrating the acceleration.

You then use the fact that the velocity is $6\mathbf{i}\,\text{m s}^{-1}$ when $t = 3$ to find the constant of integration.

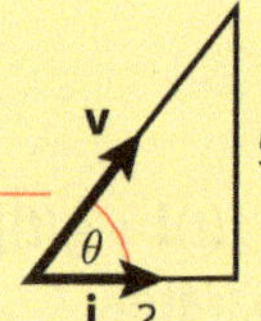

You find the angle the velocity vector makes with **i** using trigonometry.

b $\mathbf{r} = \int \mathbf{v}\,dt = \int((4t - 6)\mathbf{i} + (9 - t^2)\mathbf{j})\,dt$
$= (2t^2 - 6t)\mathbf{i} + \left(9t - \frac{t^3}{3}\right)\mathbf{j} + \mathbf{d}$
When $t = 3$, $\mathbf{r} = 20\mathbf{i} + 3\mathbf{j}$:
$20\mathbf{i} + 3\mathbf{j} = (18 - 18)\mathbf{i} + (27 - 9)\mathbf{j} + \mathbf{d}$
$= 18\mathbf{j} + \mathbf{d}$
$\mathbf{d} = 20\mathbf{i} - 15\mathbf{j}$
Hence
$\mathbf{r} = \left((2t^2 - 6t)\mathbf{i} + \left(9t - \frac{t^3}{3}\right)\mathbf{j} + 20\mathbf{i} - 15\mathbf{j}\right)$ m
When $t = 0$, $\mathbf{r} = (20\mathbf{i} - 15\mathbf{j})$ m:
$OP = |20\mathbf{i} - 15\mathbf{j}| = \sqrt{20^2 + 15^2} = 25$ m
When $t = 0$, the distance of P from O is 25 m.

You find the position vector by integrating the velocity vector. Remember to include the constant of integration.

The constant of integration is a vector. This constant is different from the constant in part **a** so you should give it a different letter.

Watch out Read the question carefully to work out whether you need to find a vector or a scalar quantity. The **distance** from O is the magnitude of the displacement vector, so use Pythagoras' Theorem.

Example 13

The velocity of a particle P at time t seconds is $((3t^2 - 8)\mathbf{i} + 5\mathbf{j})$ m s^{-1}. When $t = 0$, the position vector of P with respect to a fixed origin O is $(2\mathbf{i} - 4\mathbf{j})$ m.

a Find the position vector of P after t seconds.

A second particle Q moves with constant velocity $(8\mathbf{i} + 4\mathbf{j})$ m s^{-1}. When $t = 0$, the position vector of Q with respect to the fixed origin O is $2\mathbf{i}$ m.

b Prove that P and Q collide.

a Let the position vector of P after t seconds be $\mathbf{p}$ metres.
$\mathbf{p} = \int \mathbf{v}\,dt = \int((3t^2 - 8)\mathbf{i} + 5\mathbf{j})\,dt$
$= (t^3 - 8t)\mathbf{i} + 5t\mathbf{j} + \mathbf{c}$
When $t = 0$, $\mathbf{p} = 2\mathbf{i} - 4\mathbf{j}$:
$2\mathbf{i} - 4\mathbf{j} = 0\mathbf{i} + 0\mathbf{j} + \mathbf{c} \Rightarrow \mathbf{c} = 2\mathbf{i} - 4\mathbf{j}$
Hence
$\mathbf{p} = (t^3 - 8t)\mathbf{i} + 5t\mathbf{j} + 2\mathbf{i} - 4\mathbf{j}$
$= (t^3 - 8t + 2)\mathbf{i} + (5t - 4)\mathbf{j}$
The position vector of P after t seconds is $((t^3 - 8t + 2)\mathbf{i} + (5t - 4)\mathbf{j})$ m.

There are two position vectors in this question and to write them both as **r** m would be confusing. It is sensible to write the position vector of P as **p** m and the position vector of Q as **q** m.

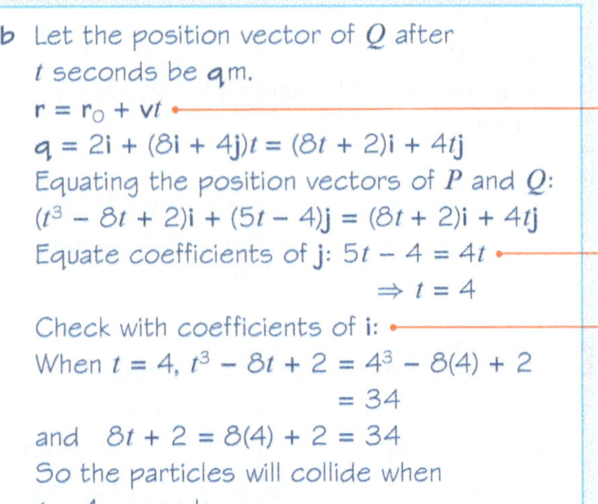

b Let the position vector of Q after t seconds be $\mathbf{q}$ m.

$\mathbf{r} = \mathbf{r}_0 + \mathbf{v}t$

$\mathbf{q} = 2\mathbf{i} + (8\mathbf{i} + 4\mathbf{j})t = (8t + 2)\mathbf{i} + 4t\mathbf{j}$

Equating the position vectors of P and Q:

$(t^3 - 8t + 2)\mathbf{i} + (5t - 4)\mathbf{j} = (8t + 2)\mathbf{i} + 4t\mathbf{j}$

Equate coefficients of $\mathbf{j}$: $5t - 4 = 4t$

$\Rightarrow t = 4$

Check with coefficients of $\mathbf{i}$:

When $t = 4$, $t^3 - 8t + 2 = 4^3 - 8(4) + 2 = 34$

and $8t + 2 = 8(4) + 2 = 34$

So the particles will collide when $t = 4$ seconds.

Use the equation for the position vector of a particle moving with constant velocity. You could also integrate $8\mathbf{i} + 4\mathbf{j}$ with the boundary condition $\mathbf{q} = 2\mathbf{i}$ when $t = 0$.

Problem-solving

Equate the position vectors for each particle. If they collide there will be a single value of t for which $\mathbf{p} = \mathbf{q}$. This means that the coefficients of $\mathbf{i}$ will be equal **and** the coefficients of $\mathbf{j}$ will be equal.

The coefficient of $\mathbf{i}$ involves a t^3 term so it is easier to start by equating the $\mathbf{j}$ components.

Now check $\mathbf{i}$ as well, as the particles only collide if **both** coefficients match.

Exercise 2E SKILLS PROBLEM-SOLVING

(E) **1** A particle P starts from rest at a fixed origin O. The acceleration of P at time t seconds (where $t \geqslant 0$) is $(6t^2\mathbf{i} + (8 - 4t^3)\mathbf{j})\,\mathrm{m\,s^{-2}}$. Find:

a the velocity of P when $t = 2$ **(3 marks)**

b the position vector of P when $t = 4$. **(3 marks)**

(E) **2** A particle P is moving in a plane with velocity $\mathbf{v}\,\mathrm{m\,s^{-1}}$ at time t seconds where

$$\mathbf{v} = (3t^2 + 2)\mathbf{i} + (6t - 4)\mathbf{j},\ t \geqslant 0$$

When $t = 2$, P has position vector $9\mathbf{j}$ m with respect to a fixed origin O. Find:

a the distance of P from O when $t = 0$ **(4 marks)**

b the acceleration of P at the instant when it is moving parallel to the vector $\mathbf{i}$. **(4 marks)**

(E) **3** At time t seconds, where $t \geqslant 0$, a particle P is moving in a plane with velocity $\mathbf{v}\,\mathrm{m\,s^{-1}}$ and acceleration $\mathbf{a}\,\mathrm{m\,s^{-2}}$, where $\mathbf{a} = (2t - 4)\mathbf{i} + 6\sin t\mathbf{j}$.

Given that P is instantaneously at rest when $t = \frac{\pi}{2}$ seconds, find:

a $\mathbf{v}$ in terms of π and t **(5 marks)**

b the exact speed of P when $t = \frac{3\pi}{2}$. **(3 marks)**

(E/P) **4** At time t seconds (where $t \geqslant 0$), a particle P is moving in a plane with acceleration $\mathbf{a}\,\mathrm{m\,s^{-2}}$, where

$$\mathbf{a} = (5t - 3)\mathbf{i} + (8 - t)\mathbf{j}$$

When $t = 0$, the velocity of P is $(2\mathbf{i} - 5\mathbf{j})\,\mathrm{m\,s^{-1}}$. Find:

a the velocity of P after t seconds **(3 marks)**

b the value of t for which P is moving parallel to $\mathbf{i} - \mathbf{j}$ **(4 marks)**

c the speed of P when it is moving parallel to $\mathbf{i} - \mathbf{j}$. **(3 marks)**

5 At time t seconds (where $t \geqslant 0$), a particle P is moving in a plane with acceleration $(2\mathbf{i} - 2t\mathbf{j})\,\text{m s}^{-2}$. When $t = 0$, the velocity of P is $2\mathbf{j}\,\text{m s}^{-1}$ and the position vector of P is $6\mathbf{i}\,\text{m}$ with respect to a fixed origin P.

a Find the position vector of P at time t seconds. **(5 marks)**

At time t seconds (where $t \geqslant 0$), a second particle Q is moving in the plane with velocity $((3t^2 - 4)\mathbf{i} - 2t\mathbf{j})\,\text{m s}^{-1}$. The particles collide when $t = 3$.

b Find the position vector of Q at time $t = 0$. **(4 marks)**

E **6** At time $t = 0$, a particle P is at rest at a point with position vector $(4\mathbf{i} - 6\mathbf{j})\,\text{m}$ with respect to a fixed origin O. The acceleration of P at time t seconds (where $t \geqslant 0$) is $((4t - 3)\mathbf{i} - 6t^2\mathbf{j})\,\text{m s}^{-2}$. Find:

a the velocity of P when $t = \frac{1}{2}$ **(5 marks)**

b the position vector of P when $t = 6$. **(5 marks)**

E **7** At time t seconds (where $t \geqslant 0$) a particle P is moving in a plane with acceleration $\mathbf{a}\,\text{m s}^{-2}$, where $\mathbf{a} = (8t^3 - 6t)\mathbf{i} + (8t - 3)\mathbf{j}$.

When $t = 2$, the velocity of P is $(16\mathbf{i} + 3\mathbf{j})\,\text{m s}^{-1}$. Find:

a the velocity of P after t seconds **(4 marks)**

b the value of t when P is moving parallel to $\mathbf{i}$. **(3 marks)**

E/P **8** At time t seconds the velocity of a particle P is $((4t - 3)\mathbf{i} + 4\mathbf{j})\,\text{m s}^{-1}$.
When $t = 0$, the position vector of P is $(\mathbf{i} + 2\mathbf{j})$ m, relative to a fixed origin O.

a Find an expression for the position vector of P at time t seconds. **(4 marks)**

A second particle Q moves with constant velocity $(5\mathbf{i} + k\mathbf{j})\,\text{m s}^{-1}$ relative to the fixed origin O. When $t = 0$, the position vector of Q is $(11\mathbf{i} + 5\mathbf{j})\,\text{m}$.

b Given that the particles P and Q collide, find:

i the value of k

ii the position vector of the point of collision. **(6 marks)**

Challenge

A particle P is moving in a plane. At time t seconds, P is moving with velocity $\mathbf{v}\,\text{m s}^{-1}$, where $\mathbf{v} = 3t\cos t\,\mathbf{i} + 5t\mathbf{j}$. Given that P is initially at the point with position vector $4\mathbf{i} + \mathbf{j}$ m relative to a fixed origin O, find the position vector of P when $t = \frac{\pi}{2}$.

2.6 Constant acceleration formulae

You can use calculus to derive the formulae for motion with constant acceleration.

Example 14 SKILLS REASONING/ARGUMENTATION

A particle moves in a straight line with constant acceleration, $a\,\text{m s}^{-2}$. Given that its initial velocity is $u\,\text{m s}^{-1}$ and its initial displacement is 0 m, show that:

a its velocity, $v\,\text{m s}^{-1}$, at time t s is given by $v = u + at$

b its displacement, s m, at time t s is given by $s = ut + \frac{1}{2}at^2$.

a $v = \int a \, dt$

$= at + c$

When $t = 0$, $v = u$,

so $u = a \times 0 + c = c$ — Use the initial condition you are given for the velocity to work out the value of c.

So $v = u + at$

b $s = \int v \, dt$

$= \int (u + at) \, dt$ — Use the equation for velocity you have just proved.

$= ut + \frac{1}{2}at^2 + c$

When $t = 0$, $s = 0$ — Use the initial condition you are given for the displacement to work out the value of c.

so $0 = u \times 0 + \frac{1}{2} \times a \times 0^2 + c$

$c = 0$

So $s = ut + \frac{1}{2}at^2$

Watch out The ***suvat* equations** can be used only when the acceleration is constant.

Exercise 2F SKILLS REASONING/ARGUMENTATION

(P) **1** A particle moves on the x-axis with constant acceleration $a\,\text{m s}^{-2}$. The particle has initial velocity 0 and initial displacement x m. After time t seconds the particle has velocity $v\,\text{m s}^{-1}$ and displacement s m.
Prove that $s = \frac{1}{2}at^2 + x$.

2 A particle moves in a straight line with constant acceleration $5\,\text{m s}^{-2}$.

a Given that its initial velocity is $12\,\text{m s}^{-1}$, use calculus to show that its velocity at time t s is given by $v = 12 + 5t$.

b Given that the initial displacement of the particle is 7 m, show that $s = 12t + 2.5t^2 + 7$.

(P) **3** A particle moves in a straight line from a point O. At time t seconds, its displacement, s m, from P is given by $s = ut + \frac{1}{2}at^2$ where u and a are constants. Prove that the particle moves with constant acceleration a.

4 Which of these equations for displacement describe constant acceleration? Explain your answers.

A $s = 2t^2 - t^3$ **B** $s = 4t + 7$ **C** $s = \frac{t^2}{4}$ **D** $s = 3t - \frac{2}{t^2}$ **E** $s = 6$

(E/P) **5** A particle moves in a straight line with constant acceleration. The initial velocity of the particle is $5\,\text{m s}^{-1}$ and after 2 seconds it is moving with velocity $13\,\text{m s}^{-1}$.

a Find the acceleration of the particle. **(3 marks)**

b Without making use of the **kinematics** formulae, show that the displacement, s m, of the particle from its starting position is given by:

$$s = pt^2 + qt + r, \; t \geqslant 0$$

where p, q and r are constants to be found. **(5 marks)**

Watch out An exam question might specify that you cannot use certain formulae or techniques. In this case you need to use calculus to find the answer to part **b**.

E/P **6** A train travels along a straight track, passing point A at time $t = 0$ and passing point B 40 seconds later. Its distance from A at time t seconds is given by:

$s = 25t - 0.2t^2, 0 \leqslant t \leqslant 40$

a Find the distance AB. **(1 mark)**

b Show that the train travels with constant acceleration. **(3 marks)**

A bird passes point B at time $t = 0$ at an initial velocity towards A of $7\,\text{m}\,\text{s}^{-1}$. It flies in a straight line towards point A with constant acceleration $0.6\,\text{m}\,\text{s}^{-2}$.

c Find the distance from A at which the bird is directly above the train. **(6 marks)**

Chapter review 2

1 A particle P moves in a horizontal straight line. At time t seconds (where $t \geqslant 0$) the velocity $v\,\text{m}\,\text{s}^{-1}$ of P is given by $v = 15 - 3t$. Find:

a the value of t when P is instantaneously at rest

b the distance travelled by P between the time when $t = 0$ and the time when P is instantaneously at rest.

2 A particle P moves along the x-axis so that, at time t seconds, the displacement of P from O is x metres and the velocity of P is $v\,\text{m}\,\text{s}^{-1}$, where:

$v = 6t + \frac{1}{2}t^3$

a Find the acceleration of P when $t = 4$.

b Given also that $x = -5$ when $t = 0$, find the distance OP when $t = 4$.

P **3** A particle P is moving along a straight line. At time $t = 0$, the particle is at a point A and is moving with velocity $8\,\text{m}\,\text{s}^{-1}$ towards a point B on the line, where $AB = 30\,\text{m}$. At time t seconds (where $t \geqslant 0$), the acceleration of P is $(2 - 2t)\,\text{m}\,\text{s}^{-2}$ in the direction $\overrightarrow{AB}$.

a Find an expression, in terms of t, for the displacement of P from A at time t seconds.

b Show that P does not reach B.

c Find the value of t when P returns to A, giving your answer to 3 significant figures.

d Find the total distance travelled by P in the interval between the two instants when it passes through A.

E **4** A particle starts from rest at a point O and moves along a straight line OP with an acceleration, a, after t seconds given by $a = (8 - 2t^2)\,\text{m}\,\text{s}^{-2}$. Find:

a the greatest speed of the particle as it moves in the direction OP **(5 marks)**

b the distance covered by the particle in the first two seconds of its motion. **(4 marks)**

5 A particle P passes through a point O and moves in a straight line. The displacement, s metres, of P from O, t seconds after passing through O is given by:

$$s = -t^3 + 11t^2 - 24t$$

a Find an expression for the velocity, $v\,\text{m s}^{-1}$, of P at time t seconds. **(2 marks)**

b Calculate the values of t at which P is instantaneously at rest. **(3 marks)**

c Find the value of t at which the acceleration is zero. **(2 marks)**

d Sketch a velocity–time graph to illustrate the motion of P in the interval $0 \leqslant t \leqslant 6$, showing on your sketch the coordinates of the points at which the graph crosses the axes. **(3 marks)**

e Calculate the values of t in the interval $0 \leqslant t \leqslant 6$ between which the speed of P is greater than $16\,\text{m s}^{-1}$. **(6 marks)**

E **6** A body moves in a straight line. Its velocity, $v\,\text{m s}^{-1}$, at time t seconds is given by $v = 3t^2 - 11t + 10$. Find:

a the values of t when the body is instantaneously at rest **(3 marks)**

b the acceleration of the body when $t = 4$ **(3 marks)**

c the total distance travelled by the body in the interval $0 \leqslant t \leqslant 4$. **(4 marks)**

E **7** A particle moves along the positive x-axis. At time $t = 0$ the particle passes through the origin with velocity $6\,\text{m s}^{-1}$. The acceleration, $a\,\text{m s}^{-2}$, of the particle at time t seconds is given by $a = 2t^3 - 8t$ for $t \geqslant 0$. Find:

a the velocity of the particle at time t seconds **(3 marks)**

b the displacement of the particle from the origin at time t seconds **(2 marks)**

c the values of t at which the particle is instantaneously at rest. **(3 marks)**

E **8** A remote-control drone flies such that its vertical height, s m, above ground level at time t seconds is given by the equation:

$$x = \frac{t^4 - 12t^3 + 28t^2 + 400}{50}, \quad 0 \leqslant t \leqslant 8$$

The diagram shows a sketch of a displacement–time graph of the drone's motion.

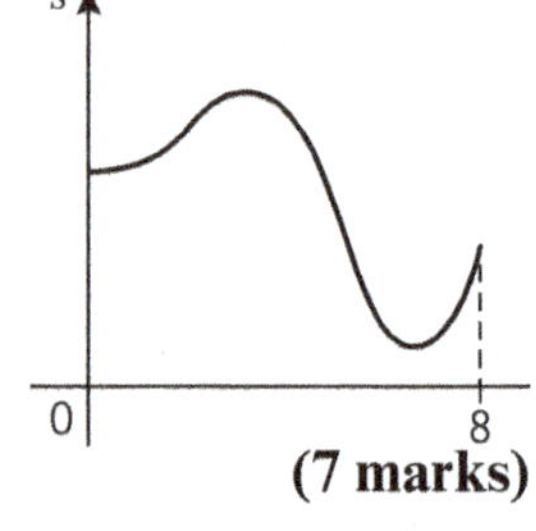

Determine the maximum and minimum height of the drone. **(7 marks)**

E/P **9** A rocket sled is used to test a parachute braking mechanism for a space capsule.

At the moment the parachute opens, the sled is 1.5 km from its launch site and is travelling away from it at a speed of $800\,\text{m s}^{-1}$. The sled comes to rest 25 seconds after the parachute opens.

The rocket sled is modelled as a particle moving in a straight horizontal line with constant acceleration. At a time t seconds after the parachute opens, its distance, s m, from the launch site is given by:

$$s = a + bt + ct^2, 0 \leqslant t \leqslant 25$$

Find the values of a, b and c in this model. **(6 marks)**

E **10** A particle P moves along the x-axis. It passes through the origin O at time $t = 0$ with speed $7\,\text{m s}^{-1}$ in the direction of x increasing.

At time t seconds the acceleration of P in the direction of x increasing is $(20 - 6t)\,\text{m s}^{-2}$.

a Show that the velocity $v\,\text{m s}^{-1}$ of P at time t seconds is given by:

$$v = 7 + 20t - 3t^2$$

(3 marks)

b Show that $v = 0$ when $t = 7$ and find the greatest speed of P in the interval $0 \leqslant t \leqslant 7$. **(4 marks)**

c Find the distance travelled by P in the interval $0 \leqslant t \leqslant 7$. **(4 marks)**

E/P **11** A particle P moves along a straight line. Initially, P is at rest at a point O on the line. At time t seconds (where $t \geqslant 0$) the acceleration of P is proportional to $(7 - t^2)$ and the displacement of P from O is s metres. When $t = 3$, the velocity of P is $6\,\text{m s}^{-1}$.
Show that $s = \frac{1}{24}t^2(42 - t^2)$. **(7 marks)**

E/P **12** A mouse leaves its hole and makes a short journey along a straight wall before returning to its hole. The mouse is modelled as a particle moving in a straight line. The distance of the mouse, s m, from its hole at time t minutes is given by:

$$s = t^4 - 10t^3 + 25t^2, \; 0 \leqslant t \leqslant 5$$

a Explain the restriction $0 \leqslant t \leqslant 5$. **(3 marks)**

b Find the greatest distance of the mouse from its hole. **(6 marks)**

P **13** At a time t seconds after launch, a space rocket can be modelled as a particle moving in a straight line with acceleration, $a\,\text{m s}^{-2}$, given by the equation:

$$a = (6.77 \times 10^{-7})t^3 - (3.98 \times 10^{-4})t^2 + 0.105t + 0.859, \quad 124 \leqslant t \leqslant 446$$

a Suggest two reasons why the space rocket might experience variable acceleration during its launch phase.

Given that the velocity of the rocket at time $t = 124$ is $974\,\text{m s}^{-1}$:

b find an expression for the velocity $v\,\text{m s}^{-1}$ of the rocket at time t. Give your coefficients to 3 significant figures.

c Hence find the velocity of the rocket at time $t = 446$, correct to 3 s.f.

From $t = 446$, the rocket maintains a constant acceleration of $28.6\,\text{m s}^{-2}$ until it reaches its escape velocity of $7.85\,\text{km s}^{-1}$. It then cuts its main engines.

d Calculate the time at which the rocket cuts its main engines.

E/P **14** Two particles P and Q move in a plane so that at time t seconds, where $t \geqslant 0$, P and Q have position vectors $\mathbf{r}_P$ metres and $\mathbf{r}_Q$ metres respectively, relative to a fixed origin O, where

$$\mathbf{r}_P = (3t^2 + 4)\mathbf{i} + \left(2t - \frac{1}{2}\right)\mathbf{j}$$

$$\mathbf{r}_Q = (t + 6)\mathbf{i} + \frac{3t^2}{2}\mathbf{j}$$

Find:

a the velocity vectors of P and Q at time t seconds **(5 marks)**

b the speed of P when $t = 2$ **(2 marks)**

c the value of t at the instant when the particles are moving parallel to one another. **(4 marks)**

d Show that the particles collide and find the position vector of their point of collision. **(6 marks)**

15 At time t seconds, a particle P has position vector $\mathbf{r}$ m with respect to a fixed origin O, where

$$\mathbf{r} = (3t^2 - 4)\mathbf{i} + (8 - 4t^2)\mathbf{j}$$

a Show that the acceleration of P is a constant.

b Find the magnitude of the acceleration of P and the size of the angle which the acceleration makes with $\mathbf{j}$.

(E) **16** At time t seconds, a particle P has position vector $\mathbf{r}$ m with respect to a fixed origin O, where

$$\mathbf{r} = 2\cos 3t\mathbf{i} - 2\sin 3t\mathbf{j}$$

a Find the velocity of P when $t = \frac{\pi}{6}$. **(5 marks)**

b Show that the magnitude of the acceleration of P is constant. **(4 marks)**

(E/P) **17** A particle of mass 0.5 kg is acted upon by a variable force $\mathbf{F}$. At time t seconds, the velocity $\mathbf{v}\,\mathrm{m\,s^{-1}}$ is given by $\mathbf{v} = (4ct - 6)\mathbf{i} + (7 - c)t^2\mathbf{j}$, where c is a constant.

a Show that $\mathbf{F} = (2c\mathbf{i} + (7 - c)t\mathbf{j})$ N. **(4 marks)**

b Given that when $t = 5$ the magnitude of $\mathbf{F}$ is 17 N, find the possible values of c. **(5 marks)**

(E) **18** At time t seconds (where $t \geqslant 0$) the particle P is moving in a plane with acceleration $\mathbf{a}\,\mathrm{m\,s^{-2}}$, where $\mathbf{a} = (8t^3 - 6t)\mathbf{i} + (8t - 3)\mathbf{j}$.

When $t = 2$, the velocity of P is $(16\mathbf{i} + 3\mathbf{j})\,\mathrm{m\,s^{-1}}$. Find:

a the velocity of P after t seconds **(3 marks)**

b the value of t when P is moving parallel to $\mathbf{i}$. **(4 marks)**

(E) **19** A particle P moves so that its acceleration $\mathbf{a}\,\mathrm{m\,s^{-2}}$ at time t seconds, where $t \geqslant 1$, is given by

$$\mathbf{a} = 4t\mathbf{i} + 5t^{-\frac{1}{2}}\mathbf{j}$$

When $t = 1$, the velocity of P is $(4\mathbf{i} + 10\mathbf{j})\,\mathrm{m\,s^{-1}}$.

Find the speed of P when $t = 5$. **(6 marks)**

(E/P) **20** In this question, $\mathbf{i}$ and $\mathbf{j}$ are horizontal unit vectors due east and due north respectively.

A clockwork train is moving on a flat, horizontal floor. At time $t = 0$, the train is at a fixed point O and is moving with velocity $(3\mathbf{i} + 13\mathbf{j})\,\mathrm{m\,s^{-1}}$. The velocity of the train at time t seconds is $\mathbf{v}\,\mathrm{m\,s^{-1}}$, and its acceleration, $\mathbf{a}\,\mathrm{m\,s^{-2}}$, is given by $\mathbf{a} = 2t\mathbf{i} + 3\mathbf{j}$.

a Find $\mathbf{v}$ in terms of t. **(3 marks)**

b Find the value of t when the train is moving in a north-east direction. **(3 marks)**

Challenge

1 A particle starts at rest and moves in a straight line. At time t seconds after the beginning of its motion, the acceleration of the particle, $a\,\mathrm{m\,s^{-2}}$, is given by:

$$a = 3t^2 - 18t + 20, \ t \geqslant 0$$

Find the distance travelled by the particle in the first 5 seconds of its motion.

2 A particle travels in a straight line with an acceleration, $a\,\mathrm{m\,s^{-2}}$, given by $a = 6t + 2$. The particle travels 50 metres in the fourth second. Find the velocity of the particle when $t = 5$ seconds.

3 A particle moves on the positive x-axis such that its displacement, s m, from O at time t seconds is given by

$$s = (20 - t^2)\sqrt{t + 1}, \ t \geqslant 0$$

a State the initial displacement of the particle.

b Show that the particle changes direction exactly once and determine the time at which this occurs.

c Find the exact speed of the particle when it crosses O.

4 Relative to a fixed origin O, the particle R has position vector $\mathbf{r}$ metres at time t seconds, where

$\mathbf{r} = (6\sin \omega t)\mathbf{i} + (4\cos \omega t)\mathbf{j}$

and ω is a positive constant.

a Find $\dot{\mathbf{r}}$ and hence show that $v^2 = 2\omega^2(13 + 5\cos 2\omega t)$, where $v\,\text{m s}^{-1}$ is the speed of R at time t seconds.

b Deduce that $4\omega \leqslant v \leqslant 6\omega$.

c At the instant when $t = \frac{\pi}{3\omega}$, find the angle between $\mathbf{r}$ and $\dot{\mathbf{r}}$, giving your answer in degrees to one decimal place.

Summary of key points

1 If the displacement, s, is expressed as a function of t, then the velocity, v, can be expressed as

$$v = \frac{ds}{dt}$$

2 If the velocity, v, is expressed as a function of t, then the acceleration, a, can be expressed as

$$a = \frac{dv}{dt} = \frac{d^2s}{dt^2}$$

3

Differentiate ↓		Integrate ↑
	displacement $= s = \int v\,dt$	
	$\frac{ds}{dt}$ = velocity $= v = \int a\,dt$	
	$\frac{dv}{dt} = \frac{d^2s}{dt^2}$ = acceleration $= a$	

4 If a particle starts from the point with position vector $\mathbf{r}_0$ and moves with constant velocity $\mathbf{v}$, then its displacement from its initial position at time t is $\mathbf{v}t$ and its position vector $\mathbf{r}$ is given by $\mathbf{r} = \mathbf{r}_0 + \mathbf{v}t$.

5 For an object moving in a plane with constant acceleration:

- $\mathbf{v} = \mathbf{u} + \mathbf{a}t$
- $\mathbf{r} = \mathbf{u}t + \frac{1}{2}\mathbf{a}t^2$

where

- $\mathbf{u}$ is the initial velocity
- $\mathbf{a}$ is the acceleration
- $\mathbf{v}$ is the velocity at time t
- $\mathbf{r}$ is the displacement at time t.

6 If $\mathbf{r} = x\mathbf{i} + y\mathbf{j}$, then $\mathbf{v} = \frac{d\mathbf{r}}{dt} = \dot{\mathbf{r}} = \dot{x}\mathbf{i} + \dot{y}\mathbf{j}$

and $\mathbf{a} = \frac{d\mathbf{v}}{dt} = \frac{d^2\mathbf{r}}{dt^2} = \ddot{\mathbf{r}} = \ddot{x}\mathbf{i} + \ddot{y}\mathbf{j}$

7 $\mathbf{v} = \int \mathbf{a}\,dt$ and $\mathbf{r} = \int \mathbf{v}\,dt$

3 CENTRES OF MASS

2.1
2.2
2.3

Learning objectives

After completing this chapter you should be able to:

Prior knowledge check

1 Work out the values of x and y:

$$8\begin{pmatrix}x\\y\end{pmatrix} = 4\begin{pmatrix}3\\1\end{pmatrix} + 6\begin{pmatrix}2\\2\end{pmatrix}$$

← International GCSE Mathematics

2 A uniform plank AB of length 6 m and mass 16 kg lies on the edge of a table. A mass of 4 kg is attached to one end of the plank at B, causing the plank to be on the point of tilting.

Find the distance AC.

← Mechanics 1 Section 8.5

3 Find the area of quadrilateral $ABCD$.

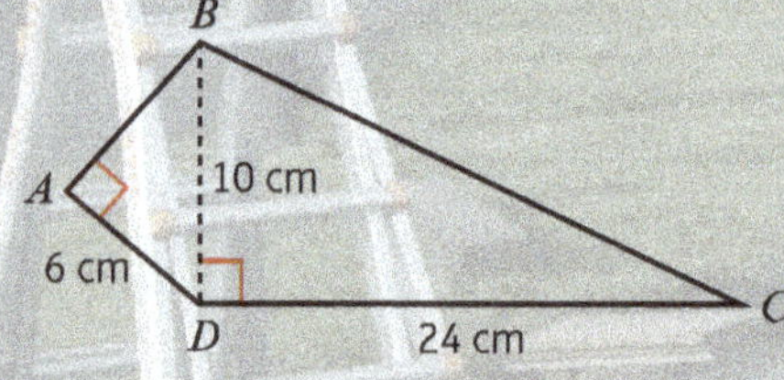

← International GCSE Mathematics

The centre of mass of large vehicles must be calculated, tested and sometimes adjusted, so that the vehicle does not topple over easily.

3.1 Centre of mass of a set of particles on a straight line

You can find the **centre of mass** of a set of particles arranged along a straight line by considering **moments**. You will use the fact that $\sum m_i x_i = \bar{x} \sum m_i$

Example 1 SKILLS PROBLEM-SOLVING

A system of three particles with masses 2 kg, 5 kg and 3 kg are placed along the x-axis at the points (3, 0), (4, 0) and (6, 0) respectively. Find the centre of mass of the system.

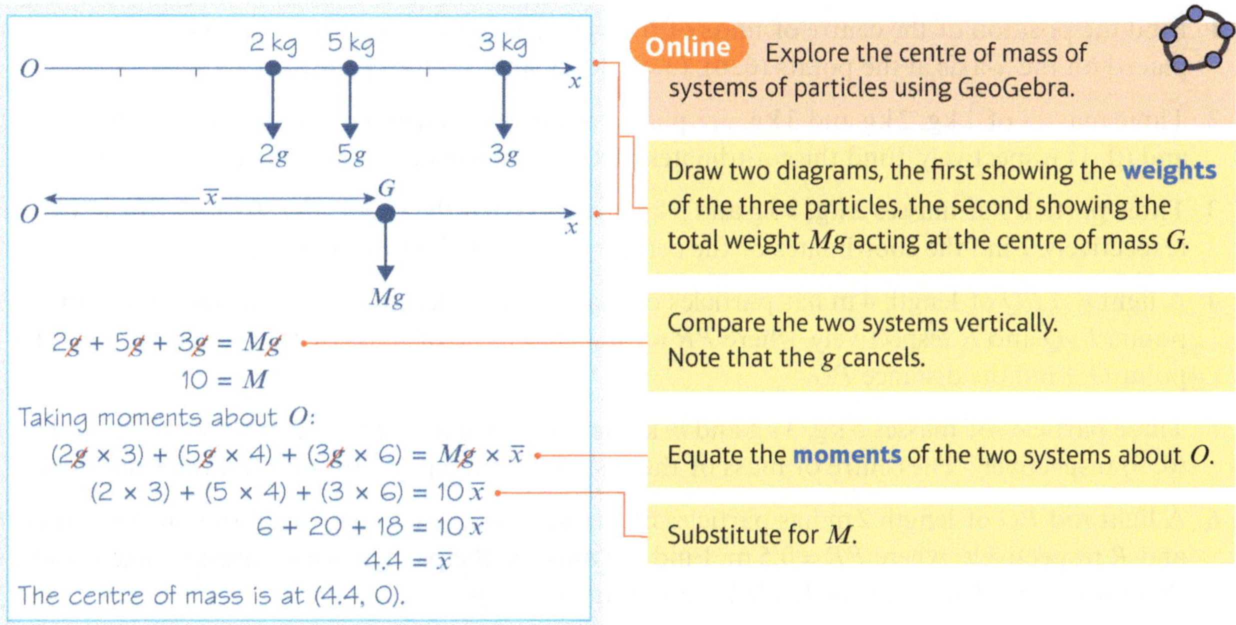

Example 2

A system of n particles with masses $m_1, m_2, \ldots, m_n$ are placed along the x-axis at the points $(x_1, 0), (x_2, 0), \ldots, (x_n, 0)$ respectively. Find the centre of mass of the system.

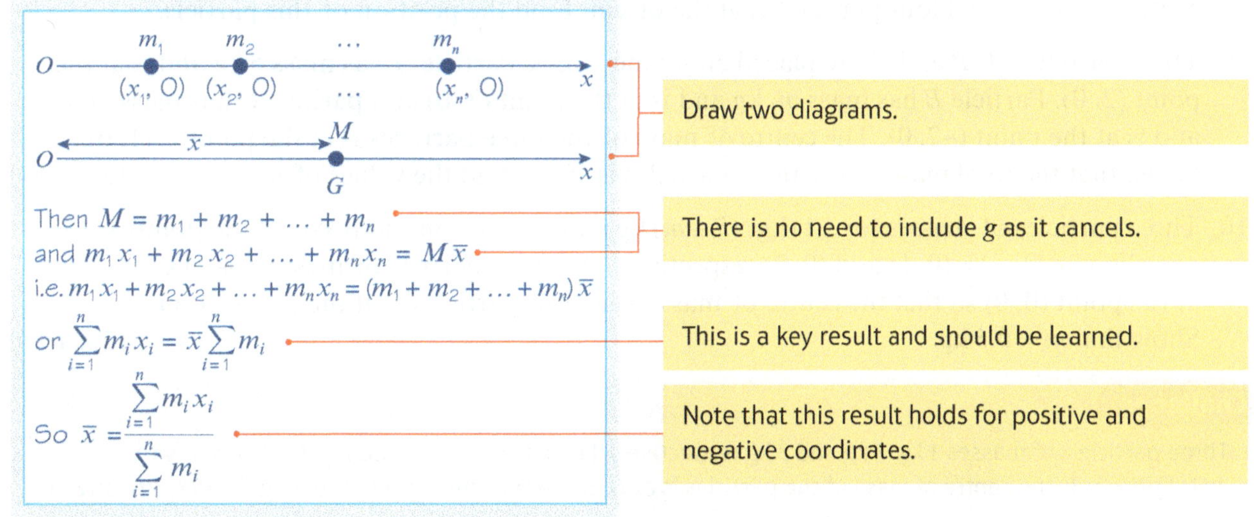

- If a system of n particles with masses $m_1, m_2, \ldots, m_n$ are placed along the x-axis at the points $(x_1, 0), (x_2, 0), \ldots, (x_n, 0)$ respectively, then:

$$\sum_{i=1}^{n} m_i x_i = \bar{x} \sum_{i=1}^{n} m_i$$

where $(\bar{x}, 0)$ is the position of the centre of mass of the system.

Notation This result could also be used for a system of particles placed along the y-axis: $\sum_{i=1}^{n} m_i y_i = \bar{y} \sum_{i=1}^{n} m_i$

Exercise 3A  SKILLS PROBLEM-SOLVING

1 Find the position of the centre of mass of four particles of masses 1 kg, 4 kg, 3 kg and 2 kg placed on the x-axis at the points (6, 0), (3, 0), (2, 0) and (4, 0) respectively.

2 Three masses of 1 kg, 2 kg and 3 kg, are placed at the points with coordinates (0, 2), (0, 5) and (0, 1) respectively. Find the coordinates of G, the centre of mass of the three masses.

3 Three particles of masses 2 kg, 3 kg and 5 kg, are placed at the points (−1, 0), (−4, 0) and (5, 0) respectively. Find the coordinates of the centre of mass of the three particles.

4 A light **rod** PQ of length 4 m has particles of masses 1 kg, 2 kg and 3 kg attached to it at the points P, Q and R respectively, where $PR = 2$ m. The centre of mass of the loaded rod is at the point G. Find the distance PG.

5 Three particles of masses 5 kg, 3 kg and m kg lie on the y-axis at the points (0, 4), (0, 2) and (0, 5) respectively. The centre of mass of the system is at the point (0, 4). Find the value of m.

(P) 6 A light rod PQ of length 2 m has particles of masses 0.4 kg and 0.6 kg fixed to it at the points P and R respectively, where $PR = 0.5$ m. Find the mass of the particle which must be fixed at Q so that the centre of mass of the loaded rod is at its **midpoint**.

(P) 7 The centre of mass of four particles of masses $2m$, $3m$, $7m$ and $8m$, which are positioned at the points $(0, a)$, (0, 2), (0, −1) and (0, 1) respectively, is the point G. Given that the coordinates of G are (0, 1), find the value of a.

(P) 8 Particles of masses 3 kg, 2 kg and 1 kg lie on the y-axis at the points with coordinates (0, −2), (0, 7) and (0, 4) respectively. Another particle of mass 6 kg is added to the system so that the centre of mass of all four particles is at the origin. Find the position of this particle.

(E/P) 9 Three particles A, B and C are placed along the x-axis. Particle A has mass 5 kg and is at the point (2, 0). Particle B has mass m_1 kg and is at the point (3, 0) and particle C has mass m_2 kg and is at the point (−2, 0). The centre of mass of the three particles is at the point G (1, 0). Given that the total mass of the three particles is 10 kg, find the values of m_1 and m_2. **(3 marks)**

(E/P) 10 Three particles of masses $(m - 1)$ kg, $(5 - m)$ kg and m kg lie on the y-axis at the points with coordinates (0, −1), (0, 1) and (0, 2) respectively. A fourth particle of mass $(m + 1)$ kg is added at the point (0, 0) so that the centre of mass of all four particles is at the point (0, 1). Show that $m = 0.5$ kg. **(3 marks)**

Challenge

Three particles, of masses 1 kg, 2 kg and 3 kg respectively, lie on the x-axis at points P, Q and R with $PQ : QR = 2 : 3$. The centre of mass of the particles is at G. Show that the ratio of the lengths $PQ : PG$ is 12 : 19.

3.2 Centre of mass of a set of particles arranged in a plane

You can use $\sum m_i x_i = \bar{x}\sum m_i$ and $\sum m_i y_i = \bar{y}\sum m_i$ to find the centre of mass of a set of point masses arranged in a plane by considering the x-coordinate and y-coordinate of the centre of mass separately.

Example 3 SKILLS PROBLEM-SOLVING

Find the coordinates of the centre of mass of the following system of particles:

2 kg at (1, 2); 3 kg at (3, 1); 5 kg at (4, 3)

Online Explore the centre of mass of particles arranged in a plane using GeoGebra.

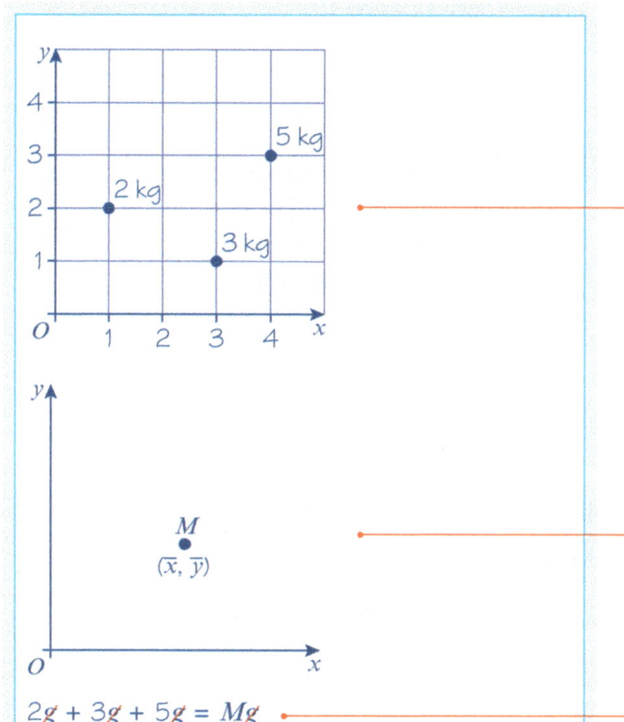

Draw two diagrams, the first showing the three particles, the second showing the total mass M placed at the centre of mass $(\bar{x}, \bar{y})$.

$2g + 3g + 5g = Mg$

$10 = M$

Equate the total weights. Note that g cancels.

Method 1

Taking moments about the y-axis:

$(2g \times 1) + (3g \times 3) + (5g \times 4) = Mg\bar{x}$

Equate the moments of the system about the y-axis.

$(2 \times 1) + (3 \times 3) + (5 \times 4) = (2 + 3 + 5)\bar{x}$

Substitute for M.

$2 + 9 + 20 = 10\bar{x}$

$3.1 = \bar{x}$

Taking moments about the x-axis:

$(2g \times 2) + (3g \times 1) + (5g \times 3) = (2 + 3 + 5)g\bar{y}$

Equate the moments of the system about the x-axis.

$2.2 = \bar{y}$

The centre of mass is at (3.1, 2.2).

Method 2

$$2\begin{pmatrix}1\\2\end{pmatrix} + 3\begin{pmatrix}3\\1\end{pmatrix} + 5\begin{pmatrix}4\\3\end{pmatrix} = (2+3+5)\begin{pmatrix}\bar{x}\\\bar{y}\end{pmatrix}$$

$$\begin{pmatrix}2\\4\end{pmatrix} + \begin{pmatrix}9\\3\end{pmatrix} + \begin{pmatrix}20\\15\end{pmatrix} = \begin{pmatrix}10\bar{x}\\10\bar{y}\end{pmatrix}$$

$$\begin{pmatrix}31\\22\end{pmatrix} = \begin{pmatrix}10\bar{x}\\10\bar{y}\end{pmatrix}$$

$$\begin{pmatrix}3.1\\2.2\end{pmatrix} = \begin{pmatrix}\bar{x}\\\bar{y}\end{pmatrix}$$

The centre of mass is at (3.1, 2.2).

Problem-solving You can reduce the working by using position vectors.

The top line is

$$\sum m_i x_i = \bar{x}\sum m_i$$

and the bottom line is

$$\sum m_i y_i = \bar{y}\sum m_i$$

Divide both sides by 10.

- If a system consists of n particles: mass m_1 with position vector $\mathbf{r}_1$, mass m_2 with position vector $\mathbf{r}_2$, …, mass m_n with position vector $\mathbf{r}_n$, then

$$\sum m_i \mathbf{r}_i = \bar{\mathbf{r}}\sum m_i$$

where $\bar{\mathbf{r}}$ is the position vector of the centre of mass of the system.

Notation The position vector of a point can be written in terms of **i** and **j** or as a column vector. For example, the position vector of the point (3, 4) is $3\mathbf{i} + 4\mathbf{j}$ or $\begin{pmatrix}3\\4\end{pmatrix}$.

Example 4

Find the coordinates of the centre of mass of the following system of particles:

4 kg at (–1, 3); 2 kg at (–2, –4); 8 kg at (4, 0); 6 kg at (1, –3)

$$4\begin{pmatrix}-1\\3\end{pmatrix} + 2\begin{pmatrix}-2\\-4\end{pmatrix} + 8\begin{pmatrix}4\\0\end{pmatrix} + 6\begin{pmatrix}1\\-3\end{pmatrix} = (4+2+8+6)\begin{pmatrix}\bar{x}\\\bar{y}\end{pmatrix}$$

$$\begin{pmatrix}-4\\12\end{pmatrix} + \begin{pmatrix}-4\\-8\end{pmatrix} + \begin{pmatrix}32\\0\end{pmatrix} + \begin{pmatrix}6\\-18\end{pmatrix} = 20\begin{pmatrix}\bar{x}\\\bar{y}\end{pmatrix}$$

$$\begin{pmatrix}30\\-14\end{pmatrix} = 20\begin{pmatrix}\bar{x}\\\bar{y}\end{pmatrix}$$

$$\begin{pmatrix}1.5\\-0.7\end{pmatrix} = \begin{pmatrix}\bar{x}\\\bar{y}\end{pmatrix}$$

Centre of mass is at (1.5, –0.7).

The result applies with positive or negative coordinates.

Simplify the **LHS**.

- If a question does not specify axes or coordinates you will need to choose your own axes and origin.

Example 5

A light rectangular plate $ABCD$ has $AB = 20$ cm and $AD = 50$ cm. Particles of masses 2 kg, 3 kg, 5 kg and 5 kg are attached to the plate at the points A, B, C and D respectively.
Find the distance of the centre of mass of the loaded plate from:

a AD **b** AB

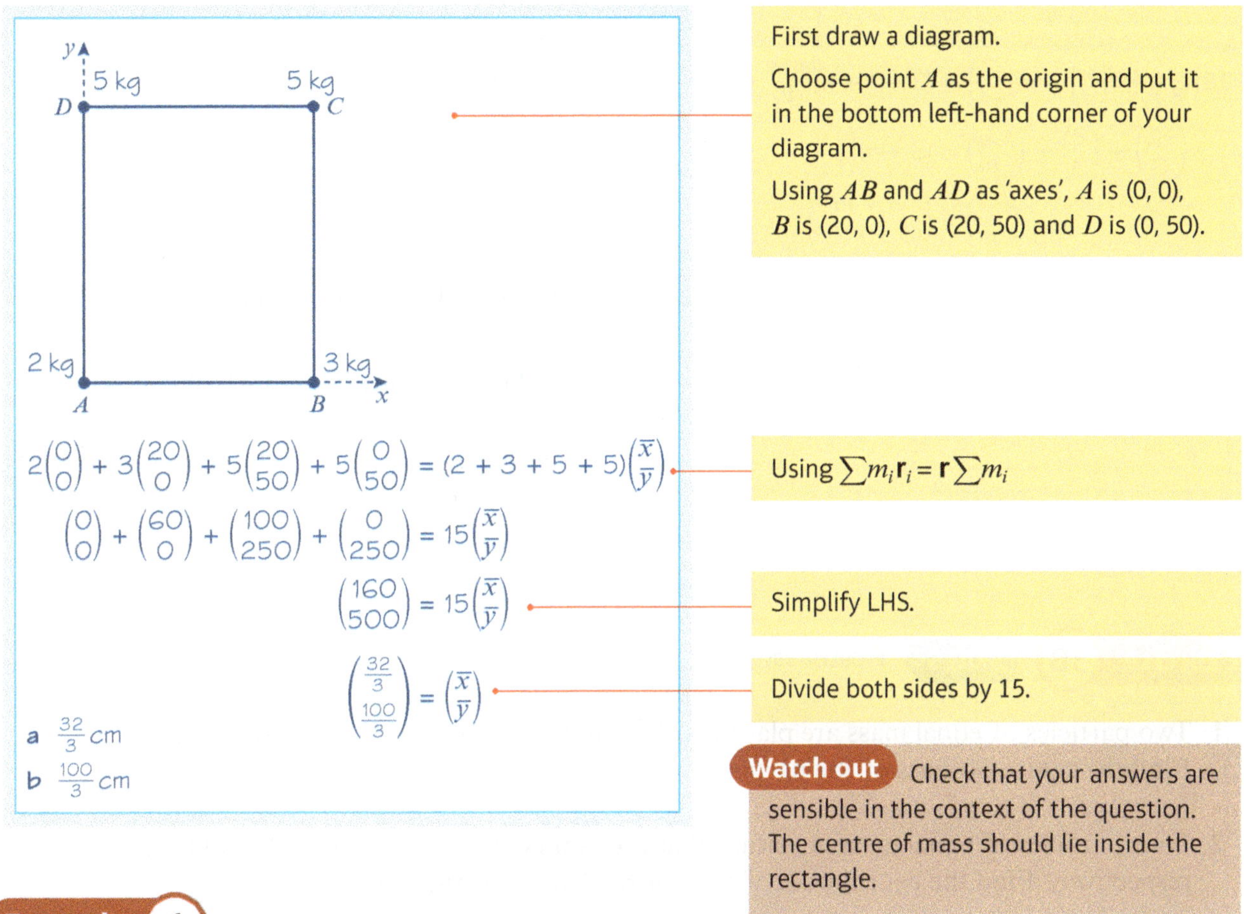

Example 6

Particles of masses 4 kg, 3 kg, 2 kg and 1 kg are placed at the points (x, y), $(3, 2)$, $(1, -5)$ and $(6, 0)$ respectively. Given that the centre of mass of the four particles is at the point $(2.5, -2)$, find the values of x and y.

$$4\begin{pmatrix}x\\y\end{pmatrix} + 3\begin{pmatrix}3\\2\end{pmatrix} + 2\begin{pmatrix}1\\-5\end{pmatrix} + 1\begin{pmatrix}6\\0\end{pmatrix} = (4+3+2+1)\begin{pmatrix}2.5\\-2\end{pmatrix}$$

Using $\sum m_i\mathbf{r}_i = \mathbf{r}\sum m_i$

$$\begin{pmatrix}4x\\4y\end{pmatrix} + \begin{pmatrix}9\\6\end{pmatrix} + \begin{pmatrix}2\\-10\end{pmatrix} + \begin{pmatrix}6\\0\end{pmatrix} = \begin{pmatrix}25\\-20\end{pmatrix}$$

$$\begin{pmatrix}4x+17\\4y-4\end{pmatrix} = \begin{pmatrix}25\\-20\end{pmatrix}$$

$$4x + 17 = 25$$
$$4y - 4 = -20$$

Equate the **i** and **j** components.

$x = 2$, $y = -4$

Solve the two equations for x and y.

Example  7

Three particles of masses 2 kg, 1 kg and m kg are situated at the points (–1, 3), (2, 9) and (2, –1) respectively. Given that the centre of mass of the three particles is at the point $(1, \bar{y})$, find:

a the value of m

b the value of $\bar{y}$.

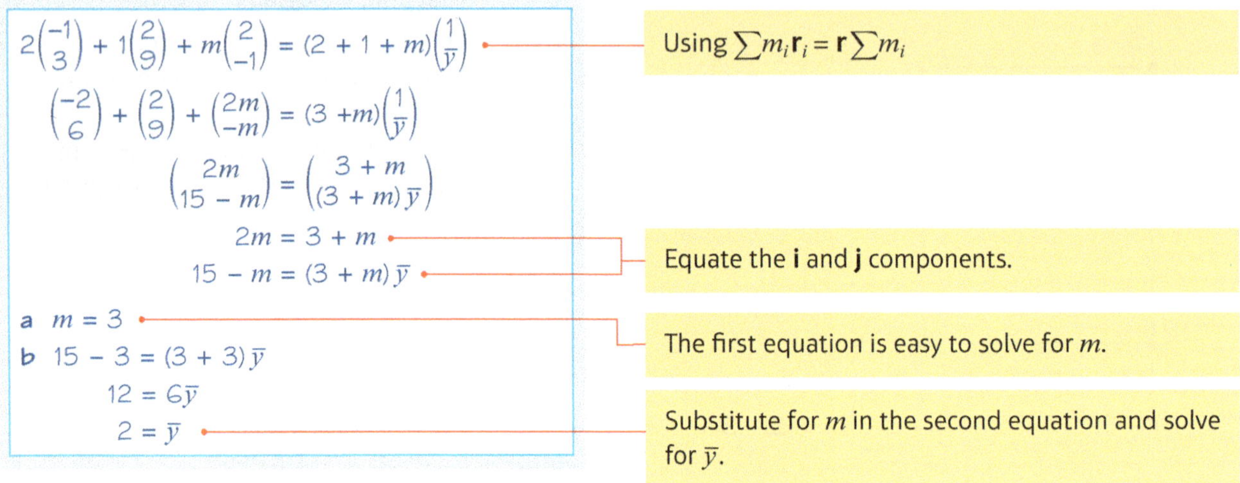

Exercise  3B SKILLS PROBLEM-SOLVING

1 Two particles of equal mass are placed at the points (1, –3) and (5, 7) respectively. Find the centre of mass of the particles.

2 Four particles of equal mass are situated at the points (2, 0), (–1, 3), (2, –4) and (–1, –2) respectively. Find the coordinates of the centre of mass of the particles.

3 A system of three particles consists of 10 kg placed at (2, 3), 15 kg placed at (4, 2) and 25 kg placed at (6, 6). Find the coordinates of the centre of mass of the system.

4 Find the position vector of the centre of mass of three particles of masses 0.5 kg, 1.5 kg and 2 kg which are situated at the points with position vectors (6**i** – 3**j**), (2**i** + 5**j**) and (3**i** + 2**j**) respectively.

5 Particles of masses m, $2m$, $5m$ and $2m$ are situated at (–1, –1), (3, 2), (4, –2) and (–2, 5) respectively. Find the coordinates of the centre of mass of the particles.

6 A light rectangular metal plate $PQRS$ has $PQ = 4$ cm and $PS = 2$ cm. Particles of masses 3 kg, 5 kg, 1 kg and 7 kg are attached respectively to the corners P, Q, R and S of the plate. Find the distance of the centre of mass of the loaded plate from:

a the side PQ **b** the side PS.

(P) 7 Three particles of masses 1 kg, 2 kg and 3 kg are positioned at the points (1, 0), (4, 3) and (p, q) respectively. Given that the centre of mass of the particles is at the point (2, 0), find the values of p and q.

(P) 8 A system consists of three particles with masses $3m$, $4m$ and $5m$. The particles are situated at the points with coordinates (−3, −4), (0.5, 4) and (0, −5) respectively. Find the coordinates of the position of a fourth particle of mass $7m$, given that the centre of mass of all four particles is at the origin.

(E) 9 A light rectangular piece of card $ABCD$ has AB = 8 cm and AD = 6 cm. Four particles of masses 300 g, 200 g, 600 g and 100 g are fixed to the rectangle at the midpoints of the sides AB, BC, CD and DE respectively. Find the distance of the centre of mass of the loaded rectangle from the sides AB and AD. **(4 marks)**

(E/P) 10 A light rectangular piece of card $ABCD$ has AB = 8 cm and AD = 6 cm. Three particles of masses 3 g, 2 g and 2 g are attached to the rectangle at the points A, B and C respectively.

a Find the mass of a particle which must be placed at the point D for the centre of mass of the whole system of four particles to lie 3 cm from the line AB. **(2 marks)**

b With this fourth particle in place, find the distance of the centre of mass of the system from the side AD. **(4 marks)**

Challenge

A light triangular piece of card ABC has sides AB = 6 cm, AC = 5 cm and BC = 5 cm. Three particles of masses m kg, 0.2 kg and 0.2 kg are fixed to the triangle at the midpoints of the sides AB, BC and AC respectively. The point P lies at the **intersection** of the lines joining each **vertex** of the triangle with the midpoint of the opposite side. Given that the centre of mass of the whole system lies at P, find the value of m.

3.3 Centres of mass of standard uniform plane laminas

You can find the positions of the centres of mass of standard **uniform** plane **laminas**, including a rectangle, a triangle and a semicircle.

- **An object which has one dimension (thickness) very small compared with its other two (length and width) is modelled as a lamina. This means that it is regarded as being two-dimensional with area, but no volume.**

Hint For example, a sheet of paper or a piece of card could be modelled as a lamina.

A lamina is **uniform** if its mass is evenly spread throughout its area.

- **If a uniform lamina has an axis of symmetry, then its centre of mass must lie on the axis of symmetry. If the lamina has more than one axis of symmetry, then it follows that the centre of mass must be at the point of intersection of the axes of symmetry.**

Uniform circular disc

Since every diameter of the disc is a line of symmetry, the centre of mass of the disc is at their intersection. This is the centre of the disc.

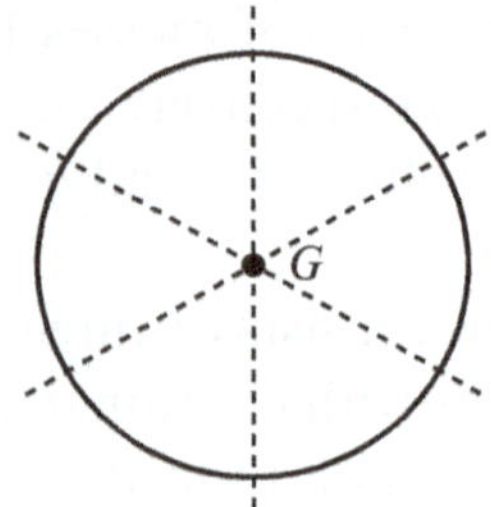

Uniform rectangular lamina

A uniform rectangular lamina has two lines of symmetry, each one joining the midpoints of a pair of opposite sides. The centre of mass is at the point where the two lines meet.

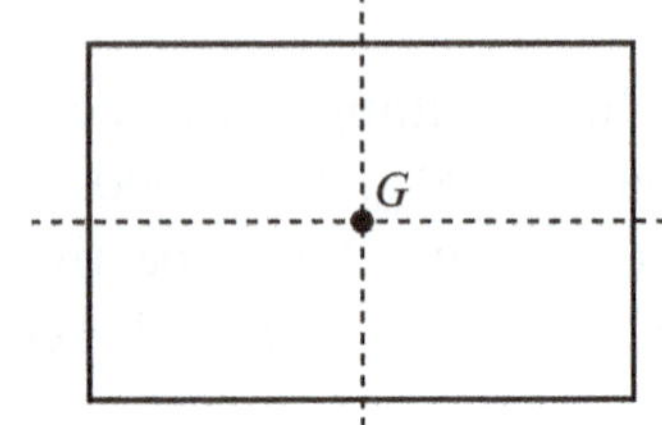

Uniform triangular lamina

A uniform triangular lamina has axes of symmetry only if it is either equilateral or isosceles.

A uniform equilateral triangle has three axes of symmetry, each one joining a vertex to the midpoint of the opposite side. These three lines are called the **medians** of the triangle.

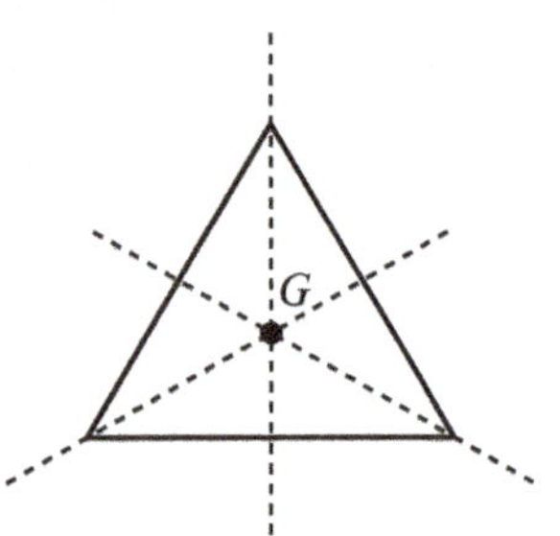

- **The centre of mass of a uniform triangular lamina is at the intersection of the medians. This point is called the centroid of the triangle.**

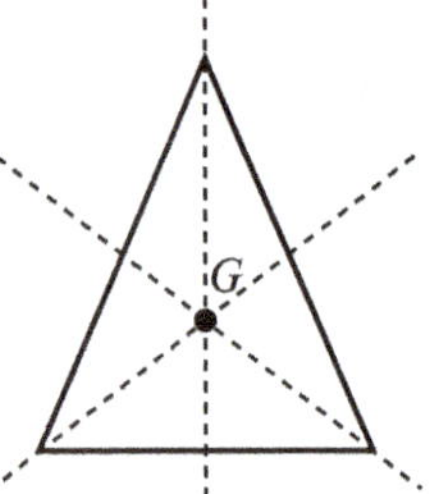

Note that the medians are not axes of symmetry of the triangle unless the triangle is equilateral (in which case all three medians are axes of symmetry) or isosceles (in which case one median is also an axis of symmetry).

Hint It can be proved that the centroid G (and therefore the centre of mass) of any triangle is two-thirds of the way down each median from each vertex:

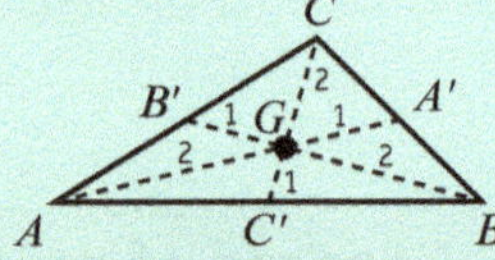

where A' is the midpoint of BC, B' is the midpoint of CA and C' is the midpoint of AB:

i.e. $\frac{AG}{GA'} = \frac{BG}{GB'} = \frac{CG}{GC'} = \frac{2}{1}$

- **If the coordinates of the three vertices of a uniform triangular lamina are (x_1, y_1), (x_2, y_2) and (x_3, y_3) then the coordinates of the centre of mass G are given by taking the average (mean) of the coordinates of the vertices:**

  G is the point $\left(\frac{x_1 + x_2 + x_3}{3}, \frac{y_1 + y_2 + y_3}{3}\right)$

Hint This is the two-dimensional version of a similar result for a uniform rod: if the ends of the rod are (x_1, y_1) and (x_2, y_2) then its centre of mass is its midpoint, $\left(\frac{x_1 + x_2}{2}, \frac{y_1 + y_2}{2}\right)$.

Uniform sector of a circle

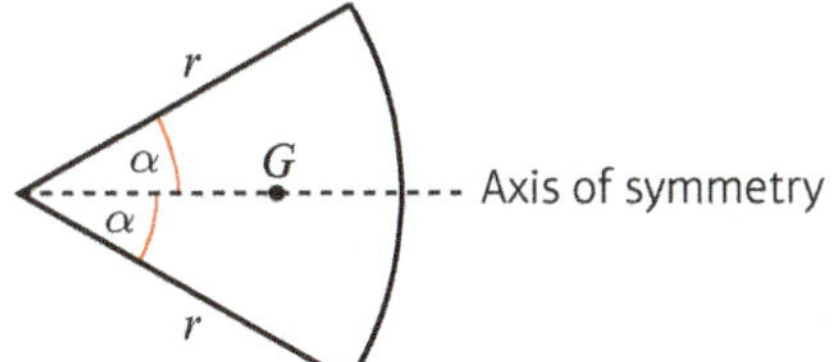

A uniform sector of a circle of radius r and centre angle 2α, where α is measured in radians, has its centre of mass on the axis of symmetry at a distance $\frac{2r\sin\alpha}{3\alpha}$ from the centre.

Example 8

A uniform triangular lamina has vertices $A(1, 4)$, $B(3, 2)$ and $C(5, 3)$. Find the coordinates of its centre of mass.

G is the point $\left(\frac{1+3+5}{3}\right), \left(\frac{4+2+3}{3}\right) = (3, 3)$

Find the mean of the vertices of the triangle. This is the **centroid** of the triangle.

Online Explore centres of mass of standard uniform plane laminas using GeoGebra.

Example 9

Find the centre of mass of the uniform triangular lamina shown.

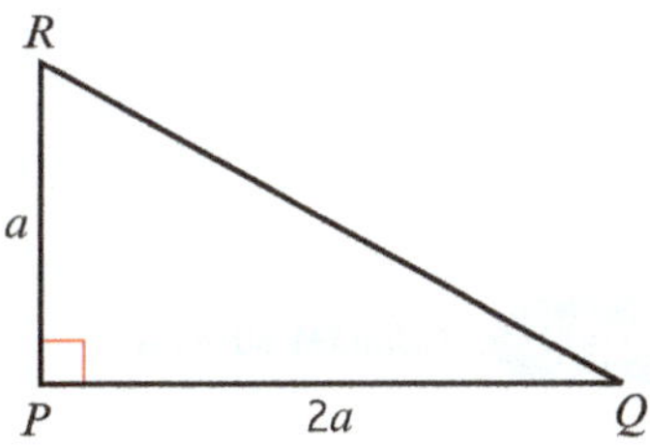

Taking P as the origin and PQ and PR as axes:
P is $(0, 0)$; Q is $(2a, 0)$; R is $(0, a)$

G is the point

$$\left(\frac{0+2a+0}{3}, \frac{0+0+a}{3}\right) = \left(\frac{2a}{3}, \frac{a}{3}\right)$$

The centre of mass is $\frac{2a}{3}$ from PR and $\frac{a}{3}$ from PQ.

Here we need to choose our own axes and origin.

Write down the coordinates of each of the three vertices.

Find the mean of the three vertices.

Watch out When you choose your own axes you must not leave your answer in coordinate form.

Example 10 SKILLS PROBLEM-SOLVING

The diagram shows a uniform semicircular lamina of radius 6 cm with centre O.
Find the centre of mass of the lamina.

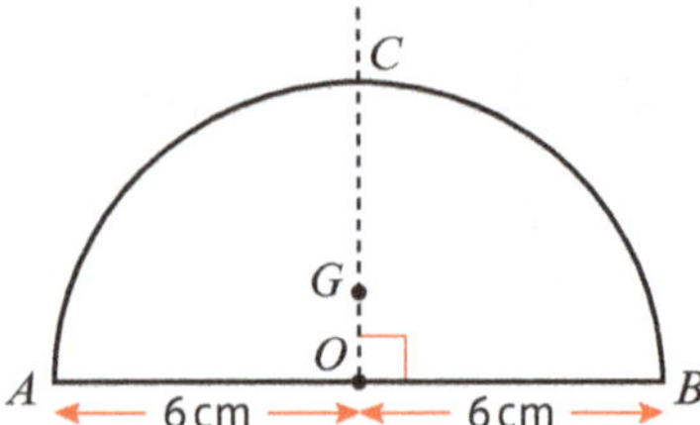

The centre of mass must lie on the line through O which is perpendicular to AB.

This is the axis of symmetry of the lamina.

Let $OG = \bar{y}$. Then:

$$\bar{y} = \frac{2 \times 6 \times \sin\frac{\pi}{2}}{\frac{3\pi}{2}}$$

Use the result for a sector which is in the formula booklet with $r = 6$ and $\alpha = \frac{\pi}{2}$.

$$= \frac{12 \times 1}{\frac{3\pi}{2}}$$

$\sin\frac{\pi}{2} = 1$
You must give the angle in radians for this formula.

$$= 12 \times \frac{2}{3\pi}$$

$$= \frac{8}{\pi}$$

Simplify.

The centre of mass of the lamina is on the line OC at a distance $\frac{8}{\pi}$ cm from O.

Exercise 3C SKILLS PROBLEM-SOLVING

1 Find the centre of mass of a uniform triangular lamina whose vertices are:

a (1, 2), (2, 6) and (3, 1) **b** (−1, 4), (3, 5) and (7, 3)

c (−3, 2), (4, 0) and (0, 1) **d** (a, a), $(3a, 2a)$ and $(4a, 6a)$

2 Find the position of the centre of mass of a uniform semicircular lamina of radius 4 cm and centre O.

(P) 3 The centre of mass of a uniform triangular lamina ABC is at the point $(2, a)$. Given that A is the point (4, 3), B is the point $(b, 1)$ and C is the point (−1, 5), find the values of a and b.

4 Find the position of the centre of mass of the following uniform triangular laminas.

a

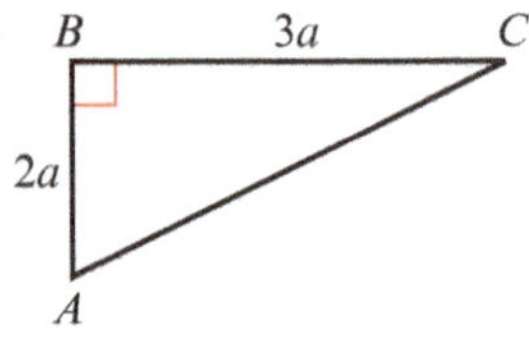

b

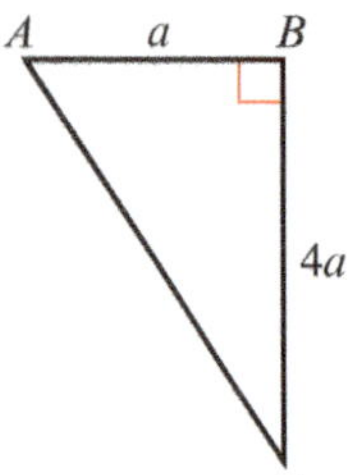

c

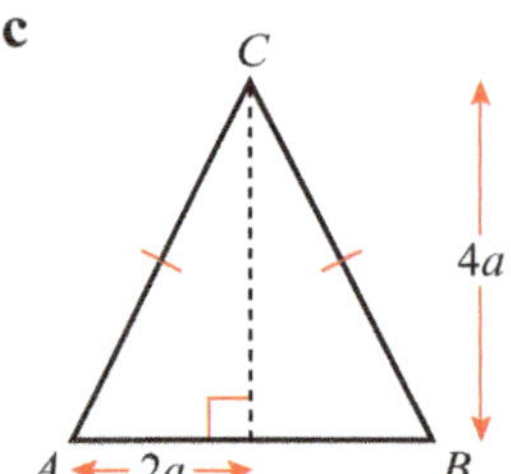

d

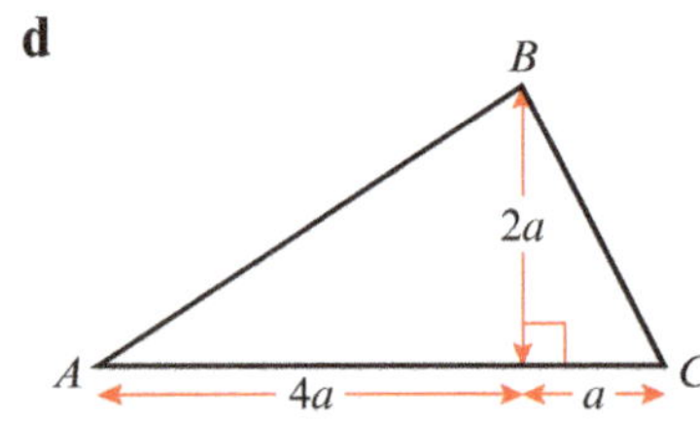

(P) **5** A uniform triangular lamina is isosceles and has the line $y = 4$ as its axis of symmetry. One of the vertices of the triangle is the point (2, 1). Given that the x-coordinate of the centre of mass of the lamina is -3, find the coordinates of the other two vertices.

(P) **6** A uniform rectangular lamina $ABCD$ is positioned such that AB lies on the line $y = 2x + 1$. Given that A is at the point (0, 1) and C is at the point (6, 7), find:

a the coordinates of the points B and D

b the coordinates of the centre of mass of the lamina.

(E/P) **7** A uniform triangular lamina ABC has coordinates $A(2, 1)$, $B(4, 1)$ and $C(x, y)$. The centre of mass of the lamina lies on the line $x = 3$. Given that the triangle ABC has an area of 4 cm^2, work out:

a the possible values of x and y **(3 marks)**

b the possible coordinates of the centre of mass. **(4 marks)**

(E/P) **8** The diagram below shows an equilateral triangle ABC where AC is 4 cm.

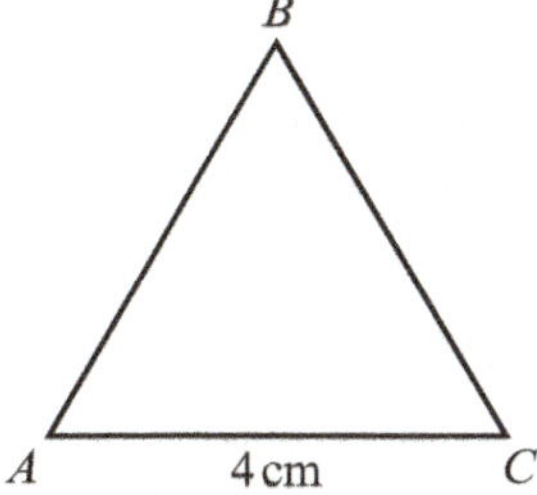

Show that the centre of mass lies $\frac{4\sqrt{3}}{3}$ cm from B. **(3 marks)**

3.4 Centre of mass of a composite lamina

A **composite uniform lamina** consists of two or more standard uniform laminas joined together. You can find the centre of mass of a composite lamina by considering each part of the lamina as a particle positioned at its centre of mass. The masses of each part of the lamina will be proportional to their areas.

Example 11 SKILLS PROBLEM-SOLVING

A uniform lamina consists of a rectangle $PQRS$ joined to an isosceles triangle QRT, as shown in the diagram.

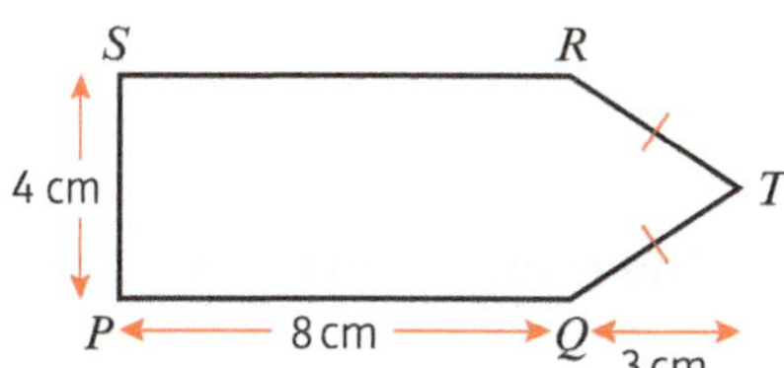

Find the distance of the centre of mass of the lamina from:

a PQ

b PS

Let the mass per unit area be m kg per cm^2.

Split the lamina along the line QR:

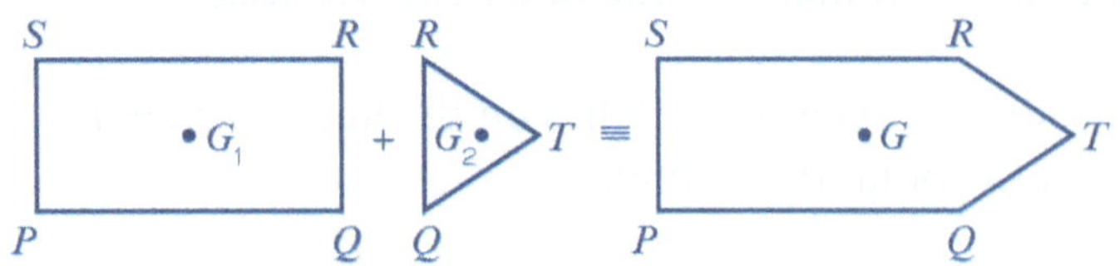

Area of $PQRS = 8 \times 4 = 32$ cm^2

So, mass of $PQRS = 32m$

Similarly, mass of $QRT = \frac{1}{2} \times 4 \times 3 \times m$

$= 6m$

So, total mass of the lamina $= 32m + 6m = 38m$

Take P as the origin, and axes along PQ and PS.

The centre of mass of $PQRS$ is at the point (4, 2).

The coordinates of Q are (8, 0).

The coordinates of R are (8, 4).

The coordinates of T are (11, 2).

The centre of mass of $\triangle QRT$ will be

$$\left(\frac{8+8+11}{3}, \frac{0+4+2}{3}\right)$$

$= (9, 2)$

Replace the lamina by two particles:

$32m$ placed at (4, 2)

and $6m$ placed at (9, 2)

$$32m\begin{pmatrix}4\\2\end{pmatrix} + 6m\begin{pmatrix}9\\2\end{pmatrix} = 38m\begin{pmatrix}\bar{x}\\\bar{y}\end{pmatrix}$$

$$\begin{pmatrix}128\\64\end{pmatrix} + \begin{pmatrix}54\\12\end{pmatrix} = 38\begin{pmatrix}\bar{x}\\\bar{y}\end{pmatrix}$$

$$\begin{pmatrix}182\\76\end{pmatrix} = 38\begin{pmatrix}\bar{x}\\\bar{y}\end{pmatrix}$$

$$\frac{91}{19} = \bar{x}$$

$$2 = \bar{y}$$

a Distance from PQ is 2 cm.

b Distance from PS is $\frac{91}{19}$ cm.

Since the lamina is uniform, the mass per unit area will be a constant.

You must always split up the lamina into standard shapes. G_1 is the centre of mass of the rectangle. G_2 is the centre of mass of the triangle.

Area of $\triangle$ is $\frac{1}{2} \times$ base $\times$ height.

It's usually a good idea to take the origin at the bottom left-hand corner of your diagram.

This is the centre of the rectangle, G_1.

Take the mean of the coordinates of the vertices of the triangle.

This is G_2.

This is the key idea behind the method.

$\begin{pmatrix}\bar{x}\\\bar{y}\end{pmatrix}$ is the position vector of the centre of mass.

Cancel the ms and simplify.

Problem-solving

Note that you could have got the answer to part **a** using the fact that the lamina has an axis of symmetry. You should always use this as it will considerably reduce the amount of working required.

- **The centre of mass of a uniform plane lamina will always lie on an axis of symmetry.**

Example 12

The diagram shows a uniform lamina.

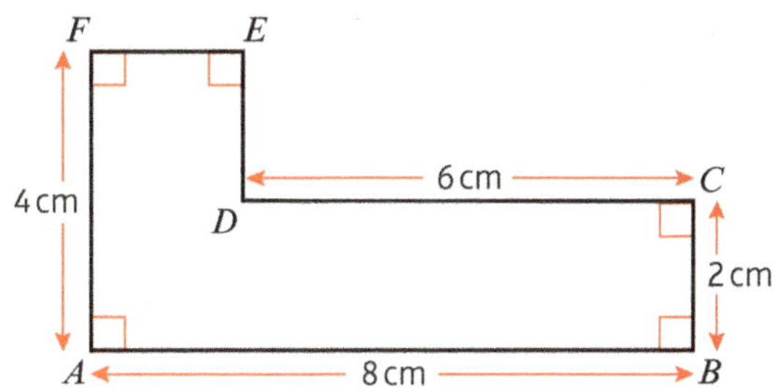

Find the distance of the centre of mass of the lamina from:

a AF **b** AB

Hint You can find the centre of mass in three different ways.

Method 1

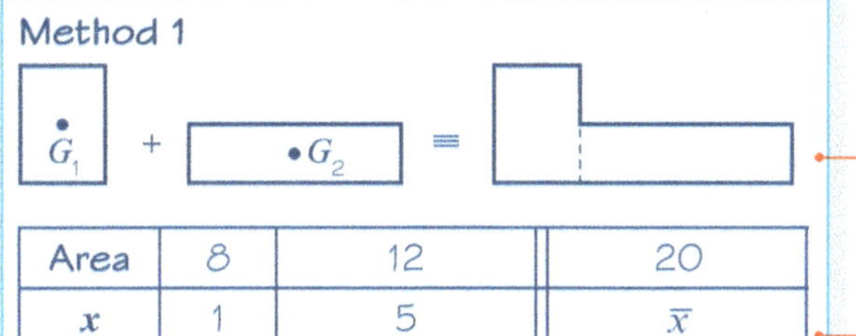

You can summarise the area of each part of the shape and the positions of G_1 and G_2 in a table. Because the lamina is uniform you only need to know the area of each piece.

Area	8	12	20
x	1	5	$\bar{x}$
y	2	1	$\bar{y}$

The centre of mass of the first rectangle is (1, 2).

The centre of mass of the second rectangle is at its centre (5, 1).

$$8\begin{pmatrix}1\\2\end{pmatrix} + 12\begin{pmatrix}5\\1\end{pmatrix} = 20\begin{pmatrix}\bar{x}\\\bar{y}\end{pmatrix}$$

Using $\sum m_i \mathbf{r}_i = \mathbf{r}\sum m_i$

$$\begin{pmatrix}8\\16\end{pmatrix} + \begin{pmatrix}60\\12\end{pmatrix} = 20\begin{pmatrix}\bar{x}\\\bar{y}\end{pmatrix}$$

$$\begin{pmatrix}68\\28\end{pmatrix} = 20\begin{pmatrix}\bar{x}\\\bar{y}\end{pmatrix}$$

Simplify.

$$3.4 = \bar{x}$$
$$1.4 = \bar{y}$$

Solve for $\bar{x}$ and $\bar{y}$.

Method 2

Split the shape using the dotted line shown.
The centre of the square is (1, 3).
The centre of the rectangle is (4, 1).

Area	4	16	20
x	1	4	$\bar{x}$
y	3	1	$\bar{y}$

$$4\begin{pmatrix}1\\3\end{pmatrix} + 16\begin{pmatrix}4\\1\end{pmatrix} = 20\begin{pmatrix}\bar{x}\\\bar{y}\end{pmatrix}$$

Using $\sum m_i \mathbf{r}_i = \mathbf{r}\sum m_i$

$$\begin{pmatrix}4\\12\end{pmatrix} + \begin{pmatrix}64\\16\end{pmatrix} = 20\begin{pmatrix}\bar{x}\\\bar{y}\end{pmatrix}$$

$$\begin{pmatrix}68\\28\end{pmatrix} = 20\begin{pmatrix}\bar{x}\\\bar{y}\end{pmatrix}$$

$$3.4 = \bar{x}$$
$$1.4 = \bar{y}$$

As before.

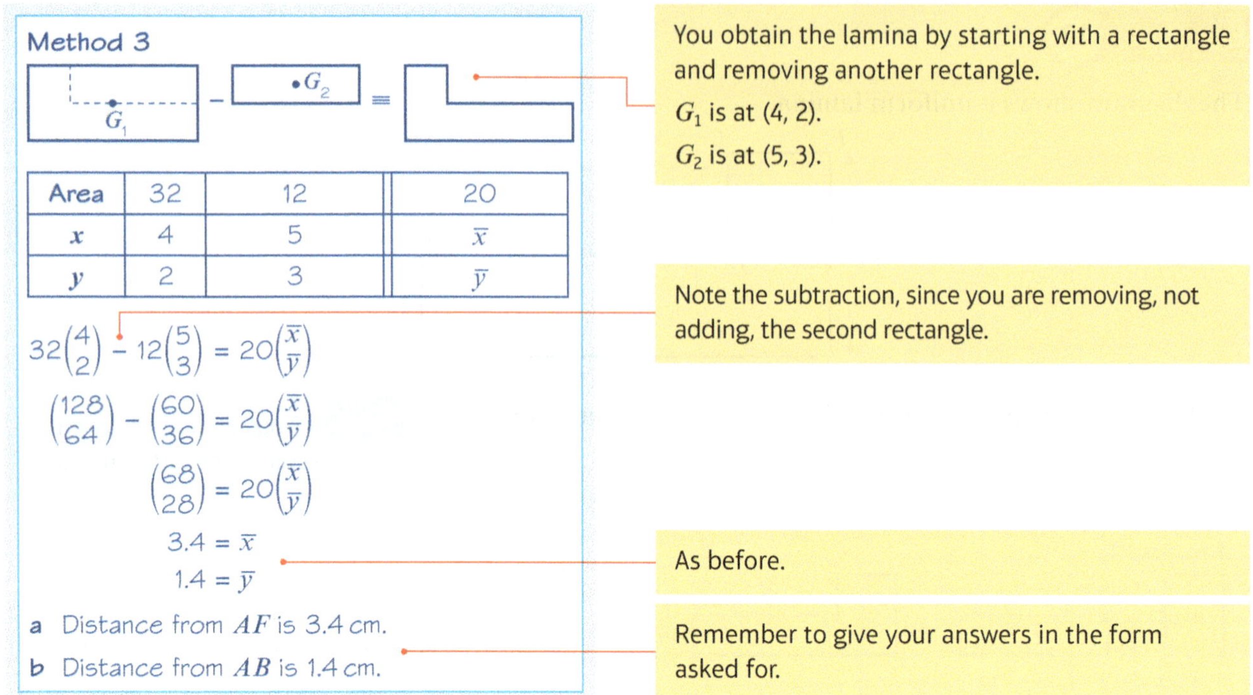

Method 3

Area	32	12	20
x	4	5	$\bar{x}$
y	2	3	$\bar{y}$

$$32\begin{pmatrix}4\\2\end{pmatrix} - 12\begin{pmatrix}5\\3\end{pmatrix} = 20\begin{pmatrix}\bar{x}\\\bar{y}\end{pmatrix}$$

$$\begin{pmatrix}128\\64\end{pmatrix} - \begin{pmatrix}60\\36\end{pmatrix} = 20\begin{pmatrix}\bar{x}\\\bar{y}\end{pmatrix}$$

$$\begin{pmatrix}68\\28\end{pmatrix} = 20\begin{pmatrix}\bar{x}\\\bar{y}\end{pmatrix}$$

$$3.4 = \bar{x}$$
$$1.4 = \bar{y}$$

a Distance from AF is 3.4 cm.

b Distance from AB is 1.4 cm.

You obtain the lamina by starting with a rectangle and removing another rectangle.
G_1 is at (4, 2).
G_2 is at (5, 3).

Note the subtraction, since you are removing, not adding, the second rectangle.

As before.

Remember to give your answers in the form asked for.

Example 13

A uniform circular disc, centre O, of radius 5 cm, has two circular holes cut in it, as shown below.

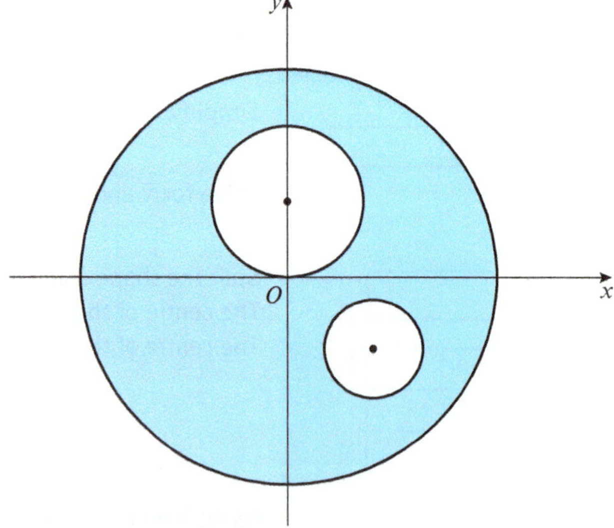

The larger hole has radius 2 cm and the smaller hole has radius 1 cm. The coordinates of the centres of the holes are (0, 2) and (2, −2) respectively. Find the coordinates of the centre of mass of the remaining lamina.

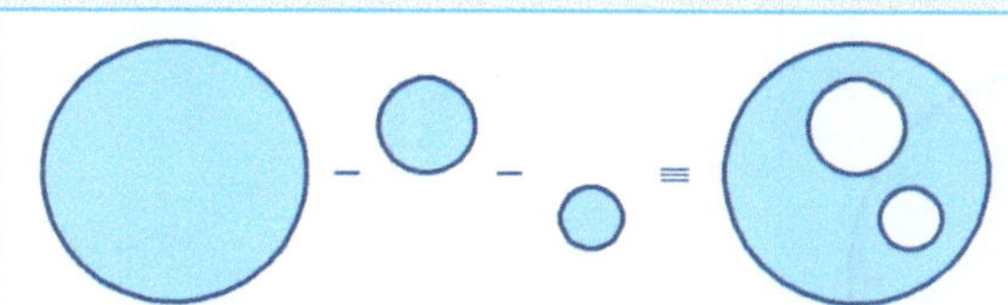

Area	$\pi \times 5^2$	$\pi \times 2^2$	$\pi \times 1^2$	$\pi(5^2 - 2^2 - 1^2)$
x	0	0	2	$\bar{x}$
y	0	2	−2	$\bar{y}$

$$\pi 5^2\begin{pmatrix}0\\0\end{pmatrix} - \pi 2^2\begin{pmatrix}0\\2\end{pmatrix} - \pi 1^2\begin{pmatrix}2\\-2\end{pmatrix} = \pi(5^2 - 2^2 - 1^2)\begin{pmatrix}\bar{x}\\\bar{y}\end{pmatrix}$$

$$\begin{pmatrix}0\\0\end{pmatrix} - \begin{pmatrix}0\\8\end{pmatrix} - \begin{pmatrix}2\\-2\end{pmatrix} = 20\begin{pmatrix}\bar{x}\\\bar{y}\end{pmatrix}$$

$$\begin{pmatrix}-2\\-6\end{pmatrix} = 20\begin{pmatrix}\bar{x}\\\bar{y}\end{pmatrix}$$

$$\begin{pmatrix}-0.1\\-0.3\end{pmatrix} = \begin{pmatrix}\bar{x}\\\bar{y}\end{pmatrix}$$

The coordinates of the centre of mass of the lamina are (−0.1, −0.3).

Problem-solving

When you remove part of a lamina, you can deal with the removed sections by adding sections with negative mass.

Setting out the key information in a table helps to clarify your working.

Note the subtraction signs for each removed section.

Cancel the πs and simplify.

Solve for $\bar{x}$ and $\bar{y}$.

Exercise 3D SKILLS PROBLEM-SOLVING

1 The following diagrams show uniform plane figures. Each one is drawn on a grid of unit squares. Find, in each case, the coordinates of the centre of mass.

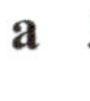

a

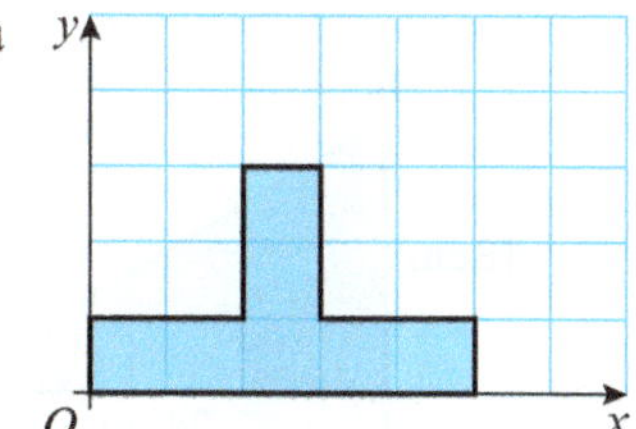

b

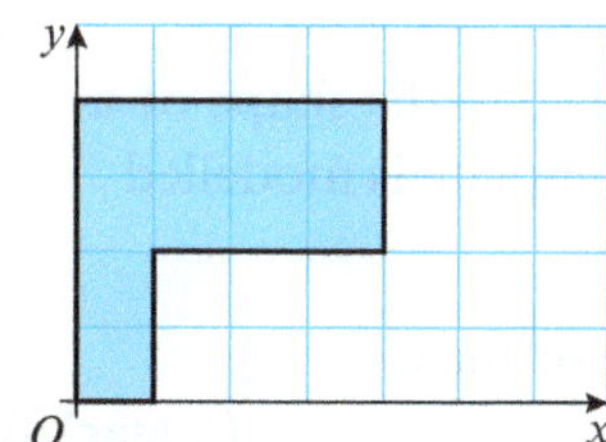

c

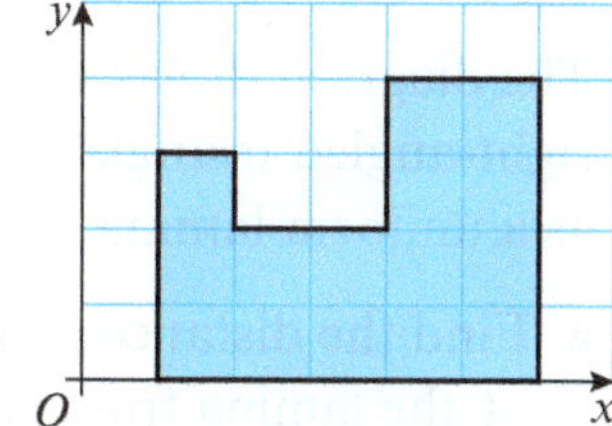

d

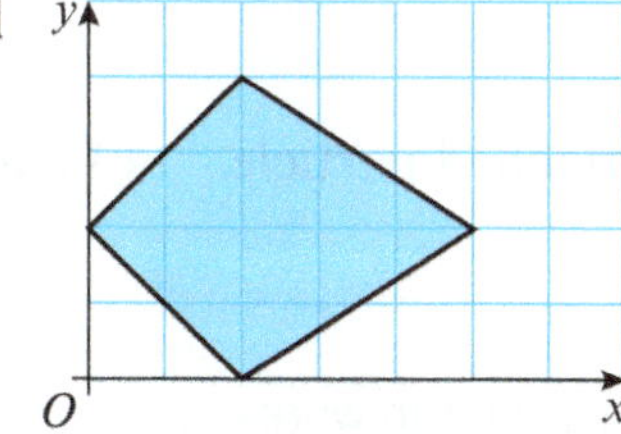

e

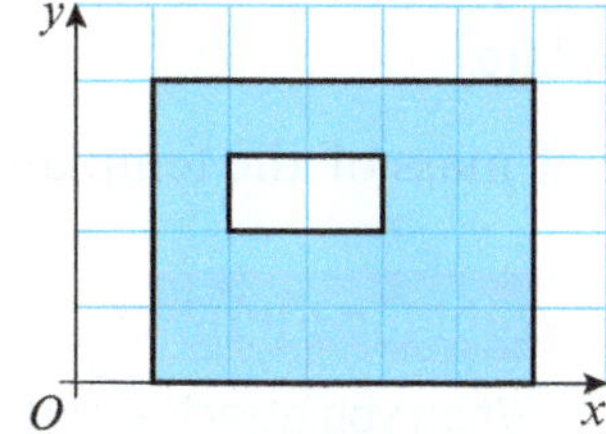

f

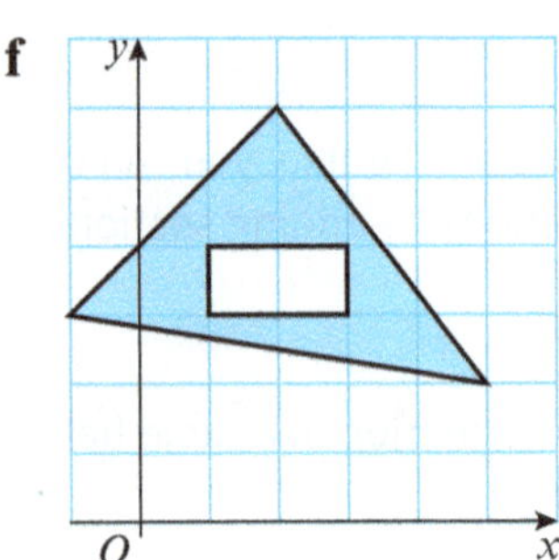

g

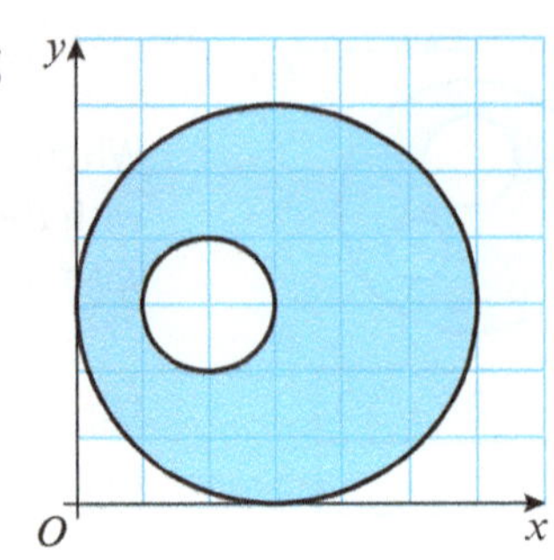

(E) **2** The uniform lamina $PQRST$ is formed by removing the triangle PQR from the rectangle $PRST$ with centre Q. The rectangle has sides of length $4a$ and $2a$. Find the distance of the centre of mass of $PQRST$ from Q. **(4 marks)**

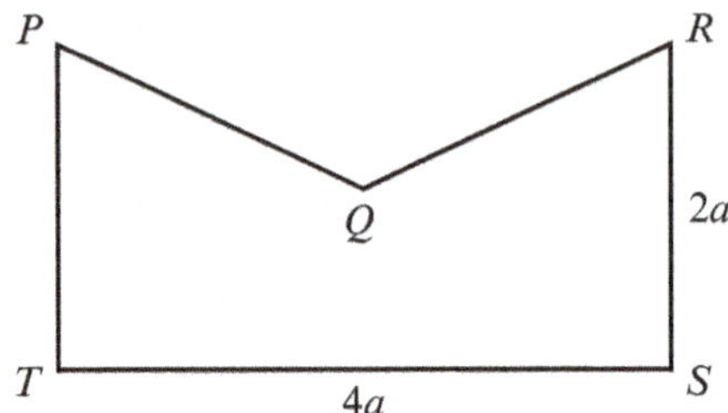

(E) **3** The uniform lamina shown in the diagram is formed by removing a square $DEFG$ of side length a from the equilateral triangle ABC of side length $5a$. The centre of mass of $DEFG$ lies on the perpendicular bisector (a line that divides something into two equal parts) of AC. Given that AC and DG are parallel and a distance a apart, work out the distance of the centre of mass of the whole lamina from B. **(6 marks)**

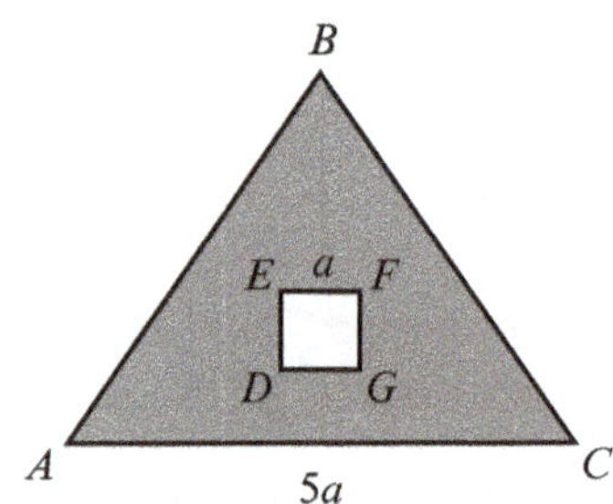

(E/P) **4** The diagram shows a metal template in the shape of a right-angled triangle ABC. The template is modelled as a uniform lamina.

B

18 cm

A 24 cm C

a Find the distance of the centre of mass of the lamina from A. **(4 marks)**

The mass of the template is 15 kg. A particle of mass 5 kg is attached to vertex C of the template.

b Find the position of the centre of mass of the template with the attached particle. **(3 marks)**

Problem-solving

When you attach a particle to a lamina, you can work out the new centre of mass by considering the lamina as a single particle whose weight acts at its centre of mass.

(E/P) **5** The diagram shows a uniform lamina formed from two rectangles. All the angles are right angles.

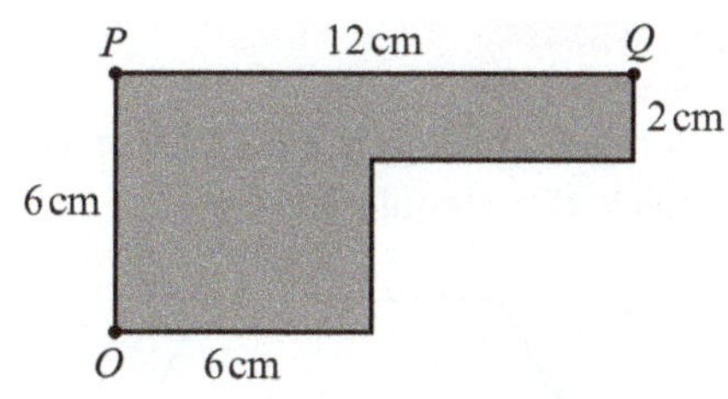

a Find the position of the centre of mass of the lamina. **(4 marks)**

The mass per unit area of the lamina is 30 g cm^{-2}. Two particles of masses 200 g and 500 g are attached to points P and Q on the lamina respectively.

b Find the new centre of mass of the lamina, giving any lengths correct to 3 significant figures. **(3 marks)**

(E/P) **6** The diagram shows a uniform piece of card $ABCDEFGH$ where $AB = 6$ cm, $BC = 10$ cm, $CD = 8$ cm, $DE = 2$ cm, $EF = 4$ cm, $FG = 6$ cm, $GH = 2$ cm, $HA = 2$ cm.

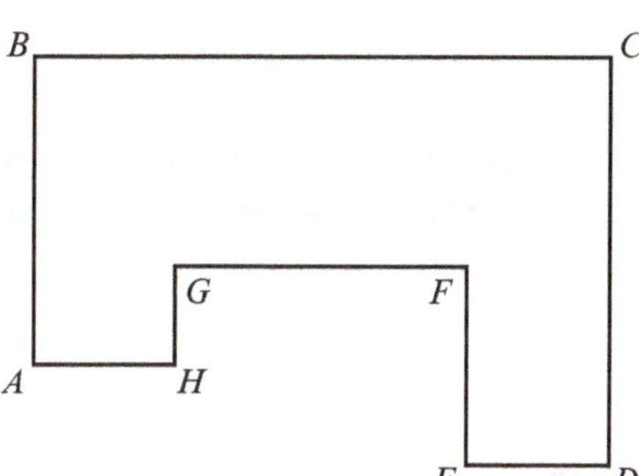

Assume that A is the origin and AH lies on the x-axis.

a Show that the centre of mass lies at the point $\left(\frac{69}{13}, \frac{41}{13}\right)$. **(7 marks)**

The template needs to be changed so that the centre of mass lies at the point $\left(\frac{69}{13}, \frac{41}{13}\right)$. To achieve this, two squares of side length 3 cm are cut out of the card. Their midpoints are $\left(a, \frac{41}{13}\right)$ and $\left(5, \frac{41}{13}\right)$.

b Explain how you can use symmetry to determine the value of a. **(2 marks)**

c Find the value of a. **(5 marks)**

(E) **7** A uniform circular disc has centre O and radius $3a$. The lines AB and CD are perpendicular diameters of the disc. A circular hole of radius x is made in the disc, with the centre of the hole at the point E on AB and the edge of the hole touching O to form the lamina shown on the right. Given that the centre of mass of the lamina lies on the line AB a distance of $\frac{23}{8}a$ from the point B, find the value of x in terms of a. **(6 marks)**

Challenge

A regular hexagon $ABCDEF$ of side length x has the triangle DEF removed to leave the irregular pentagon $ABCDF$ as shown in the diagram below.

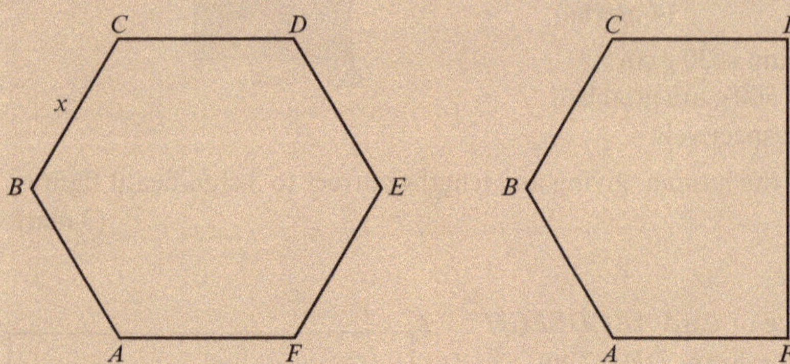

Given that the centre of mass of the hexagon is at the point M and that the centre of mass of the pentagon is at the point N, show that the length of MN is $\frac{2}{15}x$.

3.5 Centre of mass of a framework

You can find the centre of mass of a framework by using the centre of mass of each rod or wire which makes up the framework.

Links The centre of mass of a uniform straight rod is located at its midpoint.

Uniform circular arc

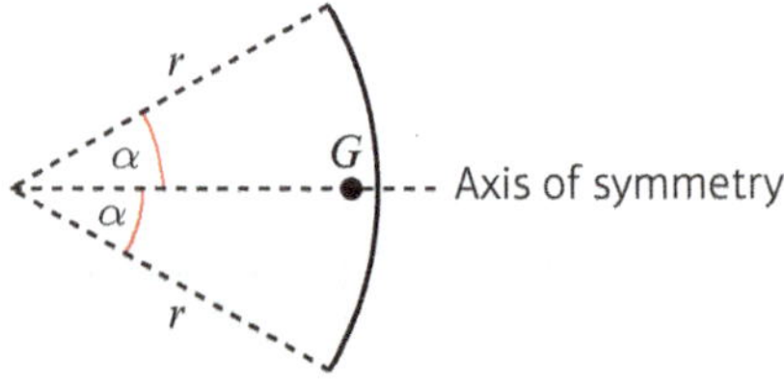

Hint This result can be found in the formula booklet.

A uniform circular arc of radius r and centre angle 2α, where α is measured in radians, has its centre of mass on the axis of symmetry at a distance $\frac{r\sin\alpha}{\alpha}$ from the centre.

- **A framework consists of a number of rods joined together or a number of pieces of wire joined together.**

Provided that you can identify the position of the centre of mass of each of the rods or pieces of wire that make up a framework you can find the position of the centre of mass of the whole framework.

Example 14

A framework consists of a uniform length of wire which has been bent into the shape of a letter L, as shown.

Find the distance of the centre of mass of the framework from AB and AF.

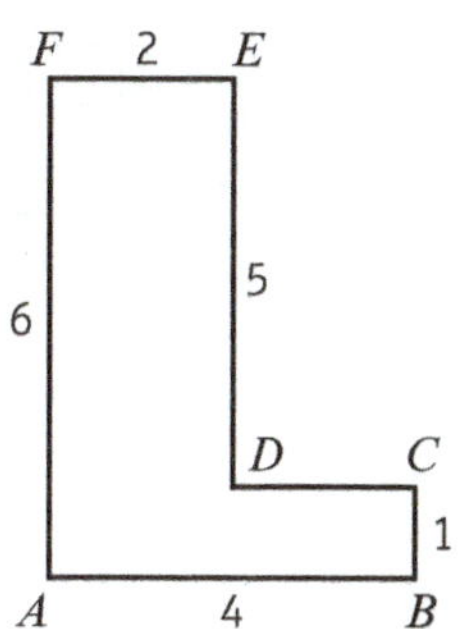

Online Explore centres of mass of a framework using GeoGebra.

Since the wire is uniform, the mass of each edge will be proportional to its length. The centre of mass of each edge will be at its midpoint.

Taking A as the origin and axes along AB and AF:

$$4\begin{pmatrix}2\\0\end{pmatrix} + 1\begin{pmatrix}4\\0.5\end{pmatrix} + 2\begin{pmatrix}3\\1\end{pmatrix} + 5\begin{pmatrix}2\\3.5\end{pmatrix} + 2\begin{pmatrix}1\\6\end{pmatrix} + 6\begin{pmatrix}0\\3\end{pmatrix} = 20\begin{pmatrix}\bar{x}\\\bar{y}\end{pmatrix}$$

Each term on the LHS consists of the length of an edge multiplied by the position vector of the midpoint of the edge.

$$\begin{pmatrix}8\\0\end{pmatrix} + \begin{pmatrix}4\\0.5\end{pmatrix} + \begin{pmatrix}6\\2\end{pmatrix} + \begin{pmatrix}10\\17.5\end{pmatrix} + \begin{pmatrix}2\\12\end{pmatrix} + \begin{pmatrix}0\\18\end{pmatrix} = 20\begin{pmatrix}\bar{x}\\\bar{y}\end{pmatrix}$$

$$\begin{pmatrix}30\\50\end{pmatrix} = 20\begin{pmatrix}\bar{x}\\\bar{y}\end{pmatrix}$$

Simplify and collect terms.

$$\begin{pmatrix}1.5\\2.5\end{pmatrix} = \begin{pmatrix}\bar{x}\\\bar{y}\end{pmatrix}$$

Distance from AF is 1.5 cm. — This is $\bar{x}$.

Distance from AB is 2.5 cm. — This is $\bar{y}$.

Example 15

Find the position of the centre of mass of a framework constructed from a uniform piece of wire bent into the shape shown,

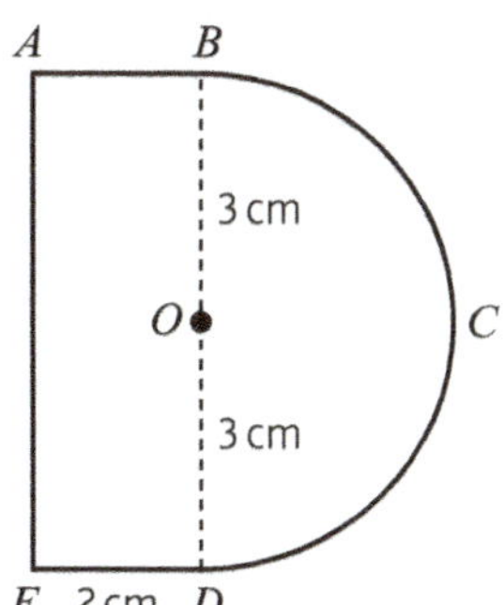

where the wire BCD is a semicircle, centre O, of radius 3 cm and wire $BAED$ forms three sides of a rectangle $ABDE$.

Take O as the origin and axes along OC and OB.

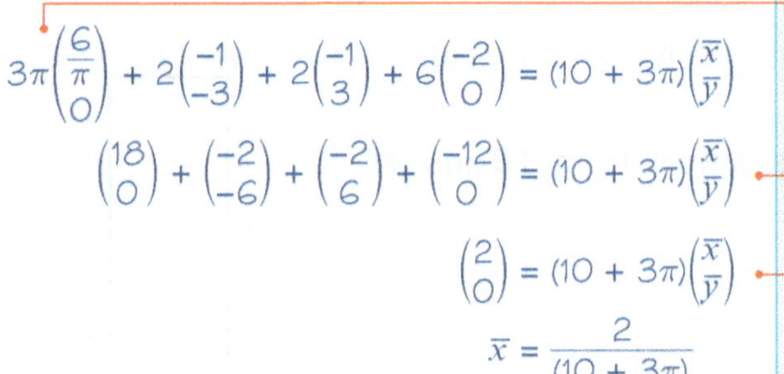

$$3\pi\begin{pmatrix}\frac{6}{\pi}\\0\end{pmatrix} + 2\begin{pmatrix}-1\\-3\end{pmatrix} + 2\begin{pmatrix}-1\\3\end{pmatrix} + 6\begin{pmatrix}-2\\0\end{pmatrix} = (10 + 3\pi)\begin{pmatrix}\bar{x}\\\bar{y}\end{pmatrix}$$

Use the result for the centre of mass of a uniform circular arc with $\alpha = \frac{\pi}{2}$ and $r = 3$ to find the centre of mass of the arc BCD. 3π is the length of the arc.

$$\begin{pmatrix}18\\0\end{pmatrix} + \begin{pmatrix}-2\\-6\end{pmatrix} + \begin{pmatrix}-2\\6\end{pmatrix} + \begin{pmatrix}-12\\0\end{pmatrix} = (10 + 3\pi)\begin{pmatrix}\bar{x}\\\bar{y}\end{pmatrix}$$

Simplify.

$$\begin{pmatrix}2\\0\end{pmatrix} = (10 + 3\pi)\begin{pmatrix}\bar{x}\\\bar{y}\end{pmatrix}$$

Collect terms.

$$\bar{x} = \frac{2}{(10 + 3\pi)}$$

$$\bar{y} = 0$$

You could have used the symmetry to deduce this without any working.

G is the centre of mass of the framework, on the axis of symmetry, a distance $\frac{2}{10 + 3\pi}$ cm to the right of O.

Exercise 3E SKILLS PROBLEM-SOLVING

1 By regarding the shapes shown below as uniform plane wire frameworks, find the coordinates of the centre of mass of each shape.

a

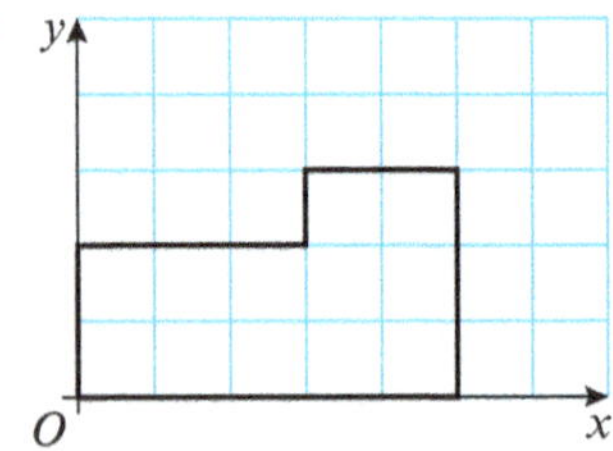

b

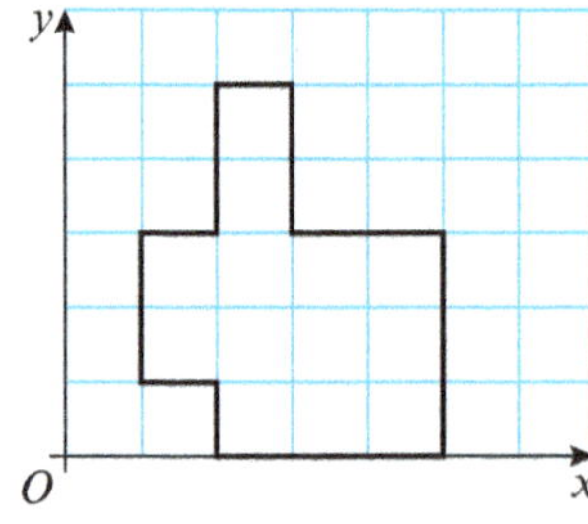

c

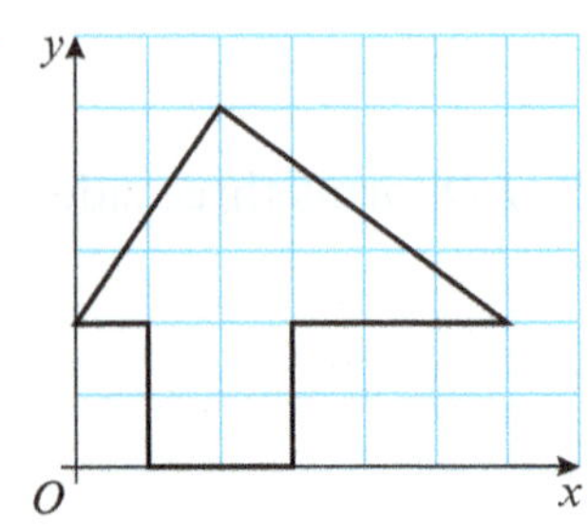

d

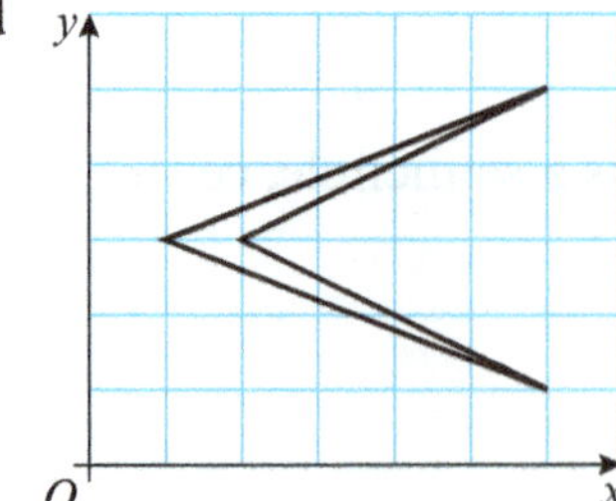

E/P **2** Find the position of the centre of mass of the framework shown in the diagram, which is formed by bending a uniform piece of wire of total length $(12 + 2\pi)$ cm to form a sector of a circle, centre O, radius 6 cm. **(4 marks)**

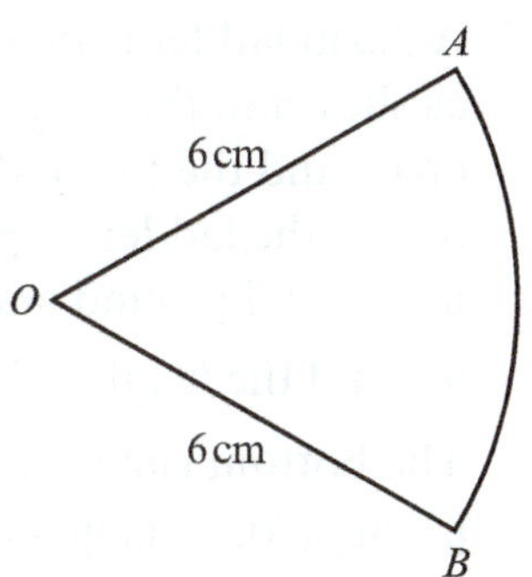

E **3** A framework consists of a uniform length of wire which has been bent into the shape of a letter T, as shown.

Find the distance of the centre of mass of the framework from AB. **(6 marks)**

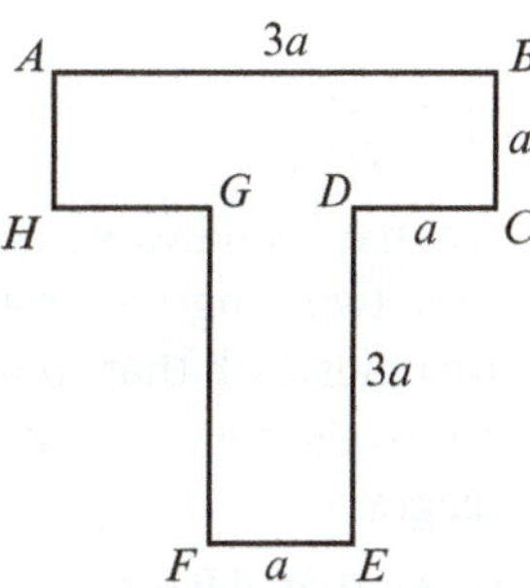

E/P **4** A uniform framework is constructed by bending wire into the shape of a semicircle and a diameter. The semicircle has radius 15 cm.

a Find the distance of the centre of mass of the framework from AB. **(4 marks)**

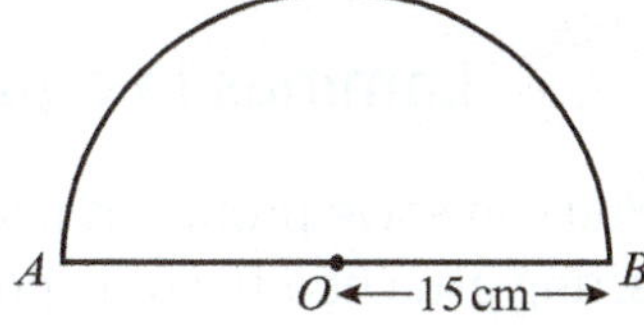

The metal used to form the framework has a mass of 8 grams per cm. Two identical particles of masses 100 g are attached to the framework, at A and B.

b Find:

i the total mass of the loaded framework

ii the distance of the centre of mass of the loaded framework from AB. **(4 marks)**

E/P **5** The diagram shows a triangular framework formed from uniform wire. Particles of masses 10 kg, 20 kg and 30 kg are attached to vertices A, B and C respectively. The mass of the unloaded framework is 15 kg.

Find the position of the centre of mass of the loaded framework. **(8 marks)**

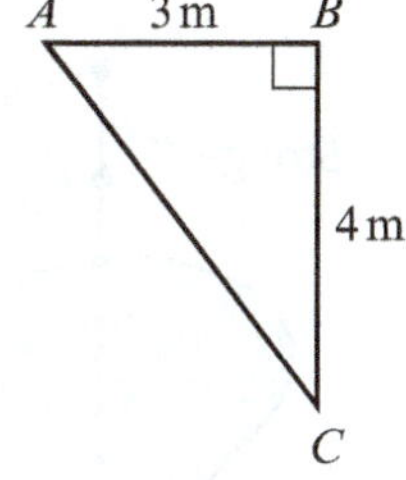

E/P **6** A uniform length of wire is bent to form the shape shown in the diagram.

ACB is a semicircle of radius 3 cm, centre O.
ADO and BEO are both semicircles of radius 1.5 cm.
Find the position of the centre of mass of the framework. **(6 marks)**

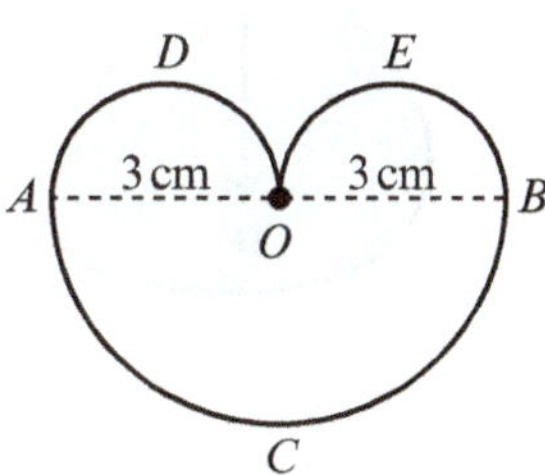

7 A 3.5 m ladder is modelled as a framework made from uniform wire as shown in the diagram. The rungs are 50 cm wide and are 50 cm apart and the top and bottom rungs are 50 cm from the base and top of the ladder respectively. The base of the ladder rests on horizontal ground and the ladder stands vertically.

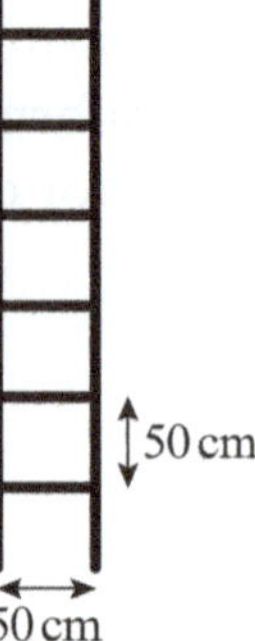

a Find the height of the centre of mass above the ground. **(2 marks)**

The bottom rung is removed from the ladder.

b Show that the height of the centre of mass of the ladder has increased by $\frac{5}{76}$ m. **(4 marks)**

Challenge

A metal framework $ABCDE$ is made from two congruent right-angled triangles such that ACD and BCE are straight lines, as shown in the diagram.

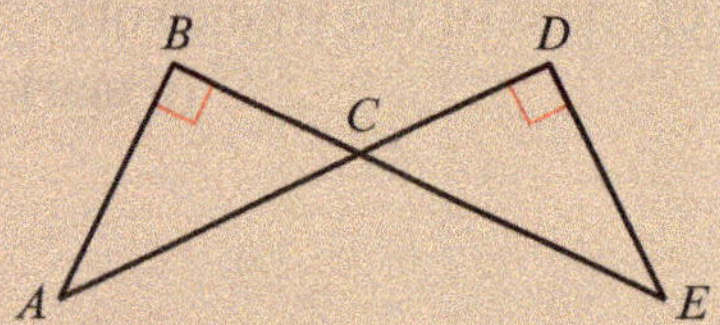

Given that $AB = 4$ cm and $CD = 3$ cm, work out the distance between C and the centre of mass of the framework.

3.6 Laminas in equilibrium

You can solve problems involving a lamina in **equilibrium**. A lamina can be **suspended** by means of a string attached to some point of the lamina, or can be allowed to **pivot** freely about a horizontal axis which passes through some point of the lamina.

- **When a lamina is suspended freely from a fixed point, or pivots freely about a horizontal axis, it will rest in equilibrium in a vertical plane with its centre of mass vertically below the point of suspension or the pivot.**

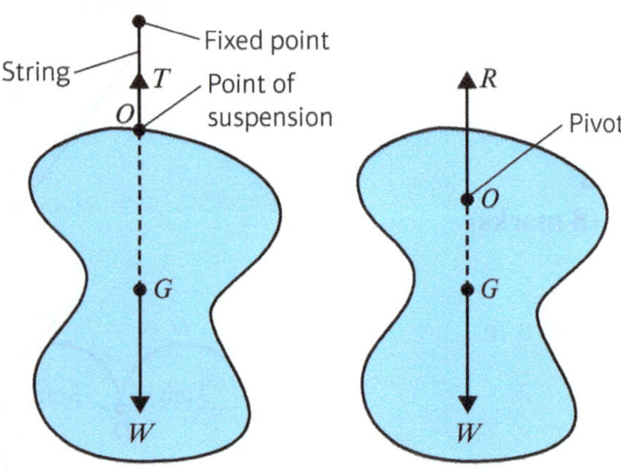

Hint The lamina on the left is suspended from a fixed point. There are only two forces acting on it: the weight of the lamina and the **tension** in the string. Both forces pass through the point of suspension.

The lamina on the right is free to rotate about a fixed horizontal pivot. There are only two forces acting on it: the weight of the lamina and the reaction of the pivot on the lamina. Both pass through the pivot.

The **resultant** of the moments about O in both laminas is zero.

Example 16 SKILLS PROBLEM-SOLVING

Find the angle that the line AB makes with the vertical if this L-shaped uniform lamina is freely suspended from:

a A

b B

c E

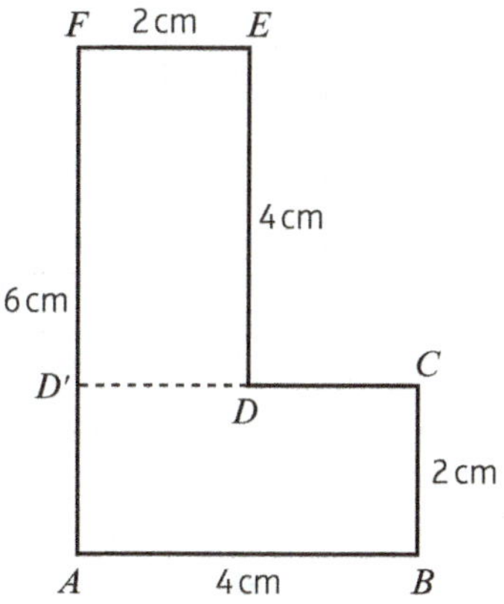

First find the centre of mass of the lamina.

Split the lamina along CD'. — Split the lamina into two rectangles.

Take A as the origin and axes along AB and AF:

$$8\begin{pmatrix}2\\1\end{pmatrix} + 8\begin{pmatrix}1\\4\end{pmatrix} = 16\begin{pmatrix}\bar{x}\\\bar{y}\end{pmatrix}$$

Area $ABCD' = 8$
Area $DEFD' = 8$

$$\begin{pmatrix}24\\40\end{pmatrix} = 16\begin{pmatrix}\bar{x}\\\bar{y}\end{pmatrix}$$

Simplify.

$\bar{x} = 1.5$

$\bar{y} = 2.5$

a

F, E, D, C, G, A, B, θ, 2.5, 1.5

Downward vertical

This is the angle between the vertical and the line AB.

$$\tan\theta = \frac{2.5}{1.5}$$

$\Rightarrow \theta = 59.0°$ (3 s.f.) — θ is the angle required.

Problem-solving

You do not need to draw the lamina hanging. Draw a line from the point of suspension to the centre of mass. Mark this in as the vertical.

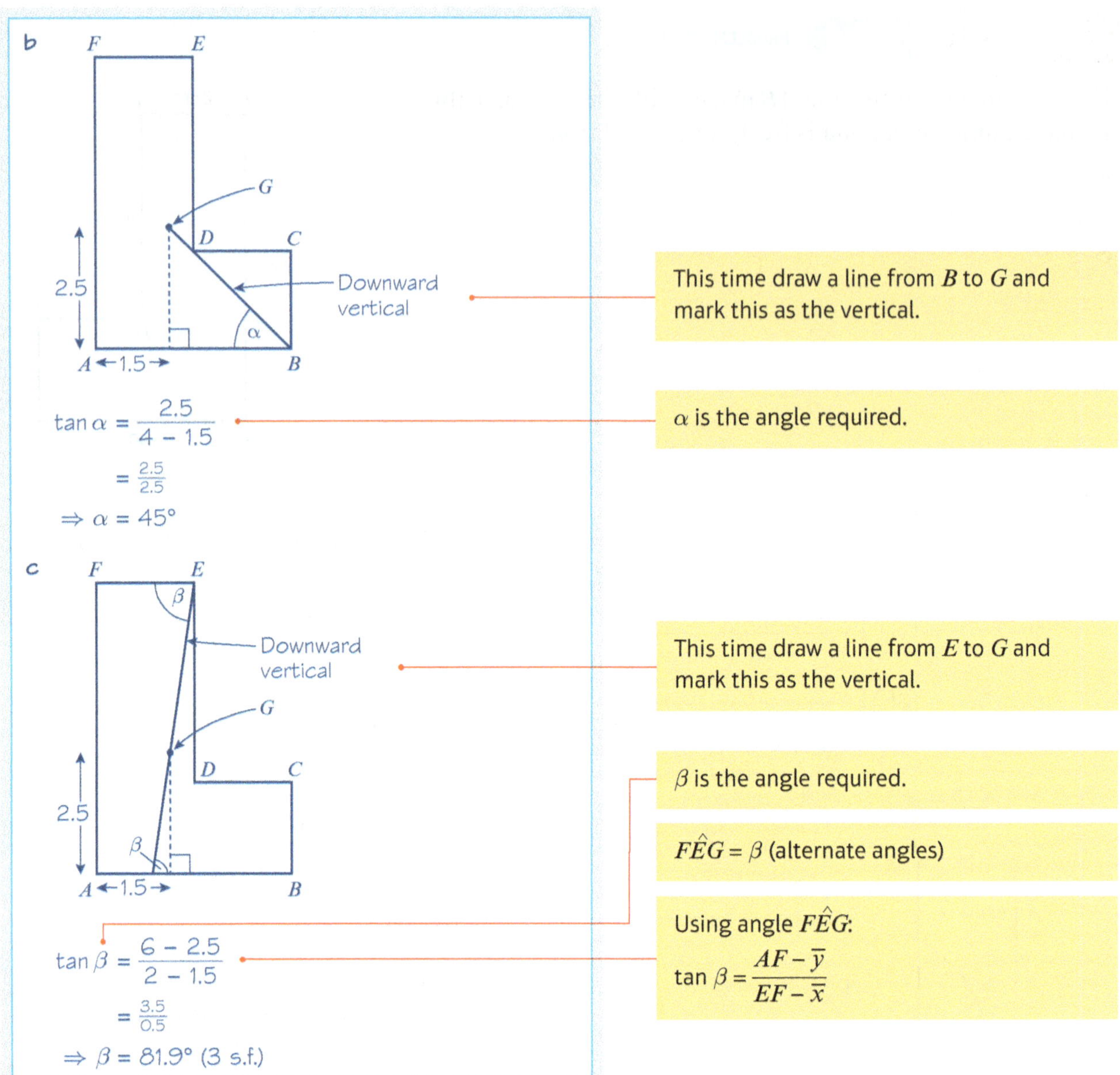

Example 17

The L-shaped lamina in Example 14 has mass M kg. Find the angle that FE makes with the vertical when a mass of $\frac{1}{10}M$ kg is attached to B and the lamina is freely suspended from F.

$$M\begin{pmatrix}1.5\\2.5\end{pmatrix} + \frac{1}{10}M\begin{pmatrix}4\\0\end{pmatrix} = \frac{11}{10}M\begin{pmatrix}\bar{x}\\\bar{y}\end{pmatrix}$$

$$\begin{pmatrix}1.9\\2.5\end{pmatrix} = \frac{11}{10}\begin{pmatrix}\bar{x}\\\bar{y}\end{pmatrix}$$

$$\begin{pmatrix}\bar{x}\\\bar{y}\end{pmatrix} = \begin{pmatrix}\frac{19}{11}\\\frac{25}{11}\end{pmatrix}$$

Problem-solving

Consider the lamina as a single particle whose weight acts at its centre of mass.

Recall from Example 14 that the centre of mass of the lamina is at (1.5, 2.5).

The centre of mass of the added mass is at (4, 0).

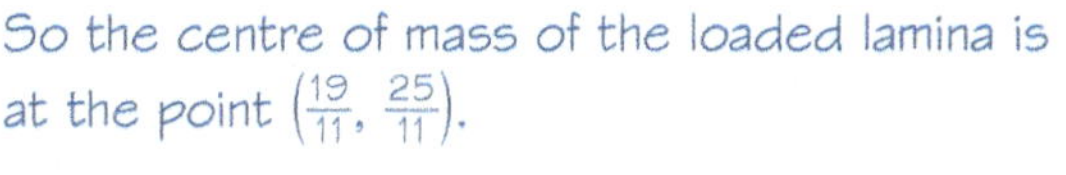

So the centre of mass of the loaded lamina is at the point $\left(\frac{19}{11}, \frac{25}{11}\right)$.

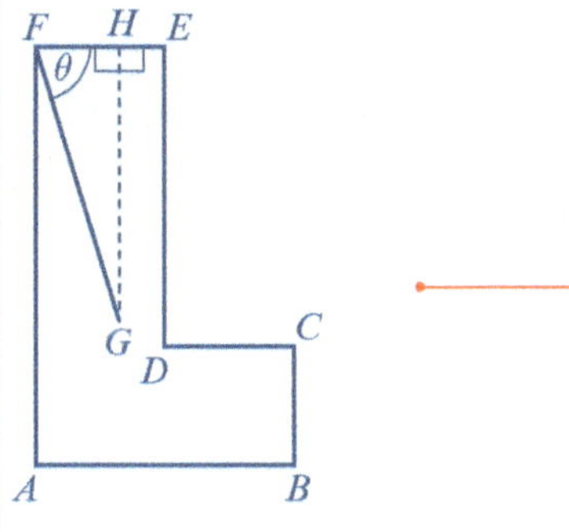

$$\tan\theta = \frac{HG}{FH} = \frac{6 - \frac{25}{11}}{\frac{19}{11}} = \frac{41}{19}$$

$$\theta = 65.1° \text{ (3 s.f.)}$$

Redraw the diagram showing the angle that you require.
G is the centre of mass and θ is the angle *FE* makes with the vertical when the lamina is suspended from *F*.

You can determine whether a body will remain in equilibrium or if equilibrium will be broken by sliding or toppling.

- **If a lamina rests in equilibrium on a rough inclined plane then the line of action of the weight of the lamina must pass through the side of the lamina *AB* which is in contact with the plane.**

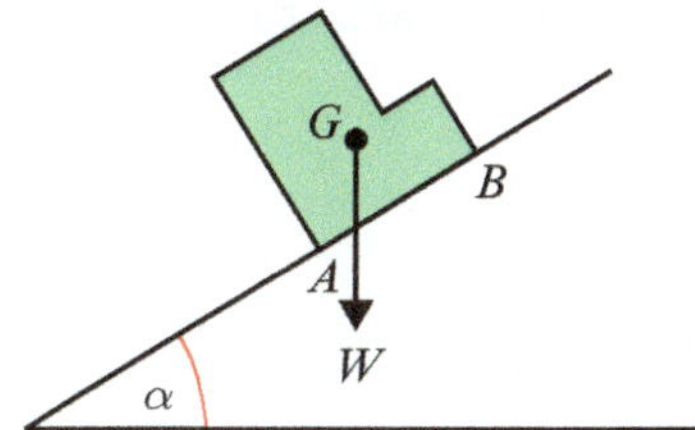

The weight of the lamina produces a clockwise moment about *A* which keeps the lamina in contact with the plane.

If the angle of the plane is increased so that the line of action of the weight passes outside the side *AB* then the weight produces an anticlockwise moment about *A*. The lamina will **topple** over.

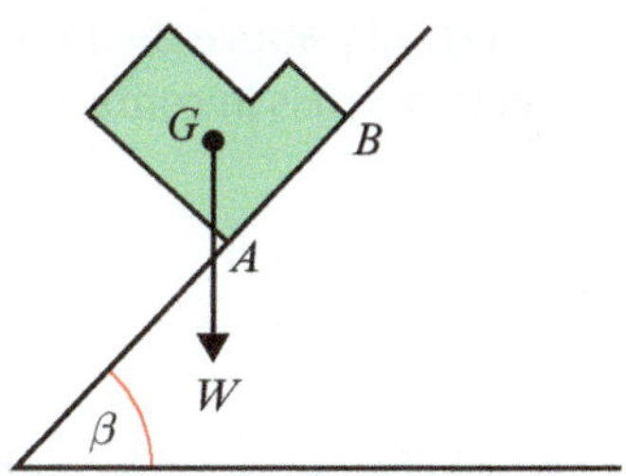

You can usually assume that the coefficient of friction between the lamina and the plane is large enough to prevent the lamina from slipping down the plane.

Example 18

The L-shaped lamina from Example 14 is placed with *AB* in contact with a rough inclined plane. The angle of the plane is gradually increased. Assuming that the lamina does not slide down the plane, find the angle that the plane makes with the horizontal when the lamina is about to topple over.

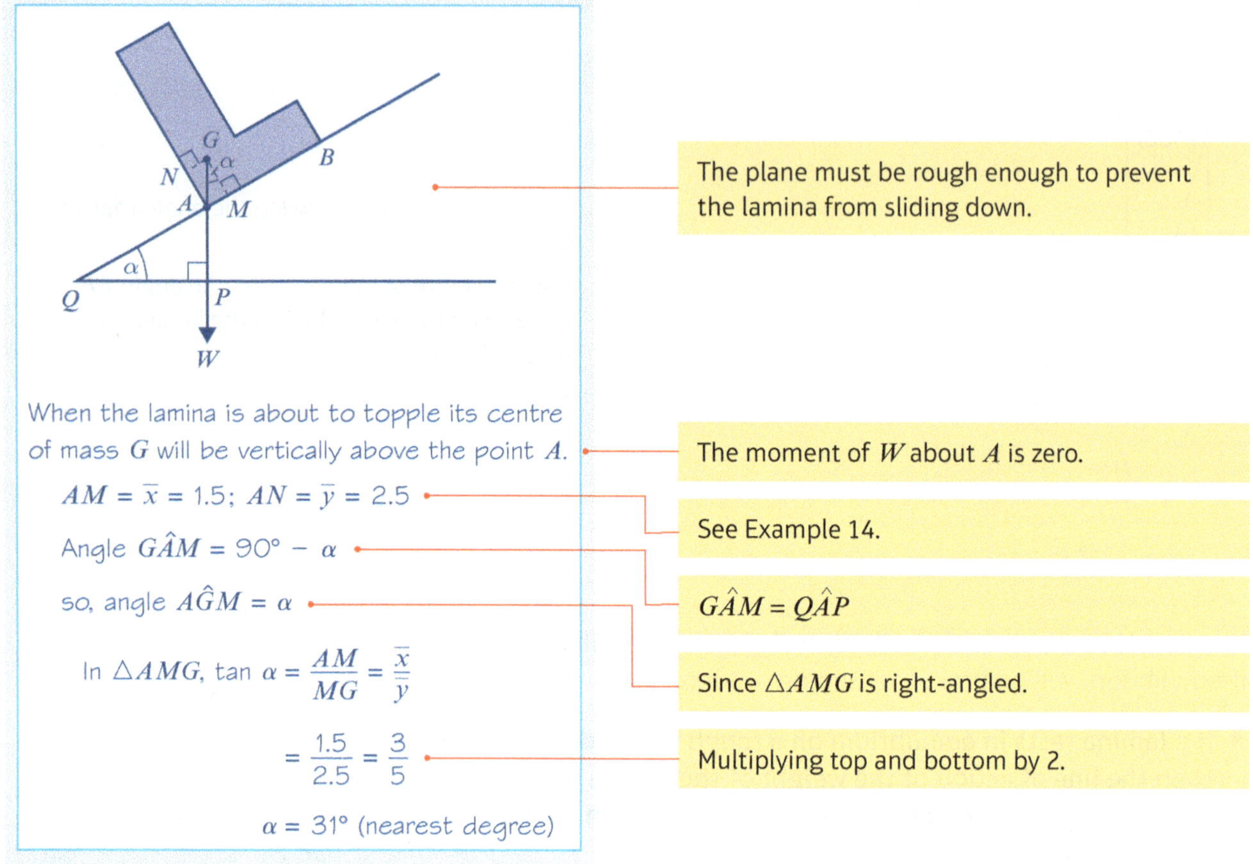

Hint When a lamina on an inclined plane is about to topple, its centre of mass will be vertically above the lowest point of the lamina which is in contact with the plane.

A similar result applies for a **rigid body**.

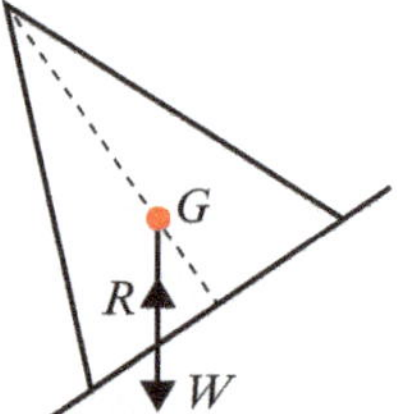

The only forces acting on the body are its weight and the total reaction between the plane and the body. As the body is in equilibrium, these forces must be equal and opposite and act in the same vertical line.

- **When a rigid body rests in equilibrium on a horizontal or rough inclined plane, then the line of action of the weight of the body must pass through the area of contact with the plane.**

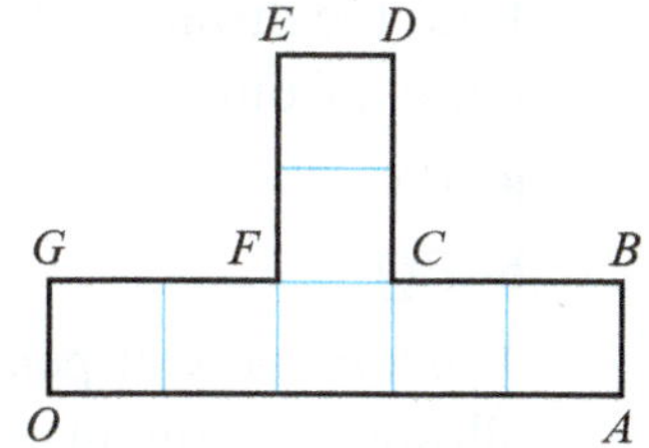

1 a The diagram shows a uniform lamina.

The lamina is freely suspended from the point O and hangs in equilibrium.
Find the angle between OA and the downward vertical.

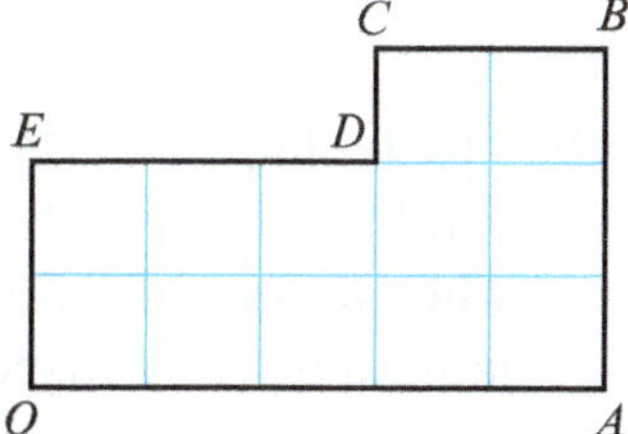

b The diagram shows a uniform lamina.

The lamina is freely suspended from the point O and hangs in equilibrium. Find the angle between OA and the downward vertical.

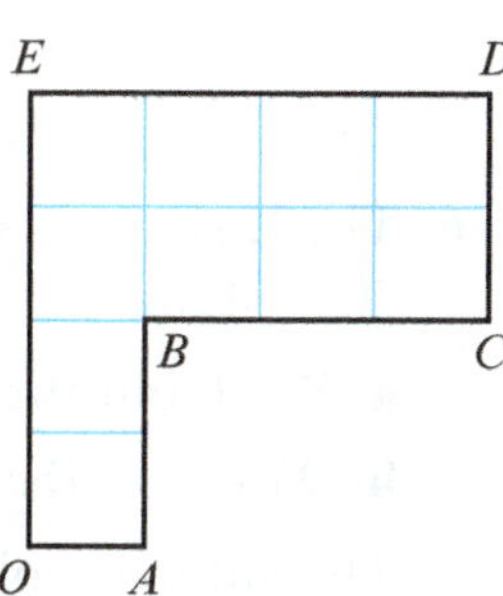

c The diagram shows a uniform lamina.

The lamina is freely suspended from the point O and hangs in equilibrium.
Find the angle between OA and the downward vertical.

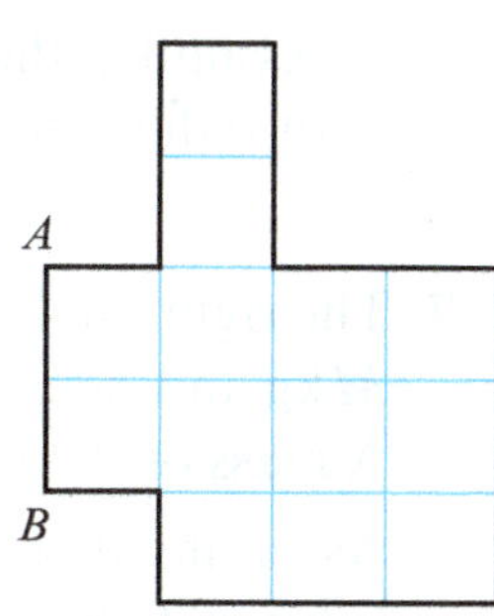

2 The diagram shows a uniform lamina.

The lamina is freely suspended from the point A and hangs in equilibrium.
Find the angle between AB and the downward vertical.

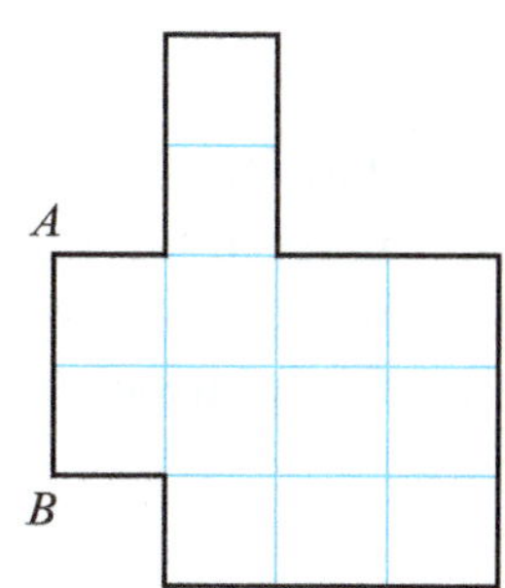

3 The diagram shows a uniform lamina.

The lamina is free to rotate about a fixed smooth horizontal axis perpendicular to the plane of the lamina, passing through the point A, and hangs in equilibrium.

Find the angle between AB and the horizontal.

E **4** $PQRS$ is a uniform lamina.

Find the distance of the centre of mass of the lamina from:

a PS **(4 marks)**

b PQ **(4 marks)**

The lamina is suspended from the point Q and allowed to hang in equilibrium.

c Find the angle that PQ makes with the vertical. **(3 marks)**

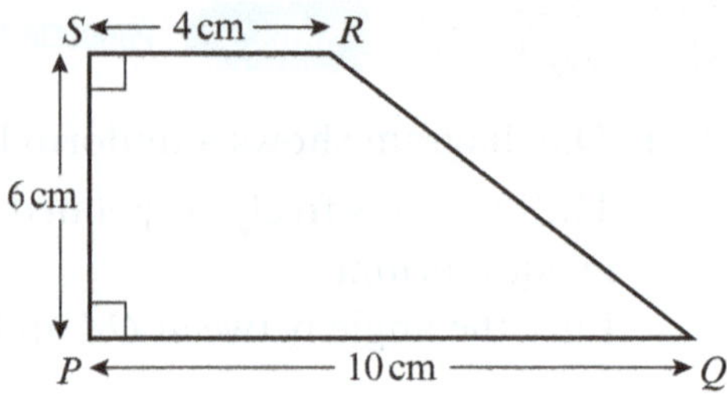

E/P **5** The L-shaped lamina shown has mass M kg. Find the angle that BC makes with the vertical when a mass of $0.2M$ kg is attached to F and the lamina is freely suspended from C. **(8 marks)**

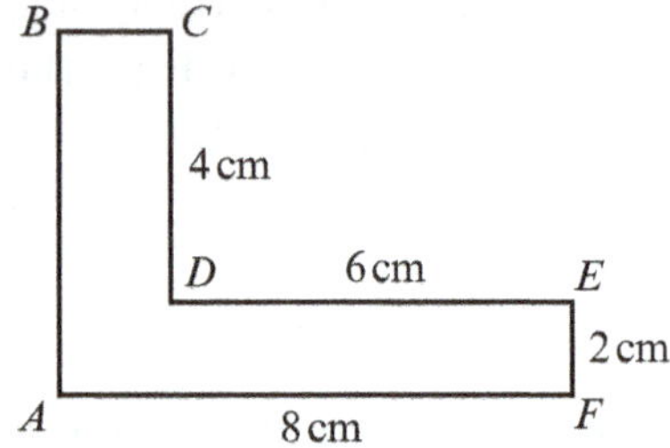

6 The L-shaped lamina $ABCDEF$ shown has sides of length $AB = 2$ cm, $BC = 4$ cm, $CD = 2$ cm, $DE = 2$ cm, $EF = 4$ cm and $FA = 6$ cm.

a Work out the horizontal distance of the centre of mass from FA.

b Work out the vertical distance of the centre of mass from EF.

The lamina is placed on a rough slope that is inclined at an angle of θ to the horizontal.

c Assuming that the lamina does not slide, work out the maximum value of θ that can be reached without the lamina toppling.

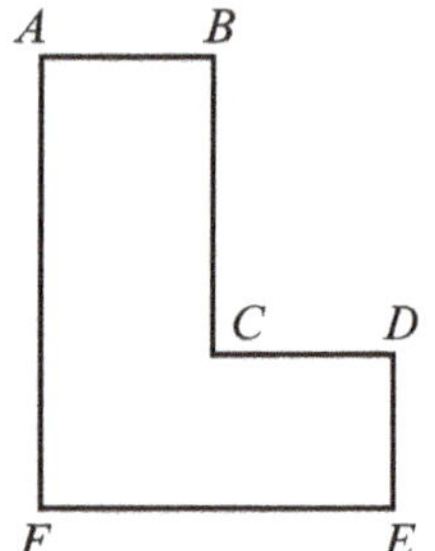

7 The right-angled triangular lamina ABC shown has mass M kg and sides of length $BC = 3$ cm and $AC = 6$ cm.
A mass of $0.2M$ kg is attached to the lamina at A.

Assuming that the lamina does not slide, work out the maximum value of θ that can be reached without the lamina toppling.

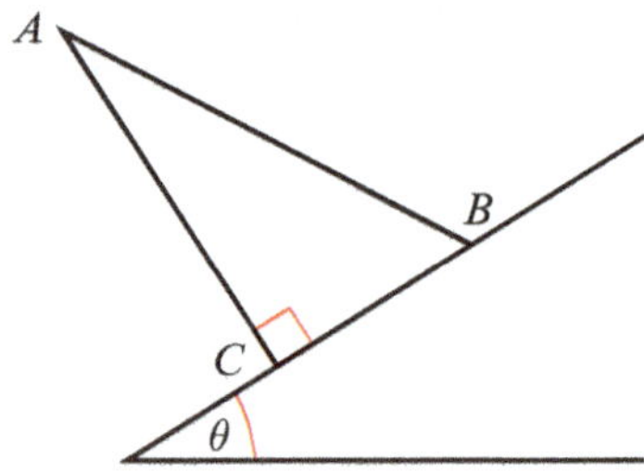

E/P **8** The lamina shown is a quarter circle with radius 4 cm and mass M kg. Find the angle that AB makes with the vertical when a mass of $0.5M$ kg is attached to C and the lamina is freely suspended from B. **(6 marks)**

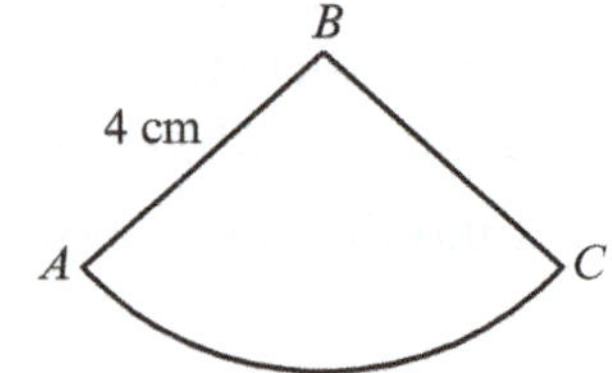

Challenge

The rectangular lamina $ABCD$ shown has mass M kg and sides of length $AB = CD = 4$ cm, $BC = DA = 6$ cm. A mass of kM kg is attached at A. The lamina is placed on a rough slope that is angled at 30° to the horizontal.

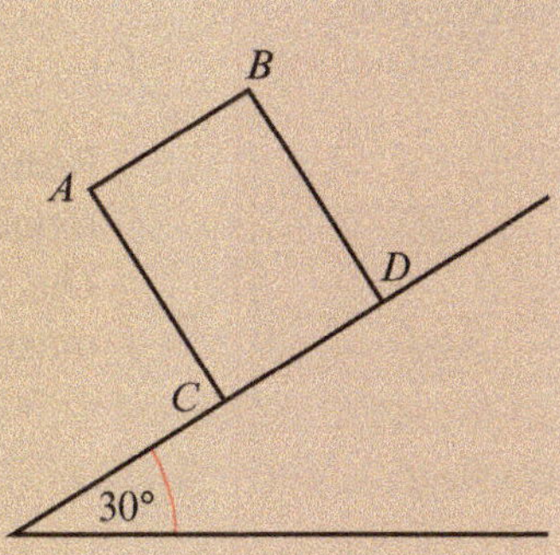

Assuming that the lamina does not slip, show that the maximum value of k that can be reached before the lamina topples is $\frac{1}{\sqrt{3}} - \frac{1}{2}$.

3.7 Frameworks in equilibrium

Problems involving frameworks in equilibrium can be solved using the same methods that are used to solve problems involving laminas in equilibrium.

Example 19

The inverted T-shaped uniform framework of weight W shown in the diagram is freely suspended from D and G by two vertical strings, so that AH is horizontal. The strings and the framework lie in the same plane.

a Find, in terms of W, the tensions T_1 and T_2.

The string at D breaks and the framework hangs freely in equilibrium from G.

b Find the angle that DE makes with the vertical when the framework comes to rest in equilibrium.

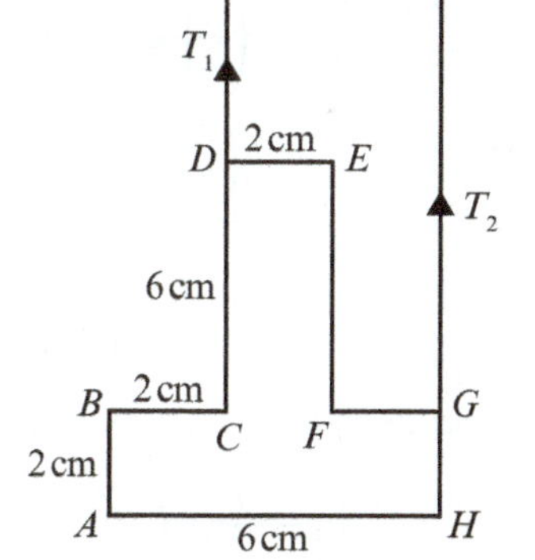

a Let AH lie on the x-axis and AB lie on the y-axis with the point A at the origin.

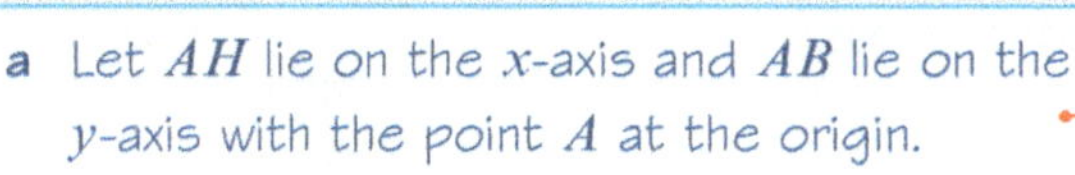

$$2\begin{pmatrix}0\\1\end{pmatrix} + 2\begin{pmatrix}1\\2\end{pmatrix} + 6\begin{pmatrix}2\\5\end{pmatrix} + 2\begin{pmatrix}3\\8\end{pmatrix} + 6\begin{pmatrix}4\\5\end{pmatrix} + 2\begin{pmatrix}5\\2\end{pmatrix} + 2\begin{pmatrix}6\\1\end{pmatrix} + 6\begin{pmatrix}3\\0\end{pmatrix} = 28\begin{pmatrix}\bar{x}\\\bar{y}\end{pmatrix}$$

$$28\begin{pmatrix}\bar{x}\\\bar{y}\end{pmatrix} = \begin{pmatrix}84\\88\end{pmatrix}$$

$$\begin{pmatrix}\bar{x}\\\bar{y}\end{pmatrix} = \begin{pmatrix}3\\\frac{22}{7}\end{pmatrix}$$

Res (↑) $T_1 + T_2 = W$

Taking moments about D gives

$W = 4T_2 \Rightarrow T_2 = \frac{1}{4}W$ and $T_1 = \frac{3}{4}W$

First find the centre of mass of the framework.

Problem-solving

The framework is in equilibrium, so the resultant moment must be 0. The centre of mass of the framework is a horizontal distance of 1 cm from D, so the clockwise moment is $1 \times W = W$.
The anticlockwise moment is due to T_2 so is equal to $4T_2$.

b The centre of mass lies $\frac{22}{7}$ cm from AH.

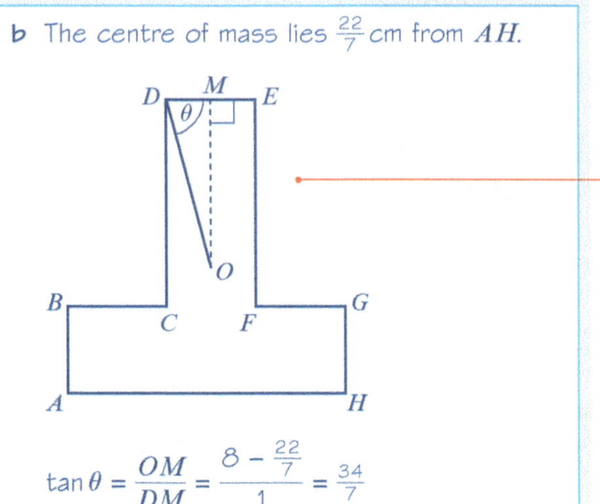

$\tan\theta = \frac{OM}{DM} = \frac{8 - \frac{22}{7}}{1} = \frac{34}{7}$

$\theta = 78.4°$ (3 s.f.)

Redraw the diagram showing the angle that you require.
O is the centre of mass and θ is the angle DE makes with the vertical when suspended from D.

Example 20

The inverted T-shaped uniform framework used in Example 19 has mass W kg.

a Find the location of the centre of mass when a mass of $\frac{1}{4}W$ kg is fixed at point H.

b Find the angle that DE makes with the vertical when the system is freely suspended from E.

a $W\begin{pmatrix}3\\ \frac{22}{7}\end{pmatrix} + \frac{1}{4}W\begin{pmatrix}6\\0\end{pmatrix} = \frac{5}{4}W\begin{pmatrix}\bar{x}\\ \bar{y}\end{pmatrix}$

$\begin{pmatrix}3\\ \frac{22}{7}\end{pmatrix} + \begin{pmatrix}\frac{3}{2}\\0\end{pmatrix} = \frac{5}{4}\begin{pmatrix}\bar{x}\\ \bar{y}\end{pmatrix}$

$\frac{5}{4}\begin{pmatrix}\bar{x}\\ \bar{y}\end{pmatrix} = \begin{pmatrix}\frac{9}{2}\\ \frac{22}{7}\end{pmatrix}$

$\begin{pmatrix}\bar{x}\\ \bar{y}\end{pmatrix} = \begin{pmatrix}\frac{18}{5}\\ \frac{88}{35}\end{pmatrix}$

So the centre of mass is at the point $\left(\frac{18}{5}, \frac{88}{35}\right)$

Recall from Example 19 that the centre of mass of the lamina is at $\left(3, \frac{22}{7}\right)$.
The centre of mass of the added mass is at (6, 0).

Find the centre of mass of the framework and the added mass.

b

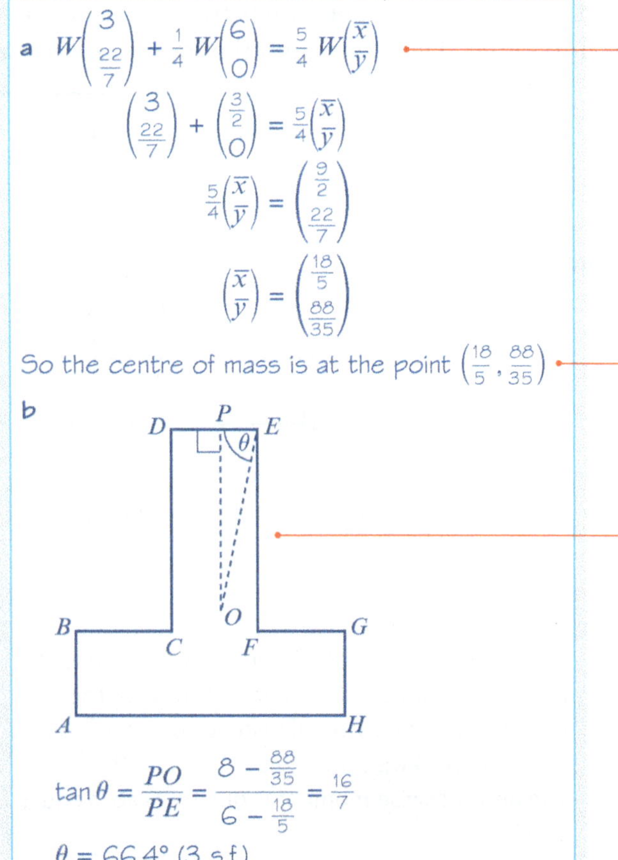

$\tan\theta = \frac{PO}{PE} = \frac{8 - \frac{88}{35}}{6 - \frac{18}{5}} = \frac{16}{7}$

$\theta = 66.4°$ (3 s.f.)

Redraw the diagram showing the angle that you require.
O is the centre of mass and θ is the angle DE makes with the vertical when the framework is suspended from E.

1 The uniform framework shown in the diagram is freely suspended from the point B and hangs in equilibrium. Find the angle between BC and the downward vertical.

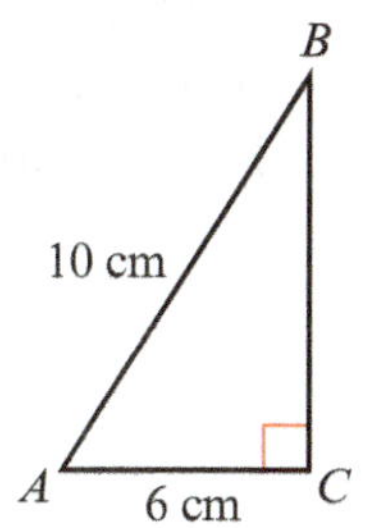

2 The uniform framework shown in the diagram is freely suspended from the point D and hangs in equilibrium. Find the angle between CD and the downward vertical.

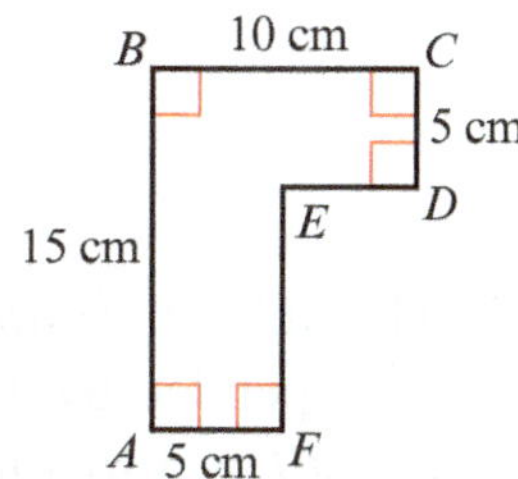

E/P 3 The uniform framework shown in the diagram is freely suspended from the point A and hangs in equilibrium.

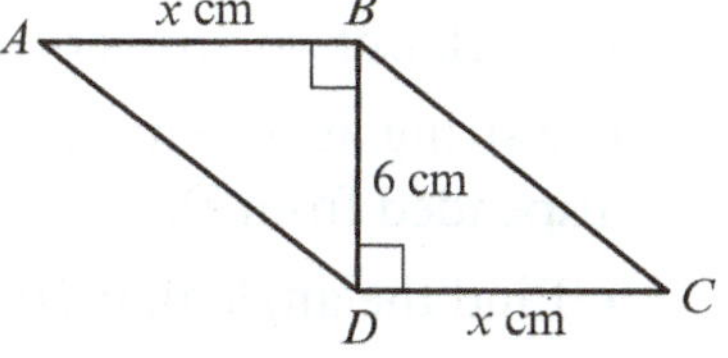

a Given that the angle between AB and the downward vertical is $\arctan \frac{3}{8}$, work out the value of x. **(4 marks)**

The framework has mass M kg.

b A particle of mass kM kg is attached to A and the framework is then freely suspended from the point B so that it hangs in equilibrium. Given that the downward vertical now makes an angle of $\arctan \frac{8}{15}$ with BD, work out the value of k. **(4 marks)**

P 4 The uniform framework shown in question 1 has mass M kg. A particle of mass $0.75M$ kg is attached to A and the framework is then freely suspended from the point B so that it hangs in equilibrium. Find the angle between BC and the downward vertical.

P 5 The uniform framework shown in question 2 has mass M kg. A particle of mass $0.15M$ kg is attached to F and the framework is then freely suspended from the point B so that it hangs in equilibrium. Find the angle between BC and the downward vertical.

E/P **6** The uniform framework shown in the diagram is made of a square of side 4 cm and two quarter circles and has mass M kg. A particle of mass $0.1M$ kg is attached to C and the framework is then freely suspended from the point A so that it hangs in equilibrium.
Find the angle between FE and the downward vertical. **(6 marks)**

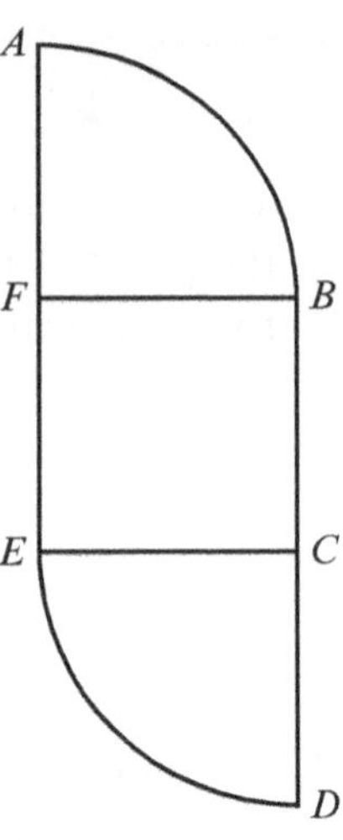

E/P **7** The uniform framework of weight W shown in the diagram is suspended by two vertical strings from A and D such that FE is horizontal. The strings and the framework lie in the same plane. The lengths shown are in metres and all the angles are right angles.

a Work out, in terms of W, the tensions T_1 and T_2. **(6 marks)**

The string at A snaps and the framework hangs in equilibrium suspended from D.

b Find the angle that DE makes with the vertical. **(3 marks)**

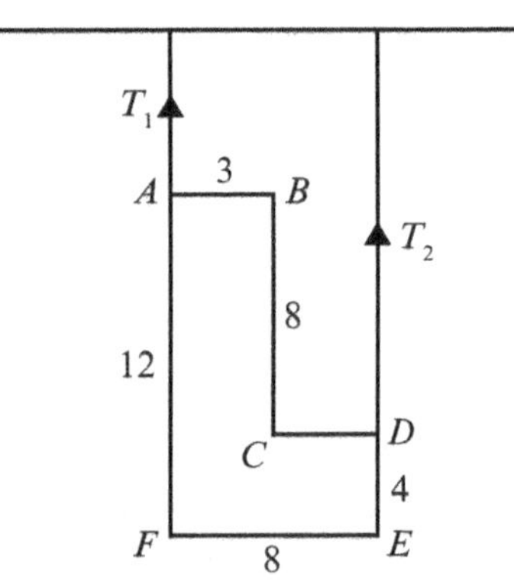

E/P **8** A uniform framework made from uniform rods is shown in the diagram. It has weight W and is suspended by one vertical string attached at A and one string angled at 60° to the vertical attached at D. The strings and the framework lie in the same plane. The framework hangs such that FE is horizontal. The lengths are $AB = 6$ cm, $AH = 4$ cm, $BC = 2$ cm, $CD = 3$ cm and $DE = 10$ cm. There is only one rod in the span BG.

a Work out, in terms of W, the tensions T_1 and T_2. **(6 marks)**

The string at D snaps.

b Find the angle that AB makes with the vertical when the framework comes to rest in equilibrium. **(3 marks)**

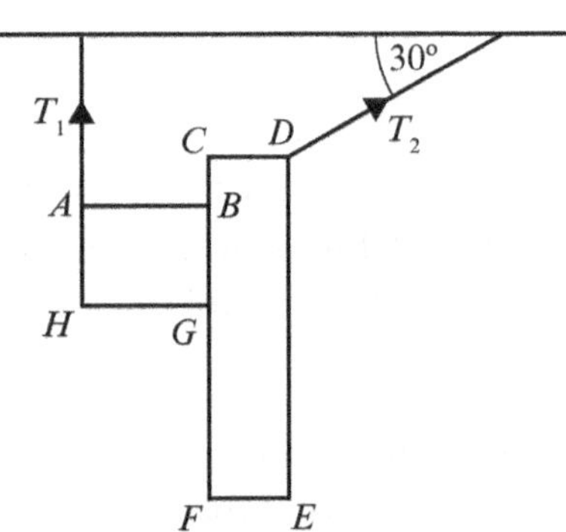

Challenge

A square uniform framework $ABCD$ is suspended by two light inextensible strings. The first string is perpendicular to AB and is attached at the point P, which lies on AB such that $AP:PB = 1:3$. The second string is attached at C and makes an angle of θ with the horizontal, as shown in the diagram.

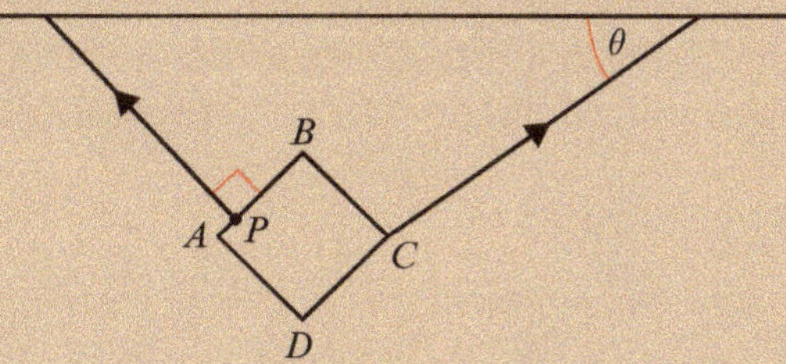

Given that the framework rests in equilibrium with AC horizontal, find θ, giving your answer in degrees, correct to one decimal place.

3.8 Non-uniform composite laminas and frameworks

You can use the methods developed in this chapter to solve problems involving non-uniform laminas and frameworks.

Example 21

The lamina shown is made from a square piece of cardboard $ABCD$ that has had the corner C folded to the centre of the square. Find the position of the centre of mass of the lamina.

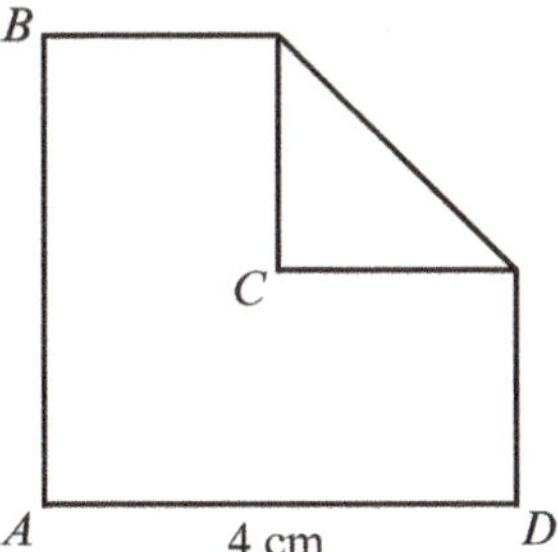

Problem-solving

You can model this situation as a composite lamina made up of three different uniform laminas. The folded section of card is modelled as a lamina with **twice** the density of the other sections.

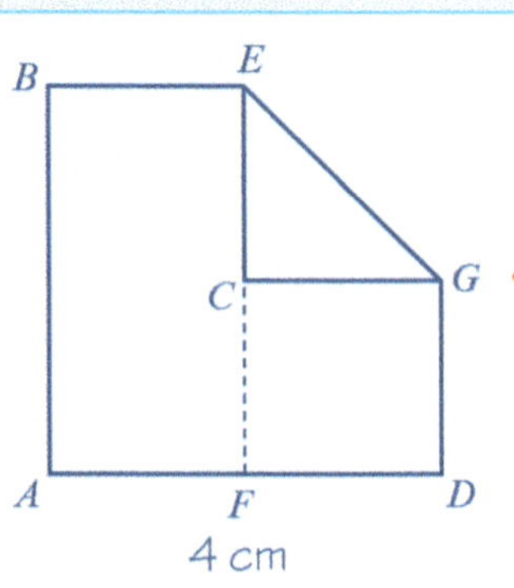

Redraw the diagram, splitting the lamina into its composite parts.

Let A be the origin so that AD lies on the x-axis and AB lies on the y-axis.

The coordinates of the centre of mass of the rectangle $ABEF$ are (1, 2).

The coordinates of the centre of mass of the square $CGDF$ are (3, 1).

The coordinates of the centre of mass of the triangle CEG are

$$\left(\frac{2+2+4}{3}, \frac{2+4+2}{3}\right) = \left(\frac{8}{3}, \frac{8}{3}\right)$$

Centre of mass of a triangle with vertices at (x_1, y_1), (x_2, y_2), (x_3, y_3) is found at: $\left(\frac{x_1 + x_2 + x_3}{3}, \frac{y_1 + y_2 + y_3}{3}\right)$

Rectangle $ABEF$ has area 8 cm^2, square $CGDF$ has area 4 cm^2. Triangle CEG has area 2 cm^2, but this material is twice the density of the other material.

Remember that the card is folded over so the triangle CEG is made of two layers.

$$8\begin{pmatrix}1\\2\end{pmatrix} + 4\begin{pmatrix}3\\1\end{pmatrix} + 4\begin{pmatrix}\frac{8}{3}\\\frac{8}{3}\end{pmatrix} = 16\begin{pmatrix}\bar{x}\\\bar{y}\end{pmatrix}$$

Use 4 for the mass of the triangular section.

$$\begin{pmatrix}\bar{x}\\\bar{y}\end{pmatrix} = \begin{pmatrix}\frac{23}{12}\\\frac{23}{12}\end{pmatrix}$$

So the centre of mass of the lamina is at $\left(\frac{23}{12}, \frac{23}{12}\right)$.

Example 22

The triangular framework shown is made from three pieces of uniform wire, AB, AC and BC of mass $2M$, $3M$ and $5M$ respectively. Find the coordinates of the centre of mass of the framework.

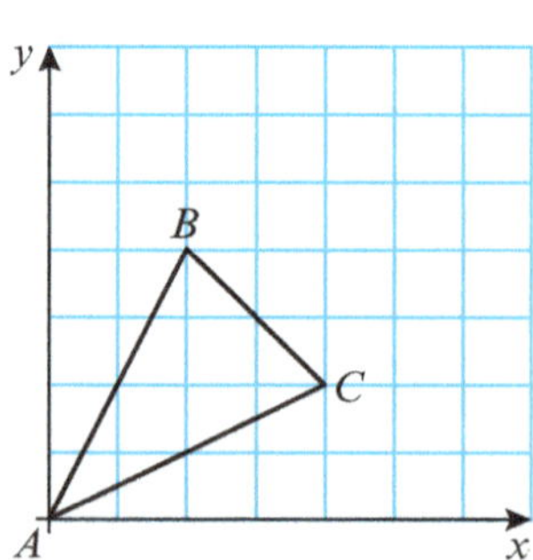

$$2M\begin{pmatrix}1\\2\end{pmatrix} + 3M\begin{pmatrix}2\\1\end{pmatrix} + 5M\begin{pmatrix}3\\3\end{pmatrix} = 10M\begin{pmatrix}\bar{x}\\\bar{y}\end{pmatrix}$$

Multiply the mass of each piece of wire by the position of its centre of mass.

$$\begin{pmatrix}\bar{x}\\\bar{y}\end{pmatrix} = \begin{pmatrix}2.3\\2.2\end{pmatrix}$$

So the coordinates of the centre of mass are at (2.3, 2.2).

Exercise 3H

(P) **1** The rectangular lamina shown is made from a rectangular piece of cardboard $ABCD$ where AB is 6 cm and BC is 4 cm.

The rectangular lamina is folded along the line EF to make the square $AEFD$.
Find the position of the centre of mass of the square lamina.

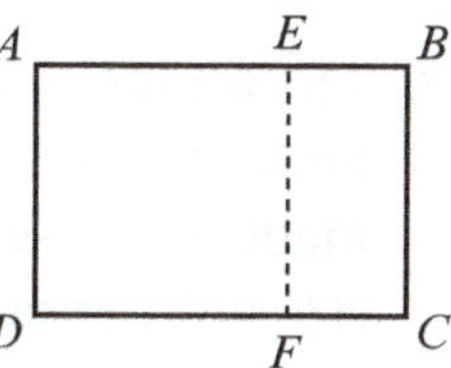

(P) **2** The rectangular lamina shown is made from a rectangular piece of cardboard $ABCD$ where AB is 10 cm and BC is 6 cm.

The lamina is folded so that the side AD lies along the side AB.
Find the position of the centre of mass of the resulting lamina.

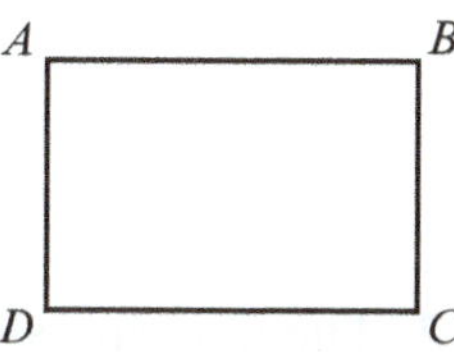

(E/P) **3** The diagram shows a rectangular lamina $ABCD$, where $AB = 10$ cm and $BC = 6$ cm. The lamina is folded so that D lies exactly on top of B. It is then suspended freely from A. Find the angle that AD makes with the vertical. **(8 marks)**

A B D C

(P) **4** The framework shown is made from four pieces of uniform wire, AB, BC, CD and DA, of masses $2M$, $3M$, M and $5M$ respectively. Find the coordinates of the centre of mass of the framework.

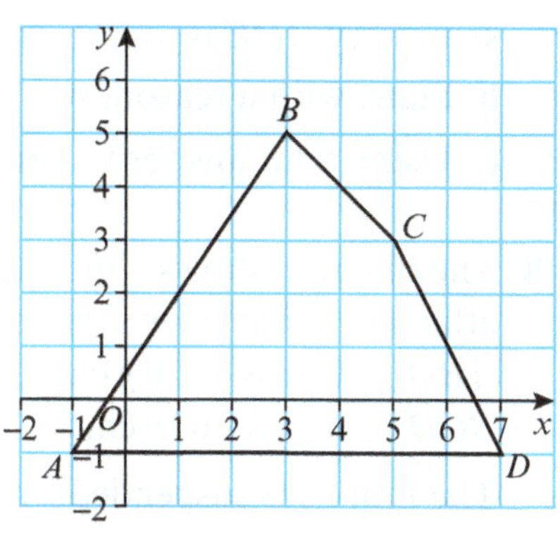

(P) **5** The framework shown is made from three pieces of uniform circular wire made from the same material. The wire used to make *BC* and *AC* is twice as thick as the wire used to make *AB*. Find the coordinates of the centre of mass of the framework.

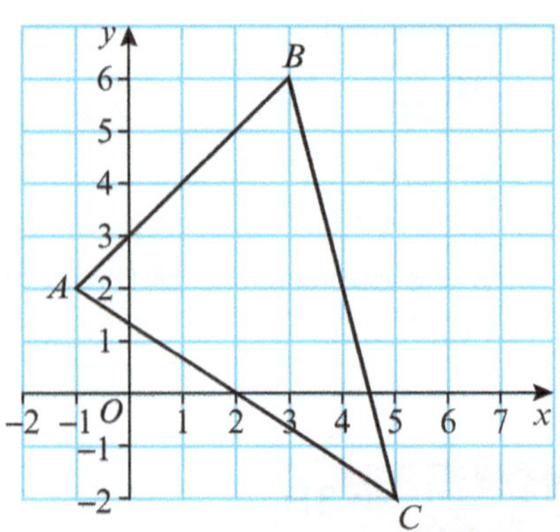

(E/P) **6** The framework shown is made from four pieces of uniform wire, *AB*, *BC*, *CD* and *AD*. *AD* and *BC* are made of the same material and *AD* has mass M. *AB* and *CD* each have mass $0.5M$. The framework is suspended freely from *B*. Find the angle that *BC* makes with the vertical. **(8 marks)**

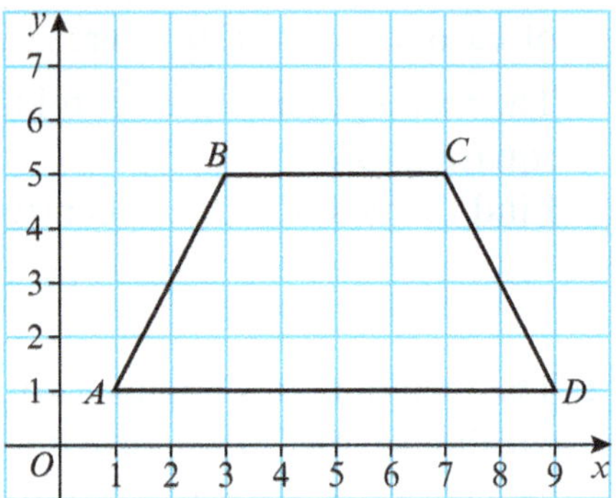

(E/P) **7** The diagram shows a sign for an ice-cream shop. The sign is made from a sheet of wood in the shape of an isosceles triangle, attached to a semicircular sheet of painted metal, as shown in the diagram. The triangle is modelled as a lamina of mass 4 kg, and the semicircle is modelled as a lamina of mass 16 kg.

a Find the distance of the centre of mass of the composite lamina from *BD*. **(4 marks)**

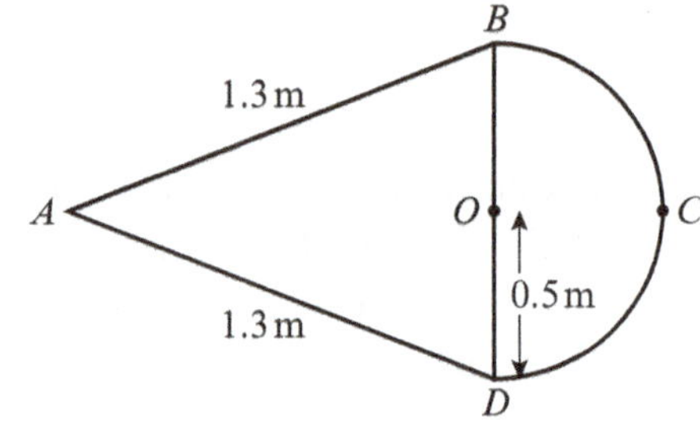

The shop owner wants to suspend the sign by two inextensible vertical wires, so that the axis of symmetry *AOC* is horizontal. She attaches one wire to point *B*.

b State, with a reason, whether she should attach the other wire to point *A* or point *C*. **(1 mark)**

c Using your answer to part **b**, find the tension in each wire. **(3 marks)**

(E/P) **8** The diagram shows a mobile made from two different flat materials attached at one edge. The mobile is modelled as a square lamina *ABDE* of density 20 g cm^{-2}, and an isosceles triangular lamina *BCD* of density 60 g cm^{-2}.

The mobile is suspended from point *B* and hangs in equilibrium.

a Find, correct to 1 d.p., the size of the acute angle that *AB* makes with the vertical. **(8 marks)**

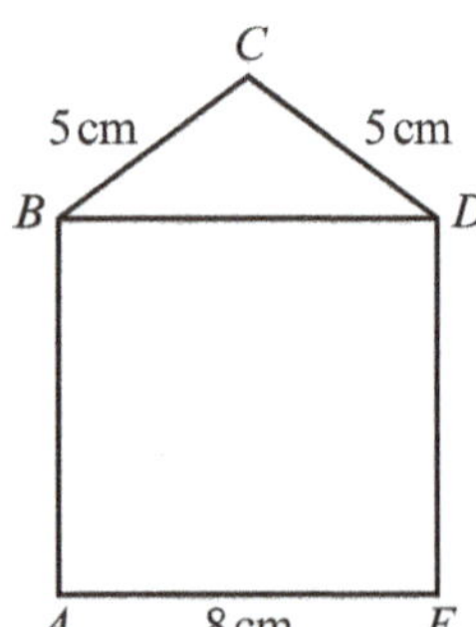

A mass of 500 g is attached to point *A*.

b Find, correct to 1 d.p., the size of the new angle that *AB* makes with the vertical. **(5 marks)**

Challenge

The first two steps in constructing a paper aeroplane from a rectangular piece of paper are as follows:

1 Fold AB and BC to the centre line of the paper.

2 Fold DE and EF to the centre line of the paper.

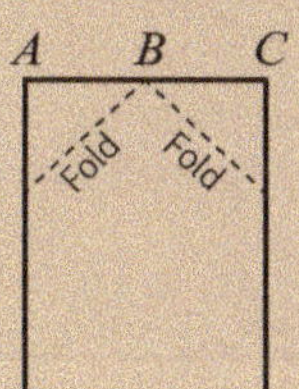

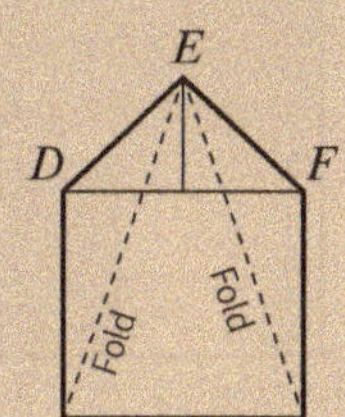

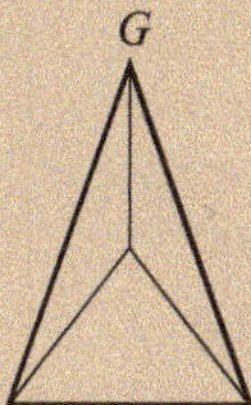

a Given that the resulting shape is an isosceles triangle, show that the sides of the original rectangle are in the ratio $2:\sqrt{2}+1$.

b Given that the width of the original rectangle $AC = x$ cm, find the position of the centre of mass of the folded isosceles triangle in terms of x.

Chapter review 3

E/P **1** The diagram shows a uniform lamina consisting of a semicircle joined to a triangle ADC.

The sides AD and DC are equal.

a Find the distance of the centre of mass of the lamina from AC. **(4 marks)**

The lamina is freely suspended from A and hangs at rest.

b Find, to the nearest degree, the angle between AC and the vertical. **(2 marks)**

The mass of the lamina is M. A particle P of mass kM is attached to the lamina at D. When suspended from A, the lamina now hangs with its axis of symmetry, BD, horizontal.

c Find, to 3 significant figures, the value of k. **(6 marks)**

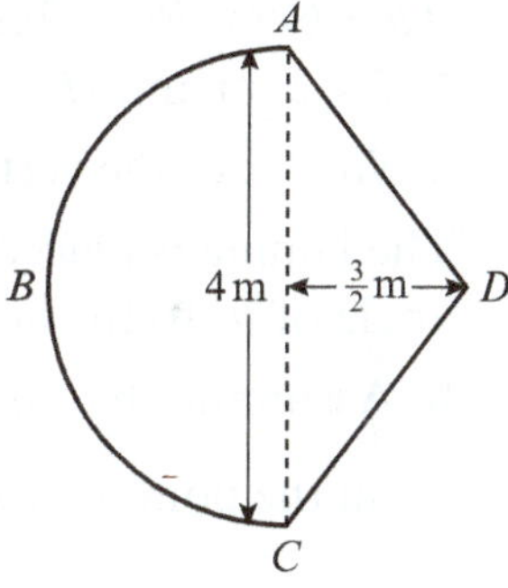

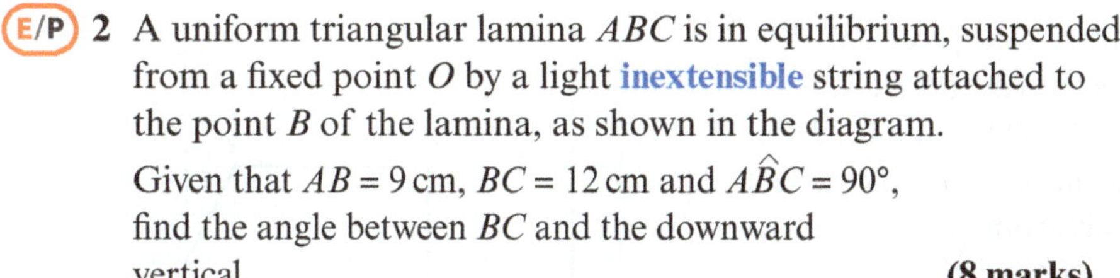

E/P **2** A uniform triangular lamina ABC is in equilibrium, suspended from a fixed point O by a light **inextensible** string attached to the point B of the lamina, as shown in the diagram.

Given that $AB = 9$ cm, $BC = 12$ cm and $A\hat{B}C = 90°$, find the angle between BC and the downward vertical. **(8 marks)**

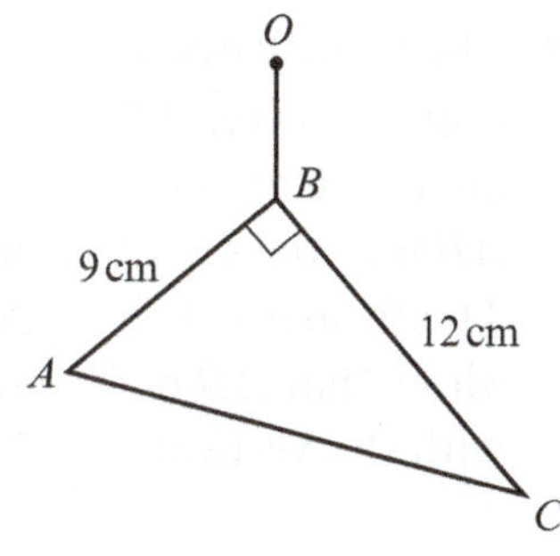

E **3** Four particles P, Q, R and S of masses 3 kg, 5 kg, 2 kg and 4 kg are placed at the points (1, 6), (−1, 5), (2, −3) and (−1, −4) respectively. Find the coordinates of the centre of mass of the particles. **(4 marks)**

E **4** A uniform lamina consists of a rectangle $ABCD$, where $AB = 3a$ and $AD = 2a$, with a square hole $EFGA$, where $EF = a$, as shown in the diagram.

Find the distance of the centre of mass of the lamina from:

a AD **(6 marks)**

b AB **(2 marks)**

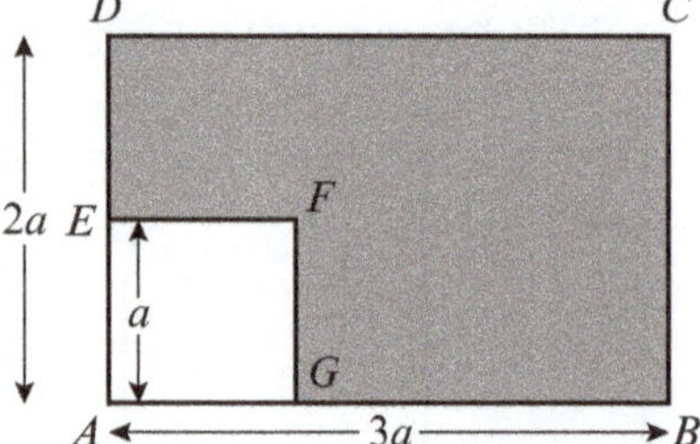

E/P **5** The lamina shown in the diagram is suspended from a point A.

a Find the angle made by AH with the vertical. **(8 marks)**

A mass of $5M$ kg is now attached at point B. Given that the lamina has a mass of $10M$:

b find the change in the angle made by AH with the vertical. **(6 marks)**

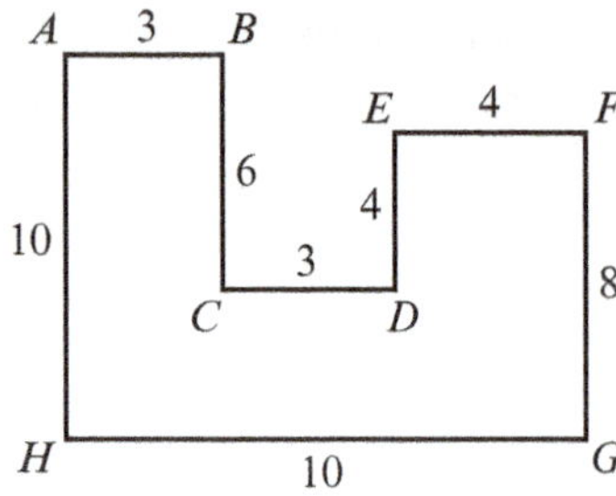

6 The T-shaped lamina $ABCDEFGH$ shown has sides of length $AB = 6$ cm, $BC = 2$ cm, $CD = 2$ cm, $DE = 2$ cm, $EF = 2$ cm, $FG = 2$ cm, $GH = 2$ cm and $HA = 2$ cm.

a Work out the vertical distance of the centre of mass from FE.

The lamina is placed on a rough slope that is inclined at an angle of $\theta°$ to the horizontal.

b Assuming that the lamina does not slide and that the lamina is on the point of toppling, show that $\theta = \tan^{-1}\left(\frac{2}{5}\right)$.

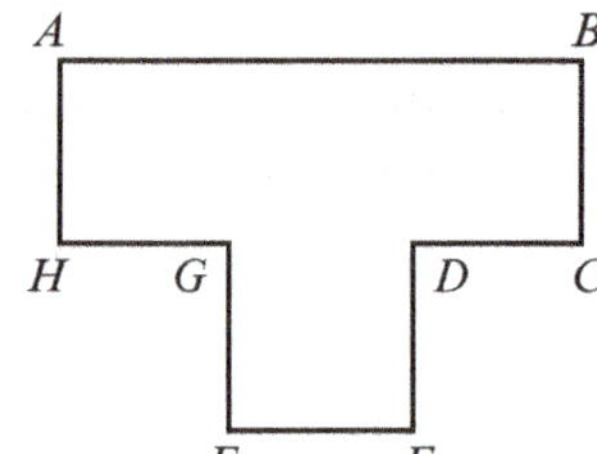

7 The T-shaped lamina $ABCDEFGH$ in question 6 is placed on an inclined plane angled at $\theta°$ to the horizontal such that AB lies on the plane. A mass equal to the mass of the lamina is attached to the lamina at E. Assuming that the lamina does not slip, show that the maximum value of θ that can be reached before the lamina topples over is $\theta = \tan^{-1}\left(\frac{6}{5}\right)$.

E/P **8** The framework shown is made from four pieces of uniform wire, AB, BC, CD and AD. AD and BC are made of the same material and AD has mass M. AB has mass $0.25M$ and CD has mass $0.5M$. The framework is suspended freely from B. Show that AB makes an angle of $\arctan\frac{125}{53}$ with the vertical. **(12 marks)**

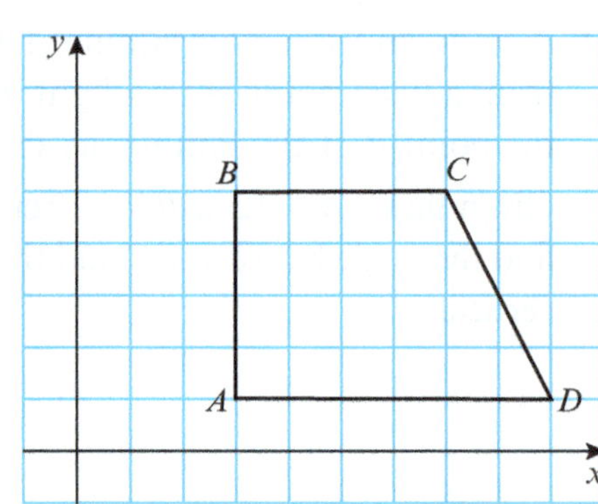

9 The rectangular lamina shown is made from two rectangular pieces of cardboard $ABCD$ and $BEFC$ where AB is 12 cm, BC is 8 cm and BE is 4 cm. The two pieces of cardboard are attached to each other along the line BC. The density of the cardboard used to make $BEFC$ is three times the density of the card used to make $ABCD$ and $ABCD$ has mass M.

The lamina is supported by two vertical strings, one attached at the midpoint of AB and one attached at E. The lamina is positioned such that AD is vertical.

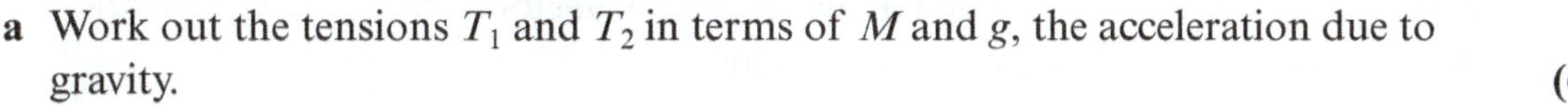

a Work out the tensions T_1 and T_2 in terms of M and g, the acceleration due to gravity. **(6 marks)**

The string at E snaps.

b Work out the angle AB makes with the vertical when the lamina has come to rest in equilibrium. **(4 marks)**

10 A piece of card is in the shape of an equilateral triangle ABC of mass $4M$ and side length 10 cm. The triangle is folded so that vertex C sits on the midpoint of AB, as shown in the diagram.

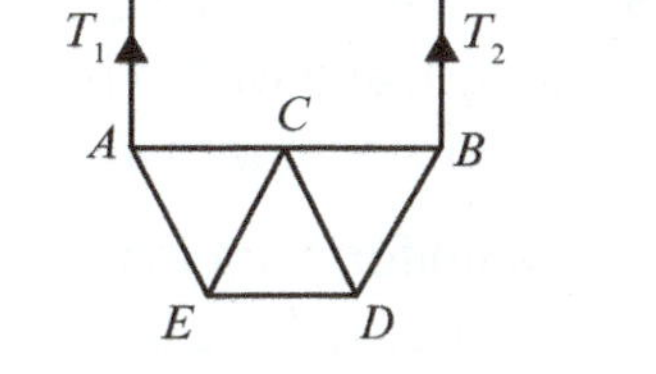

A mass of $2M$ is attached to the lamina at D. The lamina is suspended by two vertical strings attached at A and B causing AB to lie horizontally.

a Work out, in terms of M and g, the acceleration due to gravity, the values of T_1 and T_2. **(6 marks)**

The string at A snaps.

b Work out the angle AB makes with the vertical when the lamina has come to rest in equilibrium. **(4 marks)**

Challenge

The rectangular lamina $ABCD$ shown has sides of length $AB = CD = 3$ cm, $BC = DA = 6$ cm.

The lamina is folded so that AB lies along AD. The lamina is placed on a rough slope angled at 25° to the horizontal with DC resting on the slope. Assuming that the lamina does not slip, determine whether or not the lamina topples over. You must show your working clearly.

Summary of key points

1. The centre of mass of a large body is the point at which the whole mass of the body can be considered to be concentrated.
2. If a system of n particles with masses $m_1, m_2, \ldots, m_n$ are placed along the x-axis at the points $(x_1, 0), (x_2, 0), \ldots, (x_n, 0)$ respectively, then:

   $$\sum_{i=1}^{n} m_i x_i = \bar{x} \sum_{i=1}^{n} m_i$$

   where $(\bar{x}, 0)$ is the position of the centre of mass of the system.
3. If a system consists of n particles: mass m_1 with position vector $\mathbf{r}_1$, mass m_2 with position vector $\mathbf{r}_2, \ldots,$ mass m_n with position vector $\mathbf{r}_n$ then:

   $$\sum m_i \mathbf{r}_i = \bar{\mathbf{r}} \sum m_i$$

   where $\bar{\mathbf{r}}$ is the position vector of the centre of mass of the system.
4. If a question does not specify axes or coordinates you will need to choose your own axes and origin.
5. An object which has one dimension (thickness) very small compared with its other two (length and width) is modelled as a lamina. This means that it is regarded as being two-dimensional with area, but no volume.
6. If a uniform lamina has an axis of symmetry, then its centre of mass must lie on the axis of symmetry. If the lamina has more than one axis of symmetry, then it follows that the centre of mass must be at the point of intersection of the axes of symmetry.
7. The centre of mass of a uniform triangular lamina is at the intersection of the medians. This point is called the centroid of the triangle.
8. If the coordinates of the three vertices of a uniform triangular lamina are (x_1, y_1), (x_2, y_2) and (x_3, y_3) then the coordinates of the centre of mass are given by taking the average (mean) of the coordinates of the vertices:

   G is the point $\left(\dfrac{x_1 + x_2 + x_3}{3}, \dfrac{y_1 + y_2 + y_3}{3}\right)$
9. The centre of mass of a uniform plane lamina will always lie on an axis of symmetry.
10. A framework consists of a number of rods joined together or a number of pieces of wire joined together.
11. When a lamina is suspended freely from a fixed point, or pivots freely about a horizontal axis, it will rest in equilibrium in a vertical plane with its centre of mass vertically below the point of suspension or the pivot.
12. When a lamina rests on a rough inclined plane, assuming it does not slip, it will be on the point of tipping when the centre of mass lies directly above the lower vertex that is in contact with the slope.

Review exercise 1

1 The lamina $ABCDEF$ has sides of length $AB = 4$ cm, $BC = 2$ cm, $CD = 2$ cm, $DE = 4$ cm, $EF = 6$ cm and $FA = 2$ cm.

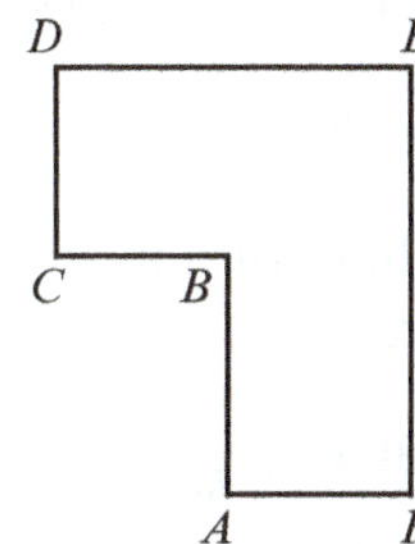

a Find the perpendicular distance of the centre of mass of the lamina from AB and AF.

The lamina is placed on a rough slope that is inclined at $\theta°$ to the horizontal.

b Assuming that the lamina does not slip, work out the maximum value of θ that can be reached before the lamina topples.

← Mechanics 2 Section 1.1

2 The lamina $ABCD$ has sides of length $AB = 6$ cm, $BC = 4$ cm, $CD = 6$ cm and $DE = 4$ cm.

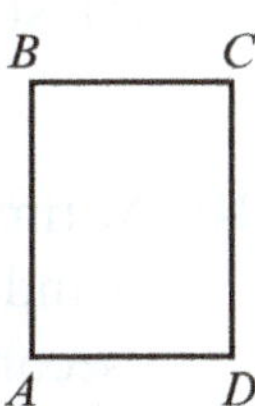

The lamina is placed on a slope angled at 10° to the horizontal. Assuming the laminas do not slip, how many such laminas can be placed on top of one another such that AD sits on top of BC, as shown, before the top lamina topples off?

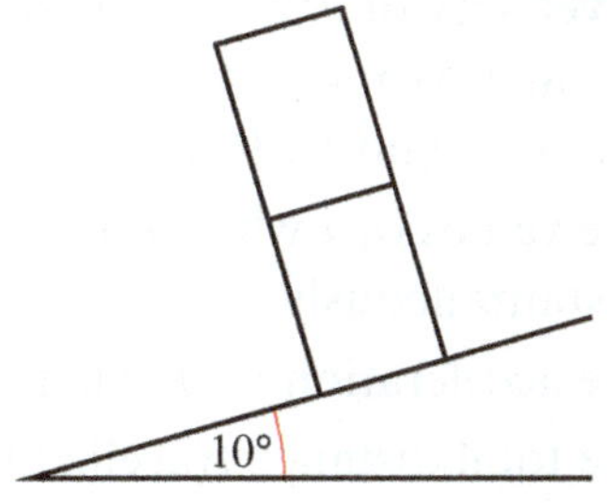

← Mechanics 2 Section 1.1

E/P 3 A ball is projected vertically upwards with a speed u m s^{-1} from a point A, which is 1.5 m above the ground. The ball moves freely under gravity until it reaches the ground. The greatest height attained by the ball is 25.6 m above A.

a Show that $u = 22.4$. **(3)**

The ball reaches the ground T seconds after it has been projected from A.

b Find, to three significant figures, the value of T. **(3)**

The ground is soft and the ball sinks 2.5 cm into the ground before coming to rest. The mass of the ball is 0.6 kg. The ground is assumed to exert a constant resistive force of magnitude F newtons.

c Find, to three significant figures, the value of F. **(4)**

d Sketch a velocity–time graph for the entire motion of the ball, showing the values of t at any points where the graph intercepts the horizontal axis. **(4)**

e State one physical factor which could be taken into account to make the model used in this question more realistic. **(1)**

← Mechanics 2 Sections 1.2, 1.3, 1.4

E 4 A ball is projected horizontally from a tabletop at a height of 0.8 m above level ground. Given that the initial velocity of the ball is 2 m s^{-1}, find:

a the time taken for the ball to reach the ground **(3)**

b the horizontal distance between the table edge and the point where the ball lands. **(2)**

← Mechanics 2 Section 1.2

E/P **5** A football is kicked horizontally off a 20 m platform and lands a distance of 40.0 m from the edge of the platform.

a Find the initial horizontal velocity of the football. **(5)**

b State two assumptions you have made in your calculations, and comment on the validity of each assumption. **(2)**

← Mechanics 2 Section 1.2

E **6** A projectile is launched from a point on horizontal ground with speed 150 m s^{-1} at an angle of 10° above the horizontal. Find:

a the time the projectile takes to reach its highest point above the ground **(4)**

b the range of the projectile. **(4)**

← Mechanics 2 Sections 1.3, 1.4

E/P **7**

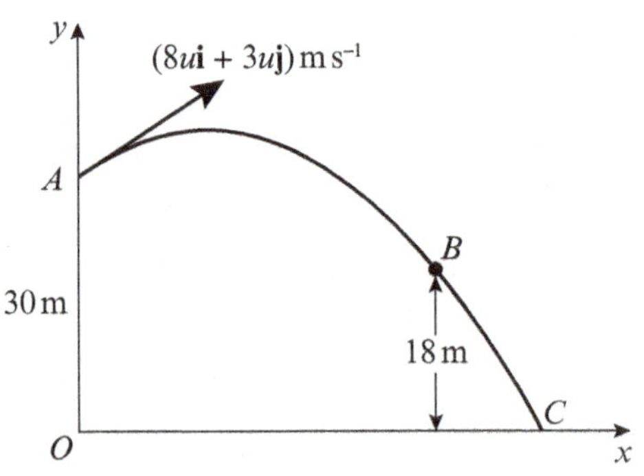

In this question, the unit vectors **i** and **j** are in a vertical plane, **i** being horizontal and **j** being vertical.

A particle P is projected from a point A with position vector 30**j** m with respect to a fixed origin O. The velocity of projection is $(8u\mathbf{i} + 3u\mathbf{j})$ m s^{-1}. The particle moves freely under gravity, passing through a point B, which has position vector $(k\mathbf{i} + 18\mathbf{j})$ m, where k is a constant, before reaching the point C on the x-axis, as shown in the figure. The particle takes 3 s to move from A to B.

Find:

a the value of u **(4)**

b the value of k **(2)**

c the angle the velocity of P makes with the x-axis as it reaches C. **(6)**

← Mechanics 2 Sections 1.3, 1.4

P **8** A projectile is launched from a point on a horizontal plane with initial speed u m s^{-1} at an angle of elevation α. The projectile moves freely under gravity until it strikes the plane. The range of the projectile is R m.

a Show that the time of flight of the projectile is $\frac{2u\sin\alpha}{g}$ seconds.

b Show that $R = \frac{u^2\sin 2\alpha}{g}$.

c Deduce that, for a fixed u, the greatest possible range is when $\alpha = 45°$.

d Given that $R = \frac{2u^2}{5g}$, find the two possible values of the angle of elevation at which the projectile could have been launched.

← Mechanics 2 Section 1.5

E/P **9** A particle P moves on the x-axis. At time t seconds, its acceleration is $(5 - 2t)$ m s^{-2}, measured in the direction of x increasing. When $t = 0$, its velocity is 6 m s^{-1} measured in the direction of x increasing. Find the time when P is instantaneously at rest in the subsequent motion. **(5)**

← Mechanics 2 Sections 2.1, 2.4

E/P **10** At time $t = 0$ a particle P leaves the origin O and moves along the x-axis. At time t seconds the velocity of P is v m s^{-1}, where $v = 6t - 2t^2$. Find:

a the maximum value of v **(4)**

b the time taken for P to return to O. **(5)**

← Mechanics 2 Sections 2.1, 2.2, 2.3, 2.4

E/P **11** A particle P moves on the positive x-axis. The velocity of P at time t seconds is $(3t^2 - 8t + 5)$ m s^{-1}. When $t = 0$, P is 12 m from the origin O. Find:

a the values of t when P is instantaneously at rest **(3)**

b the acceleration of P when $t = 4$ **(3)**

c the total distance travelled by P in the third second. **(4)**

← Mechanics 2 Sections 2.1, 2.2, 2.3, 2.4

E **12** A particle moves in a straight line and at time t seconds has velocity v m s^{-1}, where $v = 6t - 2t^{\frac{3}{2}}$, $t \geqslant 0$.

a Find an expression for the acceleration of the particle at time t. **(2)**

When $t = 0$, the particle is at the origin.

b Find an expression for the displacement of the particle from the origin at time t. **(4)**

← Mechanics 2 Sections 2.1, 2.2, 2.3, 2.4

E **13** A particle P moves in a plane such that at time t seconds, where $t \geqslant 0$, it has position vector

$$\mathbf{r} = \left(\left(\frac{1}{3}t^3 + 2t\right)\mathbf{i} + \left(\frac{1}{2}t^2 - 1\right)\mathbf{j}\right)\text{ m}$$

Find:

a the velocity vector of P at time t seconds **(2)**

b the speed of P when $t = 5$ s **(3)**

c the magnitude and direction of the acceleration of P when $t = 2$ s. **(4)**

← Mechanics 2 Sections 2.6, 2.7

E **14** A particle is acted upon by a variable force F. At time t seconds the displacement of the particle in metres relative to a fixed origin O is given by

$$\mathbf{r} = (4t^2 + 1)\mathbf{i} + (2t^2 - 3)\mathbf{j}$$

a Find the velocity of the particle when $t = 3$ s. **(3)**

b Show that the acceleration of the particle is constant. **(2)**

← Mechanics 2 Sections 2.6, 2.7

E/P **15** A particle P moves so that its velocity $\mathbf{v}$ m s^{-1} at time t seconds, where $t \geqslant 0$, is given by $\mathbf{v} = -2t\mathbf{i} + 3\sqrt{t}\mathbf{j}$. When $t = 0$, the displacement of P relative to a fixed origin is $2\mathbf{j}$ m.

Find the distance of P from O when $t = 4$ s. **(7)**

← Mechanics 2 Sections 2.6, 2.8

E **16** A particle moves in a plane with acceleration $\mathbf{a}$ m s^{-2} where

$$\mathbf{a} = t(2 - 3t^2)\mathbf{i} - 4(2t + 1)\mathbf{j},\ t \geqslant 0$$

When $t = 0$, the velocity of P is $(3\mathbf{i} + \mathbf{j})$ m s^{-1}. Find:

a the velocity of P after t s **(4)**

b the time at which P is moving in the direction of $\mathbf{i}$. **(2)**

← Mechanics 2 Sections 2.6, 2.8

E **17** In this question, $\mathbf{i}$ and $\mathbf{j}$ are horizontal unit vectors due east and due north respectively.

A wind surfer is surfing on a lake. The acceleration of the wind surfer at time t s is given by $\mathbf{a} = (-4t\mathbf{i} - 2\mathbf{j})$ m s^{-2}. At time $t = 0$ the windsurfer is moving directly east at a speed of 8 m s^{-1}.

a Find $\mathbf{v}$ in terms of t. **(4)**

b Find the value of t when the windsurfer is moving in a southerly direction. **(3)**

← Mechanics 2 Sections 2.6, 2.8

E **18** Three particles of masses $3m$, $5m$ and λm are placed at the points with coordinates (4, 0), (0, −3) and (4, 2) respectively. The centre of mass of the three particles is at (2, k).

a Show that $\lambda = 2$. **(4)**

b Calculate the value of k. **(2)**

← Mechanics 2 Section 3.2

E **19** Particles of masses $2M$, xM and yM are placed at points whose coordinates are (2, 5), (1, 3) and (3, 1) respectively. Given that the centre of mass of the three particles is at the point (2, 4), find the values of x and y. **(6)**

← Mechanics 2 Section 3.2

(E) **20** Three particles of masses 0.1 kg, 0.2 kg and 0.3 kg are placed at the points with position vectors $(2\mathbf{i} - \mathbf{j})$ m, $(2\mathbf{i} + 5\mathbf{j})$ m and $(4\mathbf{i} + 2\mathbf{j})$ m respectively. Find the position vector of the centre of mass of the particles. **(5)**

← Mechanics 2 Section 3.2

(E) **21** Three particles of mass $2M$, M and kM, where k is a constant, are placed at points with position vectors $6\mathbf{i}$ m, $4\mathbf{j}$ m and $(2\mathbf{i} - 2\mathbf{j})$ m respectively. The centre of mass of the three particles has position vector $(3\mathbf{i} + c\mathbf{j})$ m, where c is a constant.

a Show that $k = 3$. **(4)**

b Hence find the value of c. **(3)**

← Mechanics 2 Section 3.2

(E) **22**

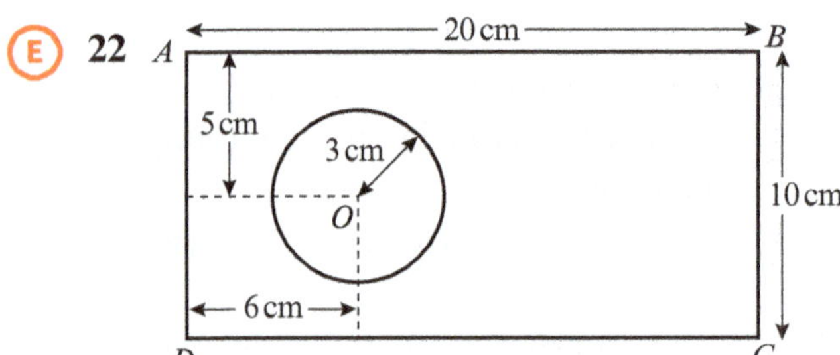

The figure shows a metal plate that is made by removing a circle of centre O and radius 3 cm from a uniform rectangular lamina $ABCD$, where $AB = 20$ cm and $BC = 10$ cm. The point O is 5 cm from both AB and CD, and is 6 cm from AD.

a Calculate, to 3 significant figures, the distance of the centre of mass of the plate from AD. **(6)**

The plate is freely suspended from A and hangs in equilibrium.

b Calculate, to the nearest degree, the angle between AB and the vertical. **(2)**

← Mechanics 2 Sections 3.4, 3.6

 **23**

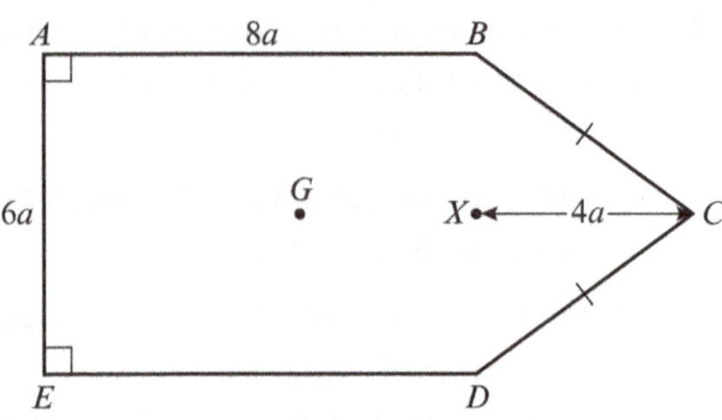

The figure shows a uniform lamina $ABCDE$ such that $ABDE$ is a rectangle, $BC = CD$, $AB = 8a$ and $AE = 6a$. The point X is the midpoint of BD and $XC = 4a$. The centre of mass of the lamina is at G.

a Show that $GX = \frac{44}{15}a$. **(6)**

The mass of the lamina is M. A particle of mass λM is attached to the lamina at C. The lamina is suspended from B and hangs freely under gravity with AB horizontal.

b Find the value of λ. **(3)**

← Mechanics 2 Sections 3.4, 3.6

(E/P) **24** A uniform square plate $ABCD$ has mass $10M$ and the length of a side of the plate is $2l$. Particles of masses M, $2M$, $3M$ and $4M$ are attached at A, B, C and D respectively. Calculate, in terms of l, the distance of the centre of mass of the loaded plate from:

a AB **(7)**

b BC **(3)**

The loaded plate is freely suspended from the vertex D and hangs in equilibrium.

c Calculate, to the nearest degree, the angle made by DA with the downward vertical.

← Mechanics 2 Sections 3.4, 3.6

E/P 25

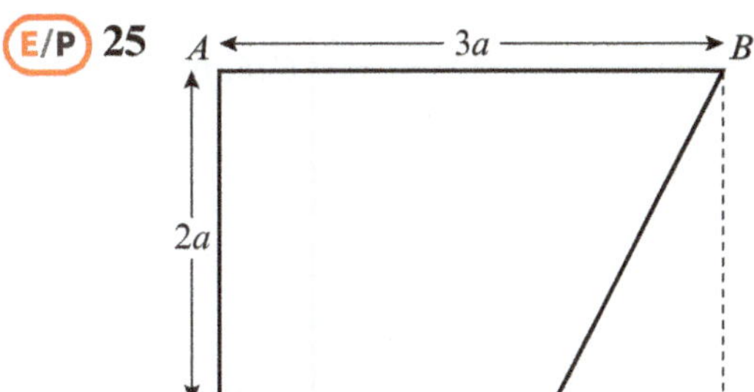

A uniform lamina $ABCD$ is made by taking a uniform sheet of metal in the form of a rectangle $ABED$, with $AB = 3a$ and $AD = 2a$, and removing the triangle BCE, where C lies on DE and $CE = a$, as shown in the figure.

a Find the distance of the centre of mass of the lamina from AD. **(5)**

The lamina has mass M. A particle of mass m is attached to the lamina at B. When the loaded lamina is freely suspended from the midpoint of AB, it hangs in equilibrium with AB horizontal.

b Find m in terms of M. **(4)**

← Mechanics 2 Sections 3.4, 3.6

E/P 26

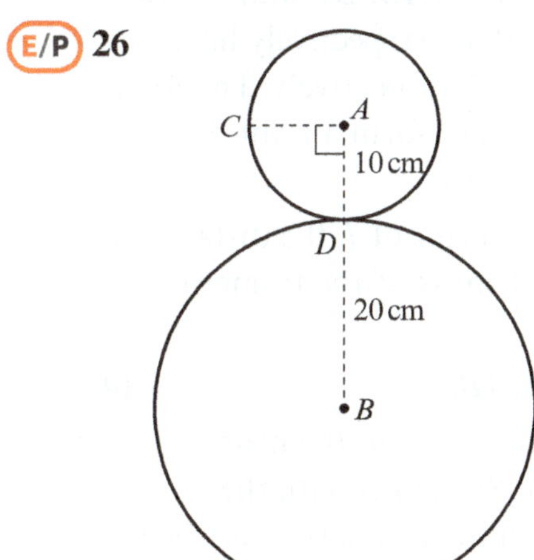

The figure shows a decoration which is made by cutting two circular discs from a sheet of uniform card. The discs are joined so that they touch at a point D on the circumference of both discs. The discs are coplanar and have centres A and B with radii 10 cm and 20 cm respectively.

a Find the distance of the centre of mass of the decoration from B. **(5)**

The point C lies on the circumference of the smaller disc and $\angle CAB$ is a right angle. The decoration is freely suspended from C and hangs in equilibrium.

b Find, in degrees to one decimal place, the angle between AB and the vertical. **(4)**

← Mechanics 2 Sections 3.4, 3.6

27

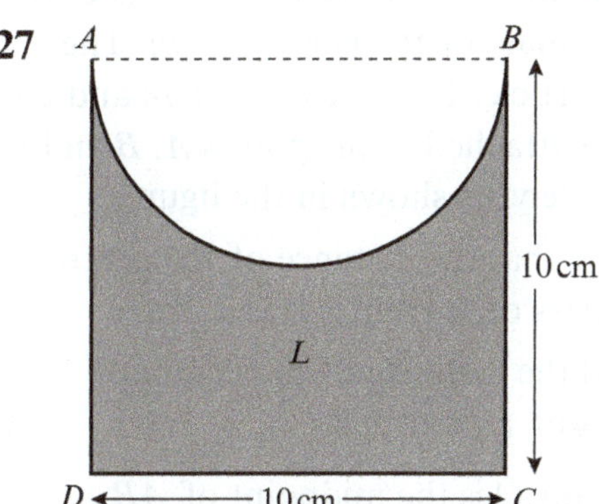

A uniform lamina L is formed by taking a uniform square sheet of material $ABCD$ of side 10 cm and removing a semicircle with diameter AB from the square, as shown in the figure.

a Find, in cm to 2 decimal places, the distance of the centre of mass of the lamina from the midpoint of AB. **(7)**

Hint The centre of mass of a uniform semicircular lamina, radius a, is at a distance $\frac{4a}{3\pi}$ from the centre of the bounding diameter.

The lamina is freely suspended from D and hangs at rest.

b Find, in degrees to one decimal place, the angle between CD and the vertical. **(4)**

← Mechanics 2 Sections 3.4, 3.6

E/P **28** 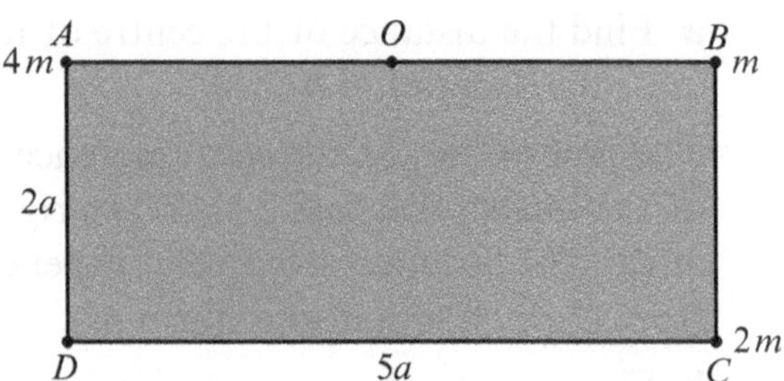

A loaded plate L is modelled as a uniform rectangular lamina $ABCD$ and three particles. The sides CD and AD of the lamina have lengths $5a$ and $2a$ respectively and the mass of the lamina is $3m$. The three particles have masses $4m$, m and $2m$ and are attached at the points A, B and C respectively, as shown in the figure.

a Show that the distance of the centre of mass of L from AD is $2.25a$. **(3)**

b Find the distance of the centre of mass of L from AB. **(2)**

The point O is the midpoint of AB. The loaded plate L is freely suspended from O and hangs at rest under gravity.

c Find, to the nearest degree, the size of the angle that AB makes with the horizontal. **(3)**

A horizontal force of magnitude P is applied at C in the direction CD. The loaded plate L remains suspended from O and rests in equilibrium with AB horizontal and C vertically below B.

d Show that $P = \frac{5}{4}mg$. **(4)**

e Find the magnitude of the force on L at O. **(4)**

← Mechanics 2 Sections 3.4, 3.6

29 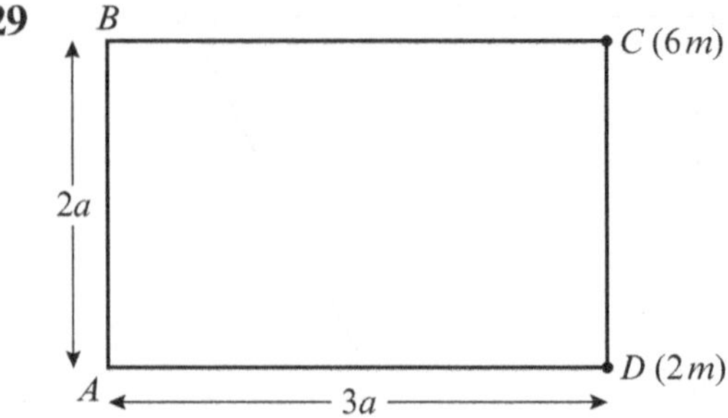

The figure shows four uniform rods joined to form a rectangular framework $ABCD$, where $AB = CD = 2a$ and $BC = AD = 3a$. Each rod has mass m. Particles of mass $6m$ and $2m$ are attached to the framework at points C and D respectively.

a Find the distance of the centre of mass of the loaded framework from

i AB **ii** AD. **(7)**

The loaded framework is freely suspended from B and hangs in equilibrium.

b Find the angle which BC makes with the vertical. **(3)**

30 Three uniform rods AB, BC and CA of mass $2m$, m and $3m$ respectively have lengths l, l and $l\sqrt{2}$ respectively. The rods are rigidly joined to form a right-angled triangular framework.

a Calculate, in terms of l, the distance of the centre of mass of the framework from

i BC **ii** AB. **(4)**

b Calculate the angle, to the nearest degree, that BC makes with the vertical when the framework is freely suspended from the point B. **(2)**

31

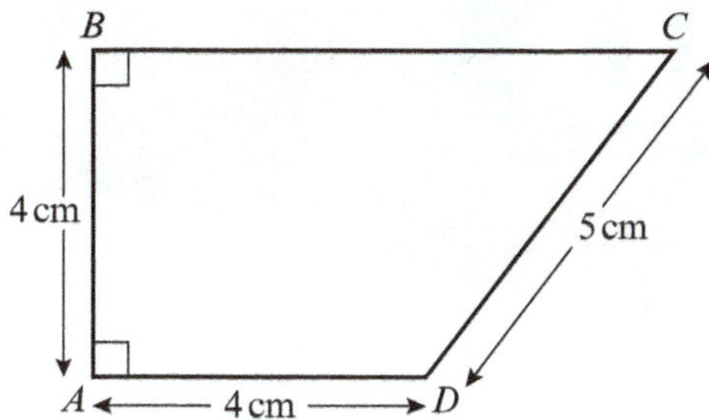

A framework is made from thin uniform wire of total length 20 cm. The framework is in the shape of a trapezium $ABCD$, where $AB = AD = 4$ cm, $CD = 5$ cm and AB is perpendicular to BC and AD as shown in the diagram.

AB, BC and AD are made from wire of mass $0.01M$ kg per cm. CD is made from wire of mass $0.015M$ kg per cm.

a Find the distance of the centre of mass of the framework from AB. **(6)**

The framework has mass M. A particle of mass kM is attached to the framework at C. When the framework is freely suspended from the midpoint of BC, the framework hangs in equilibrium with BC horizontal.

b Find the value of k. **(5)**

Challenge

1 A shop sign of weight W N can be modelled as a lamina and is shown in the diagram below. The sign is suspended by two ropes that can be modelled as light inelastic strings.

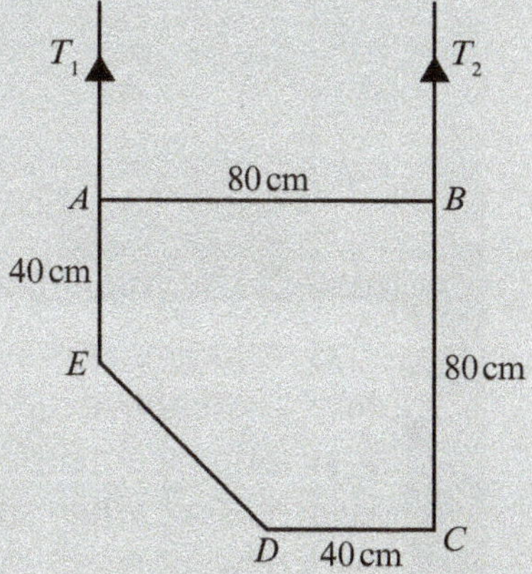

a Find the distance of the centre of mass of the lamina from AE. **(4)**

b Find the tension in each string. **(4)**

The rope attached at A will snap when the tension in it exceeds $10W$ N.

The rope attached at B will snap when the tension in it exceeds $8W$ N.

A particle of weight kW is attached to the sign at C. Given that neither rope breaks,

c find the largest possible value of k. **(4)**

← Mechanics 2 Section 3.8

2 A particle P travels in a straight line such that its velocity, v m s^{-1} at time t seconds, is given by

$$v = 3\sin kt + \cos kt, \ t \geqslant 0$$

where k is a constant and angles are measured in degrees. At time $t = 0$, the particle is at a fixed origin, O, and has acceleration 1.5 m s^{-2}.

Work out the maximum distance of the particle from the origin in its subsequent motion, and the first time at which this occurs.

← Mechanics 2 Sections 2.1, 2.3

3 A straight hill slopes upwards at an angle of θ to the horizontal, where $0 \leqslant \theta < 90°$. A projectile is launched perpendicular to the plane of the hill, with an initial velocity of u m s^{-1}, and lands a distance d m down the hill.

Show that $d = \dfrac{2u^2}{g}\tan\theta\sec\theta$.

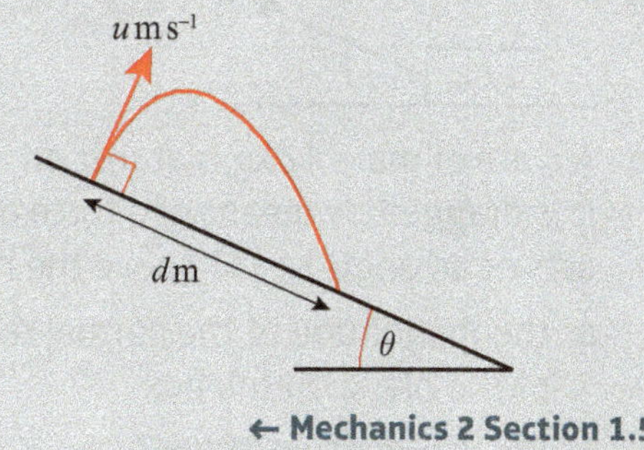

← Mechanics 2 Section 1.5

4 WORK AND ENERGY

3.1

Learning objectives

After completing this chapter you should be able to:

- Calculate the work done by a force when its point of application moves → pages 100–103
- Calculate the kinetic energy of a moving particle and the potential energy of a particle → pages 104–107
- Use the principle of conservation of mechanical energy and the work–energy principle → pages 108–112
- Calculate the power developed by an engine → pages 112–117

Prior knowledge check

1

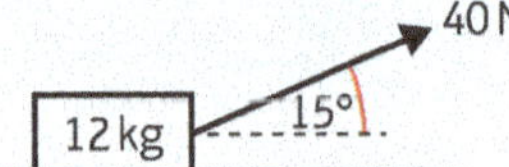

A crate of mass 12 kg is at rest on a smooth horizontal plane. It is dragged by means of a force of magnitude 40 N, which acts at an angle of 15° above the horizontal. Find:

a the magnitude of the normal reaction of the plane on the box

b the acceleration of the box

c the total distance travelled by the box in the first 5 seconds of its motion.

← Mechanics 1 Section 5.1

2 A 10 kg box rests on a rough plane inclined at 30° to the horizontal. Given that the box is on the point of slipping down the plane, find the coefficient of friction between the box and the plane. ← Mechanics 1 Section 5.3

This rock climber is increasing her height above sea-level. Her gravitational potential energy is increasing. When she abseils back down the rock face, her gravitational potential energy will be converted into kinetic energy.

4.1 Work done

If you drag an object along the ground, you have to apply a force to overcome **friction**. In order to move the object you have to do **work**. In general, work is done on an object when a force is applied to it and there is motion.

- **You can calculate the work done by a constant force when its point of application moves along a straight line using the formula:**

  work done = component of force in direction of motion × distance moved in direction of force

When the force is measured in newtons and the distance moved in metres, the work done is measured in **joules** (J).

> **Notation** If the point of application is moving in the same direction as the line of action of the constant force, this formula becomes:
> **work done = force × distance**

You can also calculate the work done against gravity when a particle is moved vertically. Work is done against gravity whenever a particle's vertical height is increased. This may be because the particle moved vertically or at an angle to the horizontal.

- **Work done against gravity = mgh, where m is the mass of the particle, g is the acceleration due to gravity and h is the vertical distance raised.**

Example 1

A box is pulled 7 m across a horizontal floor by a horizontal force of magnitude 15 N. Calculate the work done by the force.

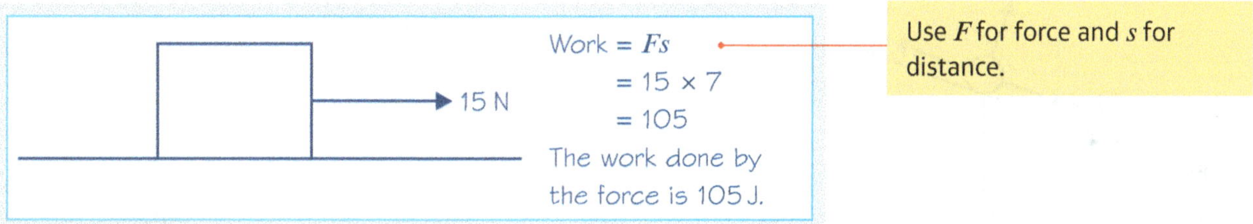

Example 2

A packing case is pulled across a horizontal floor by a horizontal rope. The case moves at a constant speed and there is a constant **resistance** to motion of magnitude R **newtons**. When the case has moved a distance of 12 m the work done is 96 J. Calculate the magnitude of the resistance.

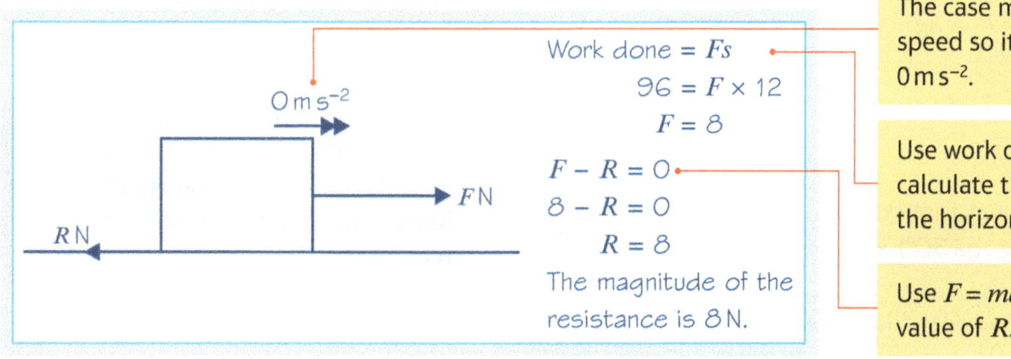

The case moves at a constant speed so its acceleration is $0\,\text{m s}^{-2}$.

Use work done = Fs to calculate the magnitude of the horizontal force.

Use $F = ma$ to calculate the value of R.

Example 3

A bricklayer raises a load of bricks with a total mass of 30 kg at a constant speed by attaching a cable to the bricks. Assuming the cable is vertical, calculate the work done when the bricks are raised a distance of 7 m.

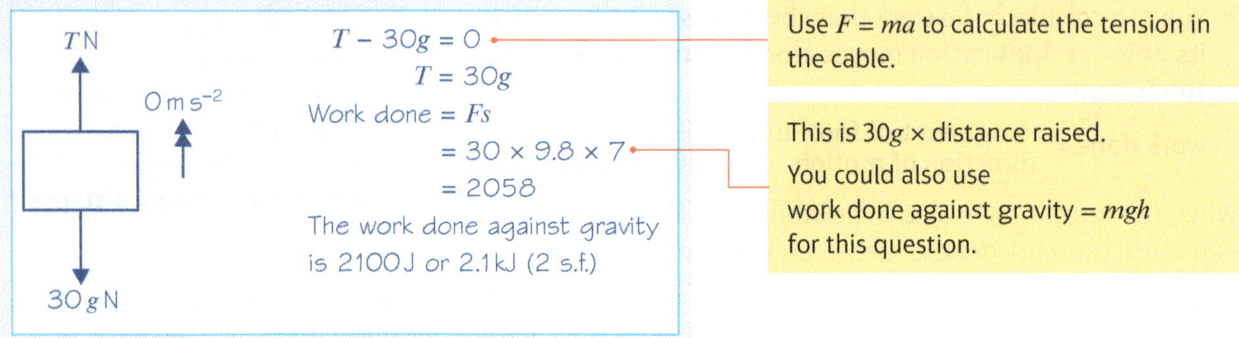

Example 4

SKILLS PROBLEM-SOLVING

A package of mass 2 kg is pulled at a constant speed up a rough plane which is inclined at 30° to the horizontal. The coefficient of friction between the package and the surface is 0.35. The package is pulled 12 m up a line of greatest slope of the plane. Calculate:

a the work done against gravity **b** the work done against friction.

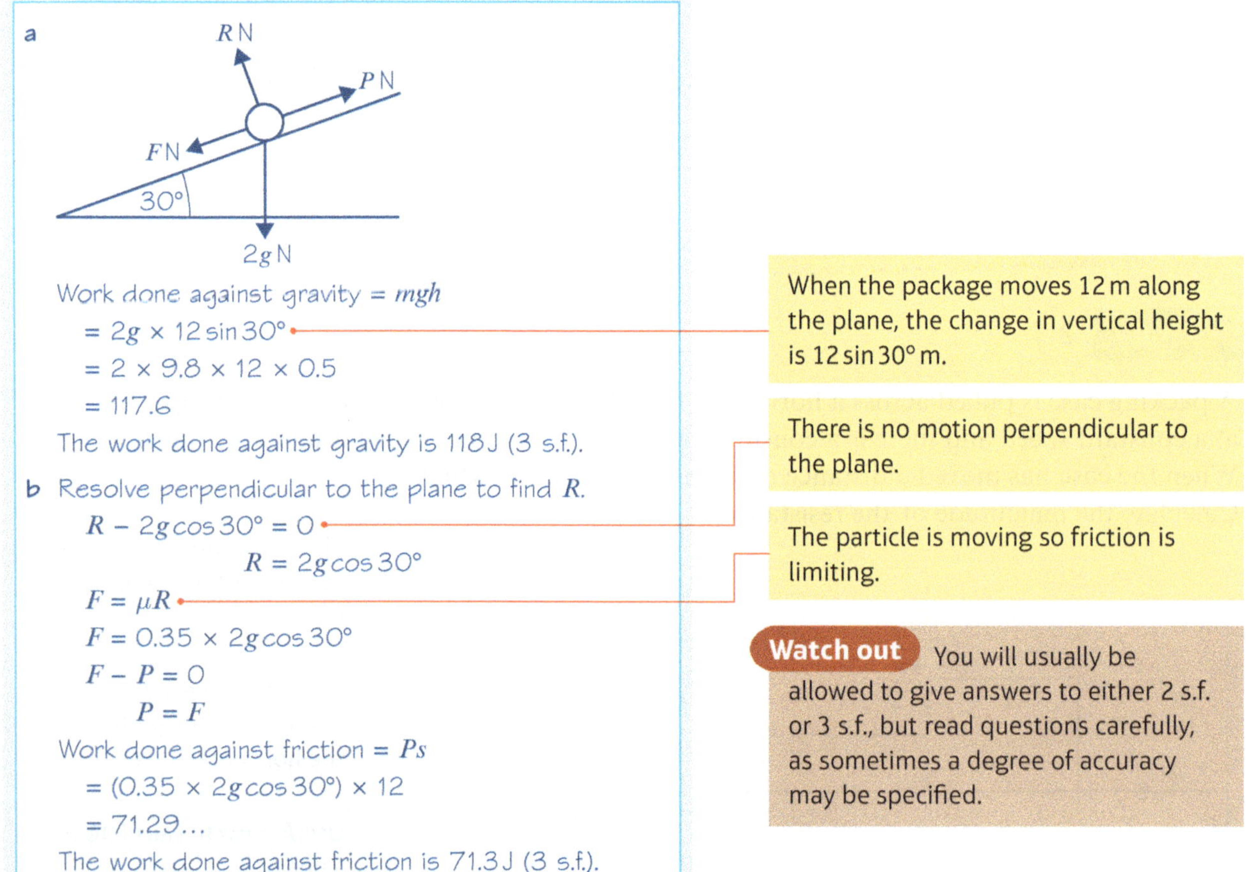

Watch out You will usually be allowed to give answers to either 2 s.f. or 3 s.f., but read questions carefully, as sometimes a degree of accuracy may be specified.

Example 5

A sledge is pulled 15 m across a smooth sheet of ice by a force of magnitude 27 N. The force is inclined at 25° to the horizontal. By modelling the sledge as a particle, calculate the work done by the force.

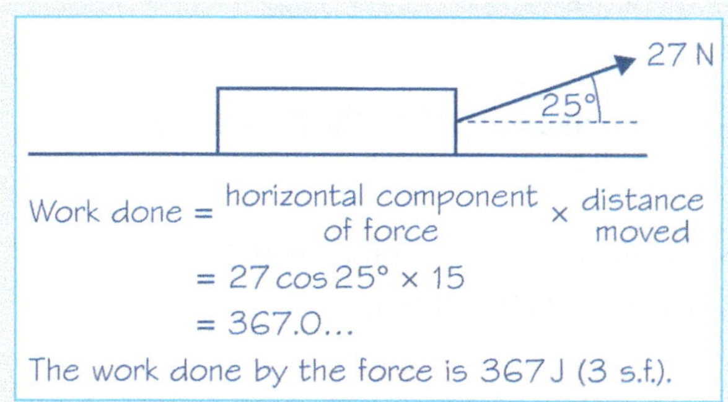

Watch out The point of application of the force is not moving in the direction of the line of action of the force. To find the work done, use the horizontal component of the force.

Exercise 4A SKILLS PROBLEM-SOLVING

Whenever a numerical value of g is required, take $g = 9.8\,\text{m s}^{-2}$, and give your answers to either 2 significant figures or 3 significant figures.

1. Calculate the work done by a horizontal force of magnitude 0.6 N which pulls a particle a distance of 4.2 m across a horizontal floor.
2. A box is pulled 12 m across a smooth horizontal floor by a constant horizontal force. The work done by the force is 102 J. Calculate the magnitude of the force.
3. Calculate the work done against gravity when a particle of mass 0.35 kg is raised a vertical distance of 7 m.
4. A crate of mass 15 kg is raised through a vertical distance of 4 m. Calculate the work done against gravity.
5. A box is pushed 15 m across a horizontal surface. The box moves at a constant speed and the resistances to motion are constant and total 22 N. Calculate the work done by the force pushing the box.
6. A ball of mass 0.5 kg falls vertically 15 m from rest. Calculate the work done by gravity.
7. A cable is attached to a crate of mass 80 kg. The crate is raised vertically at a constant speed from the ground to the top of a building. The work done in raising the crate is 30 kJ. Calculate the height of the building.
8. A sledge is pulled 14 m across a horizontal sheet of ice by a rope inclined at 25° to the horizontal. The tension in the rope is 18 N and the ice can be assumed to be a smooth surface.
   a Calculate the work done.
   b State a modelling assumption that you have made and assess its validity.
9. A parcel of mass 3 kg is pulled a distance of 4 m across a rough horizontal floor. The parcel moves at a constant speed. The work done against friction is 30 J. Calculate the coefficient of friction between the parcel and the surface.

(P) **10** A block of wood of mass 2 kg is pushed across a rough horizontal floor. The block moves at $3\,\text{m}\,\text{s}^{-1}$ and the coefficient of friction between the block and the floor is 0.55. Calculate the work done in 2 seconds.

11 A girl of mass 52 kg climbs a vertical cliff which is 46 m high. Calculate the work she does against gravity.

12 A child of mass 25 kg slides 2 m down a smooth slope inclined at 35° to the horizontal. Calculate the work done by gravity.

13 A particle of mass 0.3 kg is pulled 2 m up a line of greatest slope of a plane which is inclined at 25° to the horizontal. Calculate the work done against gravity.

(E/P) **14** A rough plane surface is inclined at an angle α to the horizontal, where $\sin\alpha = \frac{5}{13}$. A packet of mass 8 kg is pulled at a constant speed up a line of greatest slope of the plane. The coefficient of friction between the packet and the plane is 0.3.

> **Problem-solving**
> Draw a right-angled triangle with sides of length 5, 12 and 13, or use your calculator to find the exact value of $\cos\alpha$.

a Calculate the magnitude of the frictional force acting on the packet. **(5 marks)**

The packet moves a distance of 15 m up the plane. Calculate:

b the work done against friction **(3 marks)**

c the work done against gravity. **(4 marks)**

(E/P) **15** A rough surface is inclined at an angle $\arcsin\frac{7}{25}$ to the horizontal. A particle of mass 0.5 kg is pulled 3 m at a constant speed up the surface by a force acting along a line of greatest slope. The only resistances to the motion are those due to friction and gravity. The work done by the force is 12 J. Calculate the coefficient of friction between the particle and the surface. **(5 marks)**

(E) **16** A rough surface is inclined at 40° to the horizontal. A box of mass 1.5 kg is pulled at a constant speed up the surface by a force T acting along a line of greatest slope. The coefficient of friction between the particle and the surface is 0.4.

Modelling the box as a particle, calculate the work done by T when the particle travels 8 m. **(4 marks)**

(E) **17** A particle P of mass 2 kg is projected up a line of greatest slope of a rough plane which is inclined at an angle $\arcsin\frac{3}{5}$ to the horizontal. The coefficient of friction between P and the plane is 0.35.

The particle comes to instantaneous rest a distance 3 m up the line of greatest slope of the plane. Find:

a the work done by gravity **(3 marks)**

b the work done by friction **(3 marks)**

c the speed of projection. **(4 marks)**

4.2 Kinetic and potential energy

You can calculate the **kinetic energy** of a moving particle and the **potential energy** of a particle.

- **Kinetic energy (K.E.) = $\frac{1}{2}mv^2$, where m is the mass of the particle and v is its speed.**
- **Potential energy (P.E.) = mgh, where h is the height of the particle above an arbitrary fixed level.**

Notation If a particle with mass m and velocity vector $\mathbf{v}$ is moving in two dimensions then it has kinetic energy.

A particle possesses kinetic energy when it is moving. When the mass of the particle is measured in kilograms and its velocity is measured in metres per second, the kinetic energy is measured in joules.

A particle possesses potential energy whenever gravity acts on it. When the mass of the particle is measured in kilograms, the acceleration due to gravity is measured in metres per second squared. When its height above the fixed level is measured in metres, the potential energy is measured in joules.

Notation This type of potential energy is sometimes called **gravitational potential energy**.

Watch out Kinetic energy cannot be negative. Potential energy is negative if the particle is below the fixed reference level.

The work done by a force which accelerates a particle horizontally is related to the kinetic energy of that particle.

- **Work done = change in kinetic energy**

You can derive this result using formulae you already know:

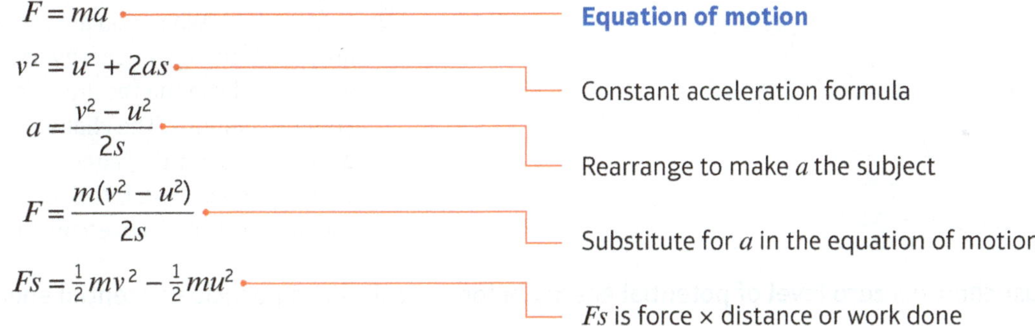

Work done = final K.E. – initial K.E.

Work done = change in K.E.

Example 6

A particle of mass 0.3 kg is moving at a speed of 9 m s^{-1}.
Calculate its kinetic energy.

K.E. $= \frac{1}{2}mv^2$

$= \frac{1}{2} \times 0.3 \times 9^2$

$= 12.15$

The K.E. of the particle is 12.2 J (3 s.f.)

Example 

A box of mass 1.5 kg is pulled across a smooth horizontal surface by a horizontal force. The initial speed of the box is $u\,\text{m s}^{-1}$ and its final speed is $3\,\text{m s}^{-1}$. The work done by the force is 1.8 J. Calculate the value of u.

$$\text{Work done} = \tfrac{1}{2}mv^2 - \tfrac{1}{2}mu^2$$
$$1.8 = \tfrac{1}{2} \times 1.5 \times 3^2 - \tfrac{1}{2} \times 1.5u^2$$
$$\tfrac{1}{2} \times 1.5u^2 = 4.95$$
$$u^2 = \frac{4.95 \times 2}{1.5}$$
$$u = 2.57 \text{ (3 s.f.)}$$

Watch out The work done by the force is equal to the **change** in kinetic energy.

Example 

SKILLS PROBLEM-SOLVING

A van of mass 2000 kg starts from rest at some traffic lights. After travelling 400 m the van's speed is $12\,\text{m s}^{-1}$. A constant resistance of 500 N acts on the van. Calculate the driving force, which can be assumed to be constant.

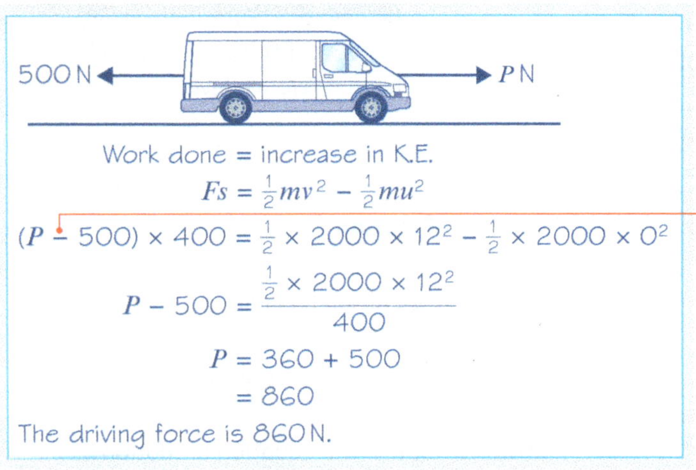

Work done = increase in K.E.

$$Fs = \tfrac{1}{2}mv^2 - \tfrac{1}{2}mu^2$$
$$(P - 500) \times 400 = \tfrac{1}{2} \times 2000 \times 12^2 - \tfrac{1}{2} \times 2000 \times 0^2$$
$$P - 500 = \frac{\tfrac{1}{2} \times 2000 \times 12^2}{400}$$
$$P = 360 + 500$$
$$= 860$$

The driving force is 860 N.

The force used to calculate the work done is the resultant force in the direction of the motion.

Watch out When you are considering an overall change in energy of a body, you should use the resultant force in your calculations. The change in energy of the van is not the same as the work done by the driving force, as some of this work is used to overcome friction.

- **You must choose a zero level of potential energy before calculating a particle's potential energy.**

If the particle moves upwards its potential energy will increase. If it moves downwards its potential energy will decrease.

Example 

A load of bricks of total mass 30 kg is lowered vertically to the ground through a distance of 15 m. Find the loss in potential energy.

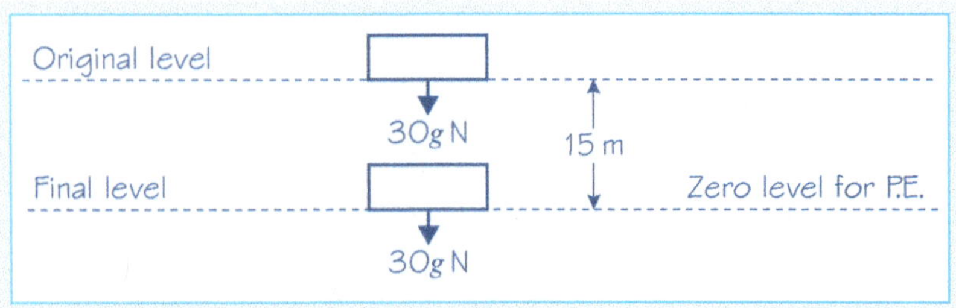

Final P.E. = 0

Initial P.E. = mgh

= 30 × 9.8 × 15

= 4410

Loss of P.E. = 4410 − 0

The loss of potential energy is 4410 J.

Take the final level to be the zero level for potential energy.

Example 10

A parcel of mass 3 kg is pulled 10 m up a plane inclined at an angle θ to the horizontal, where $\tan\theta = \frac{3}{4}$. Assuming that the parcel moves up a line of greatest slope of the plane, calculate the potential energy gained by the parcel.

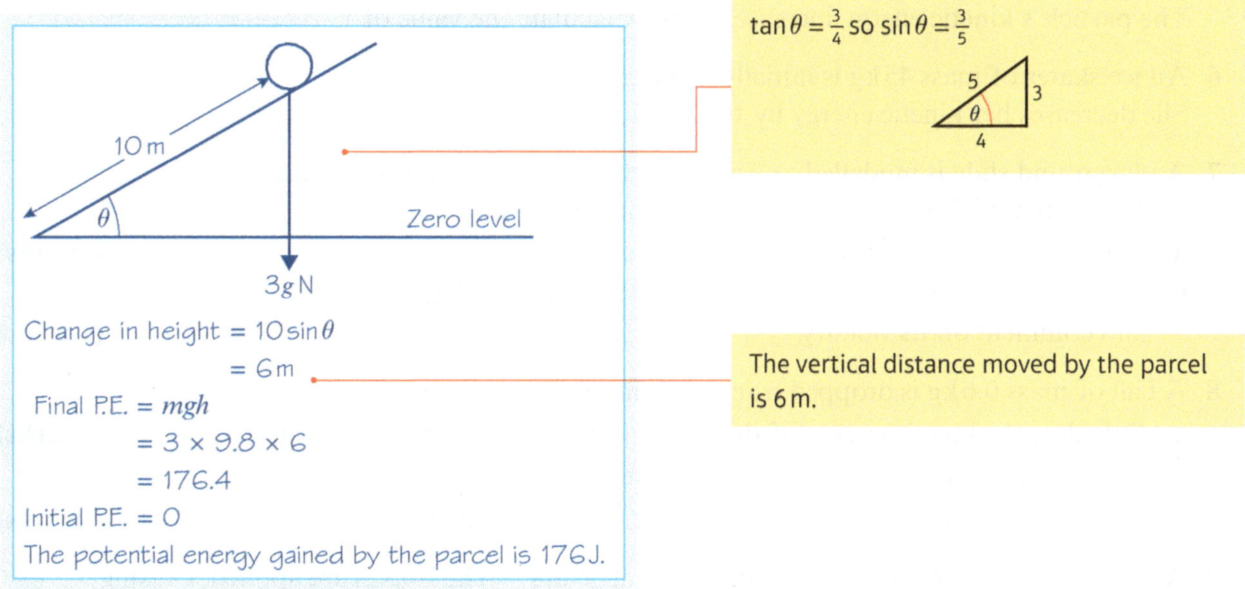

Exercise 4B

SKILLS PROBLEM-SOLVING

Whenever a numerical value of g is required, take $g = 9.8\,\text{m s}^{-2}$, and give your answers to either 2 significant figures or 3 significant figures.

1 Calculate the kinetic energy of the following objects. Put them in order, from the greatest kinetic energy to the least.

- **a** a particle of mass 0.3 kg moving at $15\,\text{m s}^{-1}$
- **b** a particle of mass 3 kg moving at $2\,\text{m s}^{-1}$
- **c** an arrow of mass 0.1 kg moving at $100\,\text{m s}^{-1}$
- **d** a boy of mass 25 kg running at $4\,\text{m s}^{-1}$
- **e** a car of mass 800 kg moving at $20\,\text{m s}^{-1}$

2 Find the change in potential energy of each of the following, stating in each case whether it is a loss or a gain.

a a particle of mass 1.5 kg raised through a vertical distance of 3 m

b a woman of mass 55 kg ascending a vertical distance of 15 m

c a man of mass 75 kg descending a vertical distance of 30 m

d a lift of mass 580 kg descending a vertical distance of 6 m.

3 A particle of mass 1.2 kg decreases its speed from 12 m s^{-1} to 4 m s^{-1}. Calculate the decrease in the particle's kinetic energy.

4 A van of mass 900 kg increases its speed from 5 m s^{-1} to 20 m s^{-1}. Calculate the increase in the van's kinetic energy.

(P) 5 A particle of mass 0.2 kg increases its speed from 2 m s^{-1} to v m s^{-1}. The particle's kinetic energy increases by 6 J. Calculate the value of v.

(P) 6 An ice skater of mass 45 kg is initially moving at 5 m s^{-1}. She decreases her kinetic energy by 100 J. Calculate her final speed.

(E) 7 A playground slide is modelled as a plane inclined at 48° to the horizontal. A child of mass 25 kg slides down the slide for 4 m.

a Calculate the potential energy lost by the child. **(4 marks)**

b State one assumption that you have made in your calculations, and comment on its validity. **(1 mark)**

(E/P) 8 A ball of mass 0.6 kg is dropped from a height of 2 m into a pond.

a Calculate the kinetic energy of the ball as it hits the surface of the water. **(3 marks)**

The ball begins to sink in the water with a speed of 4.8 m s^{-1}.

b Calculate the kinetic energy lost when the ball strikes the water. **(4 marks)**

9 A lorry of mass 2000 kg is initially travelling at 35 m s^{-1}. The brakes are applied, causing the lorry to decelerate at 1.2 m s^{-2} for 5 s. Calculate the loss of kinetic energy of the lorry.

(E) 10 A car of mass 750 kg moves along a stretch of road which can be modelled as a line of greatest slope of a plane inclined to the horizontal at 30°. As the car moves up the road for 500 m its speed reduces from 20 m s^{-1} to 15 m s^{-1}. Calculate:

a the loss of kinetic energy of the car **(3 marks)**

b the gain of potential energy of the car. **(4 marks)**

(E/P) 11 A woman of mass 80 kg climbs a vertical cliff face of height h m. Her potential energy increases by 15.7 kJ.

Find the height of the cliff. **(3 marks)**

Challenge

A 1 kg ball, initially at rest, is dropped from the top of a cliff.
The ball can be modelled as a particle falling freely under gravity.

a Find, in terms of t, the kinetic and potential energy of the ball at a time t seconds after it is dropped.

b Show that the sum of the kinetic energy and potential energy of the ball is constant.

4.3 Conservation of mechanical energy and work–energy principle

You can use the principle of conservation of mechanical energy and the work–energy principle to solve problems involving a moving particle.

- **When no external forces (other than gravity) do work on a particle during its motion, the sum of the particle's kinetic energy and potential energy remains constant.**

Notation This is called the **principle of conservation of mechanical energy**.

This is true whether the particle moves vertically or along a path inclined to the horizontal.

The total energy possessed by a particle can change during the particle's motion only if some external force is doing work on the particle. Any non-**gravitational** resistance to motion acting on the particle will reduce the total energy of the particle, as the particle will have to do work to overcome the resistance.

- **The change in the total energy of a particle is equal to the work done on the particle.**

Notation This is called the **work–energy principle**.

Example 11

A smooth plane is inclined at 30° to the horizontal. A particle of mass 0.5 kg slides down a line of greatest slope of the plane. The particle starts from rest at point A and passes point B with a speed of $6\,\text{m s}^{-1}$. Find the distance AB.

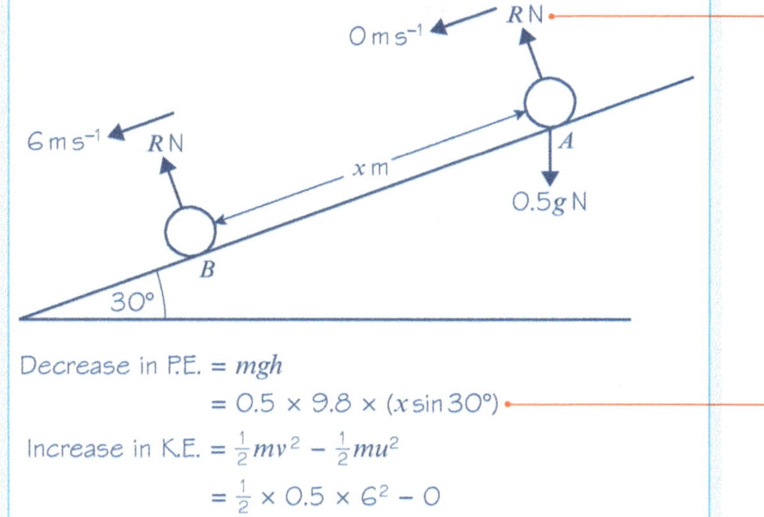

Decrease in P.E. $= mgh$

$= 0.5 \times 9.8 \times (x \sin 30°)$

Increase in K.E. $= \frac{1}{2}mv^2 - \frac{1}{2}mu^2$

$= \frac{1}{2} \times 0.5 \times 6^2 - 0$

Decrease in P.E. = increase in K.E.

$0.5 \times 9.8 \times (x \sin 30°) = \frac{1}{2} \times 0.5 \times 6^2$

$$x = \frac{\frac{1}{2} \times 0.5 \times 6^2}{0.5 \times 9.8 \times \sin 30°}$$

$x = 3.673\ldots$

The distance AB is 3.67 m (3 s.f.)

Note that the **normal reaction** R does no work on the particle as it is always perpendicular to the motion.

The vertical distance moved by the particle is $x \sin 30°$, where x is the distance AB.

Problem-solving The only force acting on the particle that is doing work is gravity so you can apply the principle of conservation of mechanical energy.

Example 12

A particle of mass 2 kg is projected with speed $8\,\text{m}\,\text{s}^{-1}$ up a line of greatest slope of a rough plane inclined at 45° to the horizontal. The coefficient of friction between the particle and the plane is 0.4. Calculate the distance the particle travels up the plane before coming to instantaneous rest.

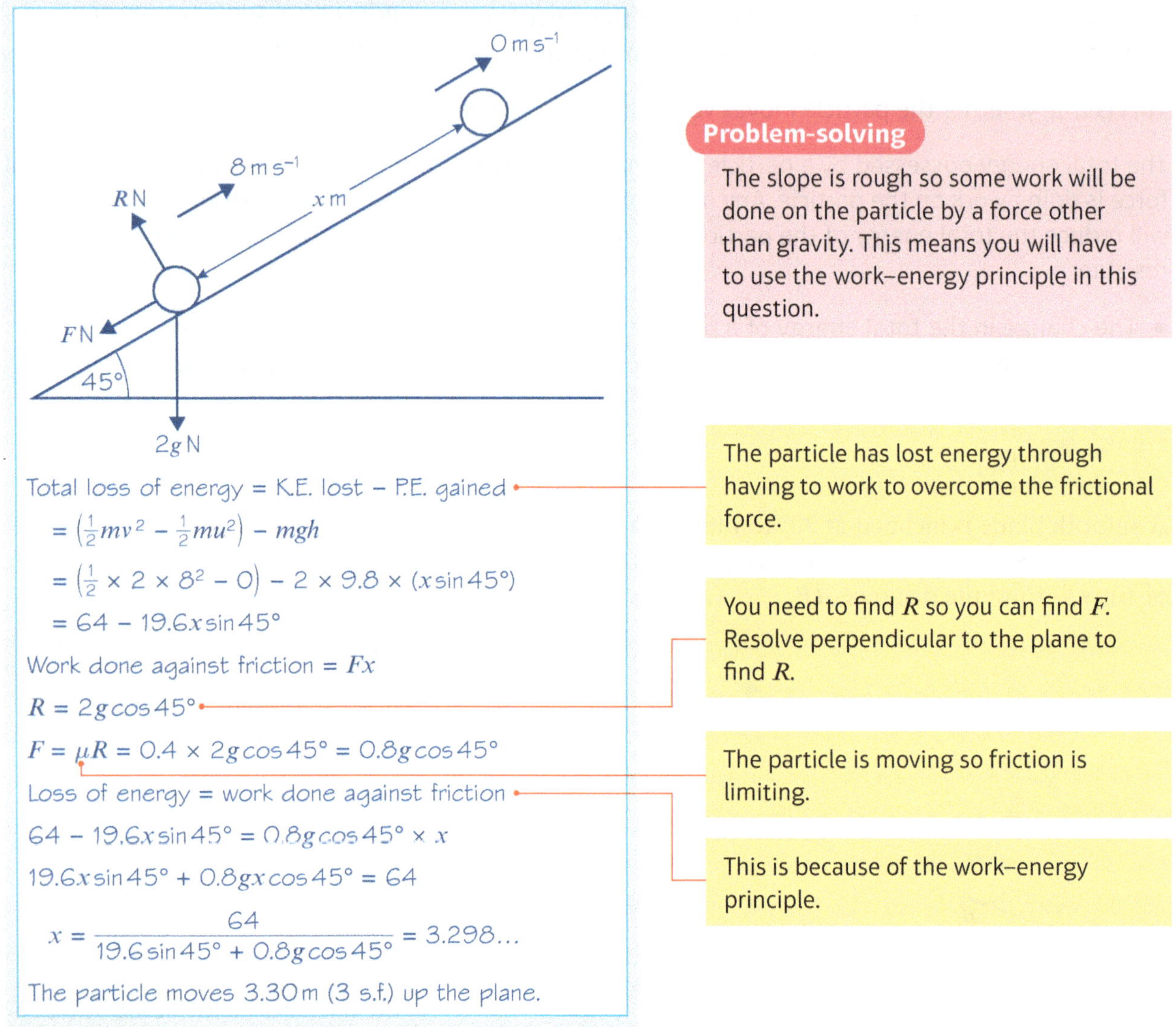

Total loss of energy = K.E. lost − P.E. gained

$= \left(\frac{1}{2}mv^2 - \frac{1}{2}mu^2\right) - mgh$

$= \left(\frac{1}{2} \times 2 \times 8^2 - 0\right) - 2 \times 9.8 \times (x \sin 45°)$

$= 64 - 19.6x \sin 45°$

Work done against friction $= Fx$

$R = 2g\cos 45°$

$F = \mu R = 0.4 \times 2g\cos 45° = 0.8g\cos 45°$

Loss of energy = work done against friction

$64 - 19.6x\sin 45° = 0.8g\cos 45° \times x$

$19.6x\sin 45° + 0.8gx\cos 45° = 64$

$$x = \frac{64}{19.6\sin 45° + 0.8g\cos 45°} = 3.298\ldots$$

The particle moves 3.30 m (3 s.f.) up the plane.

Problem-solving

The slope is rough so some work will be done on the particle by a force other than gravity. This means you will have to use the work–energy principle in this question.

The particle has lost energy through having to work to overcome the frictional force.

You need to find R so you can find F. Resolve perpendicular to the plane to find R.

The particle is moving so friction is limiting.

This is because of the work–energy principle.

Example 13

SKILLS PROBLEM-SOLVING

A skier moving downhill passes point A on a ski run at $6\,\text{m}\,\text{s}^{-1}$. After descending 50 m vertically the run begins to ascend. When the skier has ascended 25 m to point B her speed is $4\,\text{m}\,\text{s}^{-1}$.

The skier and her skis have a combined mass of 55 kg. The total distance she travels from A to B is 1400 m. The non-gravitational resistances to motion are constant and have a total magnitude of 12 N. Calculate the work done by the skier.

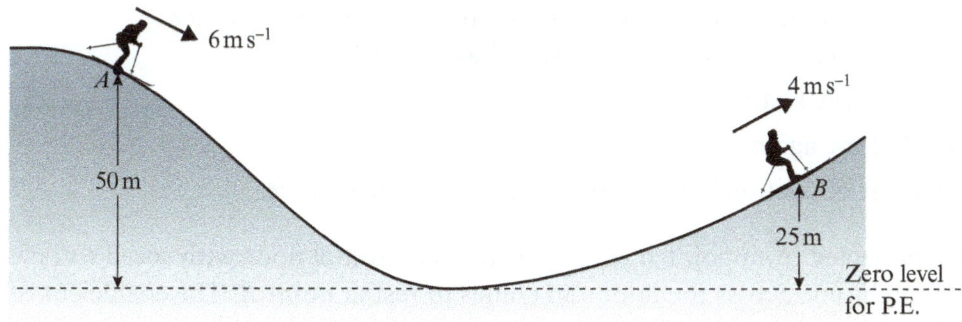

Loss of K.E. $= \frac{1}{2}mu^2 - \frac{1}{2}mv^2$
$= \frac{1}{2} \times 55 \times 6^2 - \frac{1}{2} \times 55 \times 4^2$
$= 550\,\text{J}$

Her final speed is less than her initial speed so K.E. is lost.

Loss of P.E. $= mgh$
$= 55 \times 9.8 \times (50 - 25)$
$= 13\,475\,\text{J}$

B is lower than *A* so P.E. is lost.

Total loss of energy $= 550 + 13475 = 14\,025\,\text{J}$

Use work done $= Fs$ to calculate the work done against the resistances.

Work done against resistances $= 12 \times 1400 = 16\,800\,\text{J}$
Work done by skier $= 16\,800 - 14\,025 = 2775\,\text{J}$

The skier's loss of energy provides some of the work needed to overcome the resistances. The remainder is provided by the skier doing work.

The work done by the skier is 2780 J (3 s.f.)

Exercise 4C

SKILLS PROBLEM-SOLVING

Whenever a numerical value of g is required, take $g = 9.8\,\text{m s}^{-2}$, and give your answers to either 2 significant figures or 3 significant figures.

Hint Where possible, you should use the principle of conservation of mechanical energy and the work–energy principle to answer these questions rather than the *suvat* formulae.

1 A particle of mass 0.4 kg falls a vertical distance of 7 m from rest.
- **a** Calculate the potential energy lost.
- **b** By assuming that air resistance can be neglected, calculate the final speed of the particle.

2 A stone of mass 0.5 kg is dropped from the top of a tower and falls vertically to the ground. It hits the ground with a speed of $12\,\text{m s}^{-1}$. Find:
- **a** the kinetic energy gained by the stone
- **b** the potential energy lost by the stone
- **c** the height of the tower.

3 A box of mass 6 kg is pulled in a straight line across a smooth horizontal floor by a constant horizontal force of magnitude 10 N. The box has speed $2.5\,\text{m s}^{-1}$ when it passes through point P and speed $5\,\text{m s}^{-1}$ when it passes through point Q.
- **a** Find the increase in kinetic energy of the box.
- **b** Write down the work done by the force.
- **c** Find the distance PQ.

4 A particle of mass 0.4 kg moves in a straight line across a rough horizontal surface. The speed of the particle decreases from $8\,\text{m}\,\text{s}^{-1}$ to $4\,\text{m}\,\text{s}^{-1}$ as it travels 7 m.

a Calculate the kinetic energy lost by the particle.

b Write down the work done against friction.

c Calculate the coefficient of friction between the particle and the surface.

(E) 5 A box of mass 3 kg is projected from point A across a rough horizontal floor with speed $6\,\text{m}\,\text{s}^{-1}$. The box moves in a straight line across the floor and comes to rest at point B. The coefficient of friction between the box and the floor is 0.4.

a Calculate the kinetic energy lost by the box. **(2 marks)**

b Write down the work done against friction. **(1 mark)**

c Calculate the distance AB. **(5 marks)**

6 A particle of mass 0.8 kg falls a vertical distance of 5 m from rest. By considering energy, find the speed of the particle as it hits the ground.

7 A stone of mass 0.3 kg is dropped from the top of a vertical cliff and falls freely under gravity. It hits the ground below with a speed of $20\,\text{m}\,\text{s}^{-1}$. Air resistance is negligible (too small to be significant). Use energy considerations to calculate the height of the cliff.

8 A particle of mass 0.3 kg is projected vertically upwards and moves freely under gravity. The initial speed of the particle is $u\,\text{m}\,\text{s}^{-1}$. When the particle is 5 m above the point of projection its kinetic energy is 2.1 J. Neglecting air resistance, calculate the value of u.

9 A package of mass 5 kg is released from rest and slides 2 m down a line of greatest slope of a smooth plane inclined at 35° to the horizontal.

a Calculate the potential energy lost by the package.

b Write down the kinetic energy gained by the package.

c Calculate the final speed of the package.

10 A particle of mass 0.5 kg is released from rest and slides down a line of greatest slope of a smooth plane inclined at 30° to the horizontal. When the particle has moved a distance x m, its speed is $2\,\text{m}\,\text{s}^{-1}$. Find the value of x.

11 A particle of mass 0.2 kg is projected with speed $9\,\text{m}\,\text{s}^{-1}$ up a line of greatest slope of a smooth plane inclined at 30° to the horizontal. The particle travels a distance x m before first coming to rest. By considering energy, calculate the value of x.

12 A particle of mass 0.6 kg is projected up a line of greatest slope of a smooth plane inclined at 40° to the horizontal. The particle travels 5 m before first coming to rest. Use energy considerations to calculate the speed of projection.

(E/P) 13 A box of mass 2 kg is projected with speed $6\,\text{m}\,\text{s}^{-1}$ up a line of greatest slope of a rough plane inclined at 30° to the horizontal. The coefficient of friction between the box and the plane is $\frac{1}{3}$. Use the work–energy principle to calculate the distance the box travels up the plane before coming to rest. **(5 marks)**

 **14** A skier of mass 80 kg skis down a straight hill inclined at 30° to the horizontal. The speed of the skier increases from $3\,\text{m s}^{-1}$ to $12\,\text{m s}^{-1}$.

The total resistances to motion of the skier due to friction and air resistance are modelled as a constant force of magnitude R N.

a Given that the skier travels a total distance of 50 m, find the value of R. **(5 marks)**

b Suggest one way in which the model could be refined. **(1 mark)**

 **15** A box of mass 70 kg starts at rest and slides in a straight line down a surface inclined at 20° to the horizontal. After the box has travelled a distance of 60 m, the surface becomes horizontal, and the box slides a further 50 m before coming to rest. The total resistance due to friction and air resistance is modelled as a force of constant magnitude R N which acts so as to oppose the direction of motion of the box. Find the value of R. **(7 marks)**

 **16** A girl and her sledge have a combined mass of 40 kg. She starts from rest and descends a slope which is inclined at 25° to the horizontal. At the bottom of the slope the ground becomes horizontal for 15 m before rising at 6° to the horizontal. The girl travels 25 m up the slope before coming to rest once more. There is a constant resistance to motion of magnitude 18 N. Calculate the distance the girl travels down the slope initially. **(7 marks)**

Challenge

The temperature of a gas is related to the average kinetic energy of its molecules by the formula:

average K.E. $= \frac{3}{2}kT$

where $k = 1.38 \times 10^{-23}\,\text{J K}^{-1}$ and T is the temperature in kelvin (K).

The mass of an oxygen molecule is 8 times greater than the mass of a hydrogen molecule. Two containers, one containing hydrogen and the other containing oxygen, have been in contact and able to exchange heat for a very long time, so that molecules of both gases are at the same temperature. The average speed of the oxygen molecules is $400\,\text{m s}^{-1}$. Find the average speed of the hydrogen molecules.

4.4 Power

You can calculate the **power** developed by an engine and solve problems about moving vehicles.

- **Power is the rate of doing work.**

The power developed by the engine of a moving vehicle is calculated using the following formula.

- **Power = Fv, where F is the driving force produced by the engine and v is the speed of the vehicle.**

If the driving force is measured in newtons and the speed is measured in m s^{-1}, power is measured in **watts** (W), where **1 watt is 1 joule per second**. The power of an engine is often given in kilowatts (kW).

Example 14

A truck is being pulled up a slope at a constant speed of $8\,\mathrm{m\,s^{-1}}$ by a force of magnitude 2000 N acting parallel to the direction of motion of the truck. Calculate, in kilowatts, the power developed.

Work done per second $= 2000 \times 8 = 16000\,\mathrm{J}$

Power = rate of doing work = 16000 W

The power developed is 16 kW.

> Work done per second = force × distance moved per second

> 1 kW = 1000 W

Example 15 SKILLS PROBLEM-SOLVING

A van of mass 1250 kg is travelling along a horizontal road. The van's engine is working at 24 kW. The constant resistance to motion has magnitude 600 N. Calculate:

a the acceleration of the van when it is travelling at $6\,\mathrm{m\,s^{-1}}$

b the maximum speed of the van.

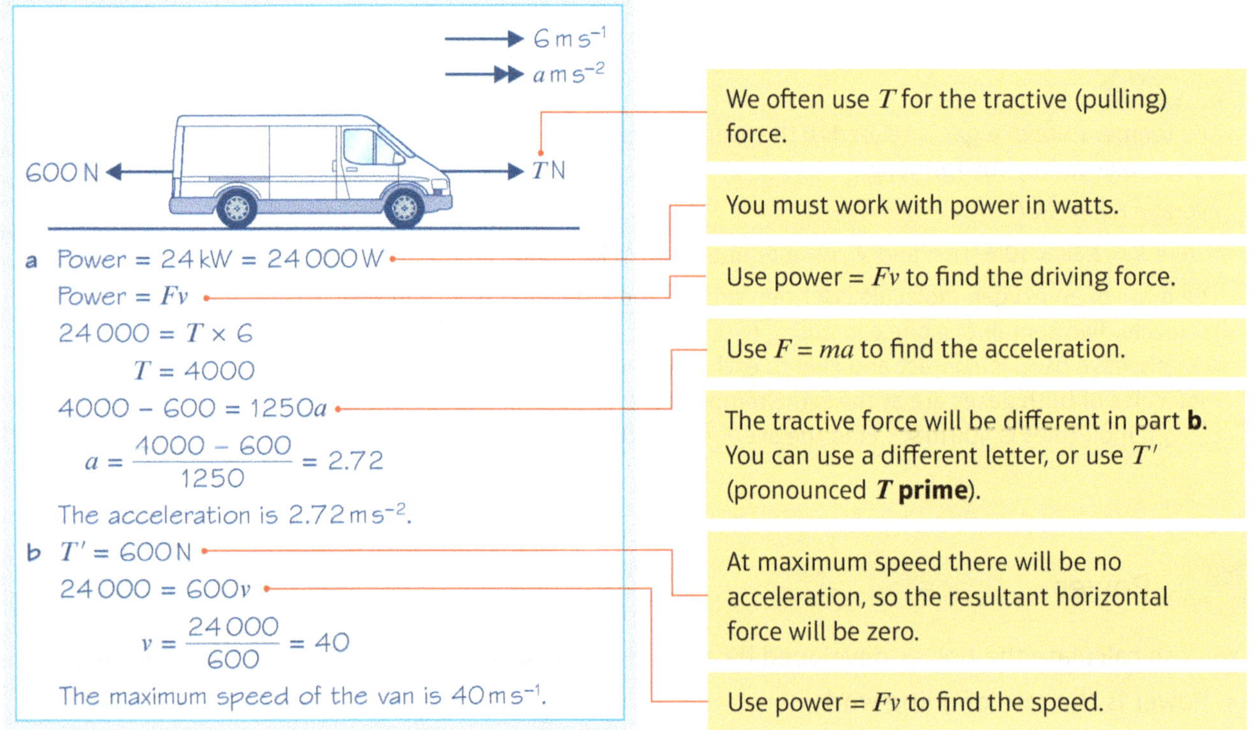

Example 16

A car of mass 1100 kg is travelling at a constant speed of $15\,\text{m s}^{-1}$ along a straight road which is inclined at 7° to the horizontal. The engine is working at a rate of 24 kW.

a Calculate the magnitude of the non-gravitational resistance to motion.

The rate of working of the engine is now increased to 28 kW.

Assuming the resistances to motion are unchanged,

b calculate the initial acceleration of the car.

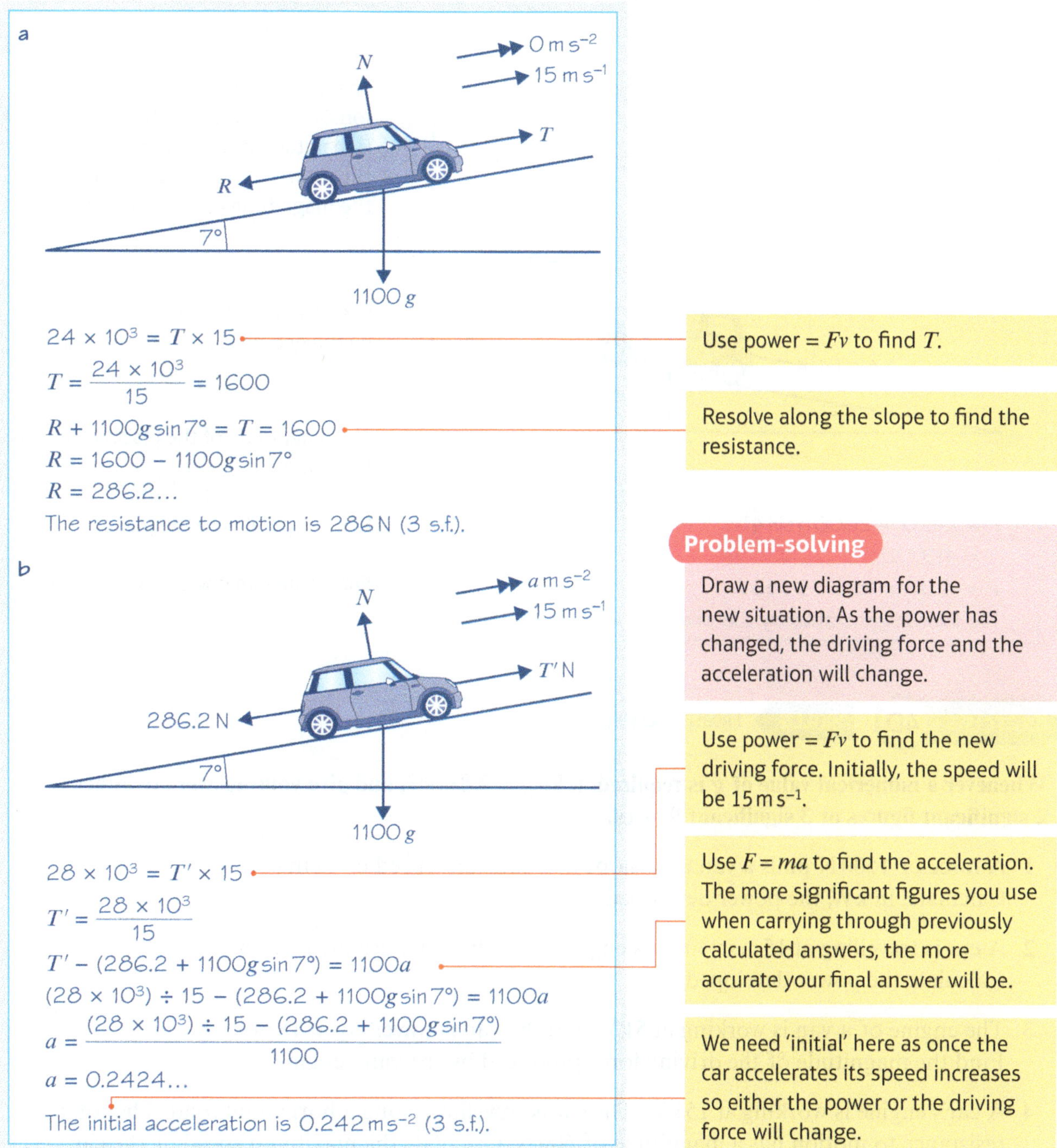

a

$24 \times 10^3 = T \times 15$

$T = \dfrac{24 \times 10^3}{15} = 1600$

$R + 1100g\sin 7° = T = 1600$

$R = 1600 - 1100g\sin 7°$

$R = 286.2\ldots$

The resistance to motion is 286 N (3 s.f.).

Use power = Fv to find T.

Resolve along the slope to find the resistance.

Problem-solving

Draw a new diagram for the new situation. As the power has changed, the driving force and the acceleration will change.

b

$28 \times 10^3 = T' \times 15$

$T' = \dfrac{28 \times 10^3}{15}$

$T' - (286.2 + 1100g\sin 7°) = 1100a$

$(28 \times 10^3) \div 15 - (286.2 + 1100g\sin 7°) = 1100a$

$a = \dfrac{(28 \times 10^3) \div 15 - (286.2 + 1100g\sin 7°)}{1100}$

$a = 0.2424\ldots$

The initial acceleration is $0.242\,\text{m s}^{-2}$ (3 s.f.).

Use power = Fv to find the new driving force. Initially, the speed will be $15\,\text{m s}^{-1}$.

Use $F = ma$ to find the acceleration. The more significant figures you use when carrying through previously calculated answers, the more accurate your final answer will be.

We need 'initial' here as once the car accelerates its speed increases so either the power or the driving force will change.

Example 17

A car of mass 2600 kg is travelling in a straight line. At the instant when the speed of the car is $v\,\text{m s}^{-1}$, the total resistances to motion are modelled as a variable force of magnitude $(800 + 5v^2)$ N. The car has a cruise control feature which adjusts the power generated by the engine to maintain a constant speed of $18\,\text{m s}^{-1}$.

Find the power generated by the engine when:

a the car is travelling on a horizontal road

b the car is travelling up a road that is inclined at an angle 4° to the horizontal.

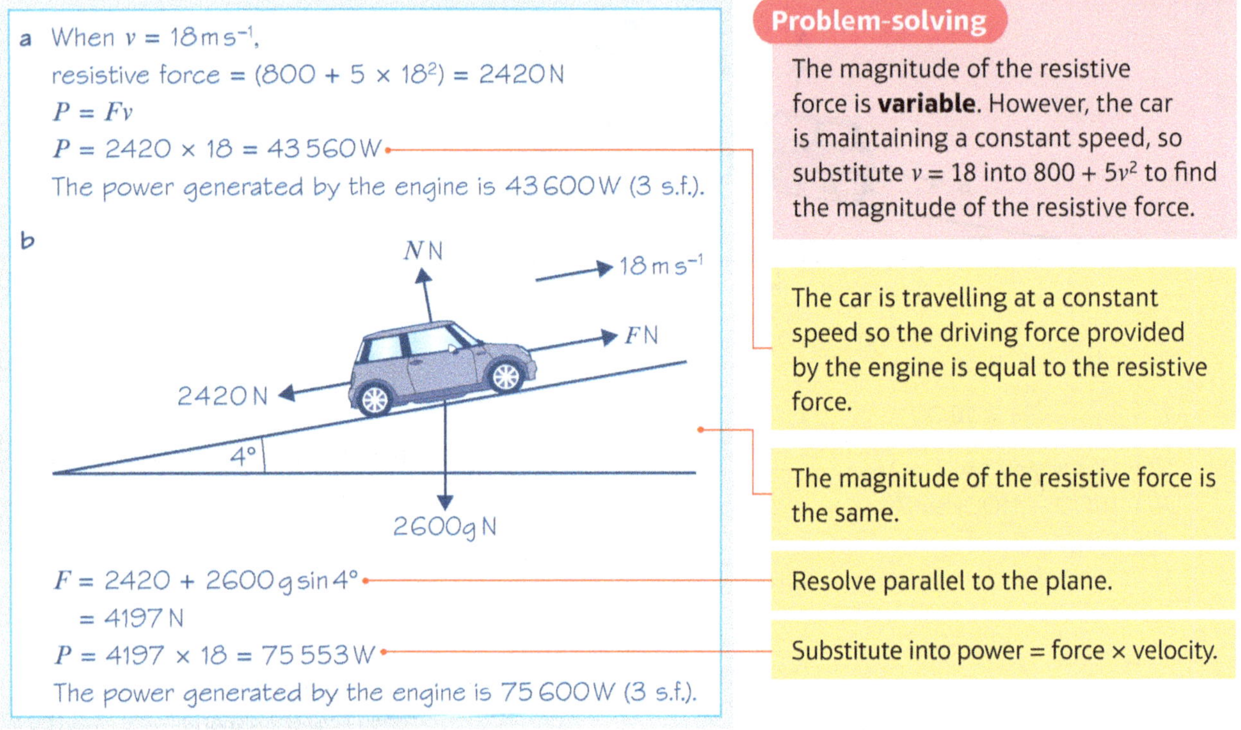

a When $v = 18\,\text{m s}^{-1}$,

resistive force $= (800 + 5 \times 18^2) = 2420$ N

$P = Fv$

$P = 2420 \times 18 = 43\,560$ W

The power generated by the engine is 43 600 W (3 s.f.).

b

$F = 2420 + 2600g \sin 4°$

$= 4197$ N

$P = 4197 \times 18 = 75\,553$ W

The power generated by the engine is 75 600 W (3 s.f.).

Problem-solving

The magnitude of the resistive force is **variable**. However, the car is maintaining a constant speed, so substitute $v = 18$ into $800 + 5v^2$ to find the magnitude of the resistive force.

The car is travelling at a constant speed so the driving force provided by the engine is equal to the resistive force.

The magnitude of the resistive force is the same.

Resolve parallel to the plane.

Substitute into power = force × velocity.

Exercise 4D  

SKILLS PROBLEM-SOLVING

Whenever a numerical value of g is required, take $g = 9.8\,\text{m s}^{-2}$, and give your answers to either 2 significant figures or 3 significant figures.

1 A force of 1500 N pulls a van up a slope at a constant speed of $12\,\text{m s}^{-1}$.
Calculate, in kW, the power developed.

2 A car is travelling at $15\,\text{m s}^{-1}$ and its engine is producing a driving force of 1000 N.
Calculate the power developed.

3 The engine of a van is working at 5 kW and the van is travelling at $18\,\text{m s}^{-1}$.
Find the magnitude of the driving force produced by the van's engine.

4 A car's engine is working at 15 kW. The car is travelling along a horizontal road. The total resistance to motion has a magnitude of 600 N. Calculate the maximum speed of the car.

5 A car has a maximum speed of $40\,\text{m}\,\text{s}^{-1}$ when travelling along a horizontal road. The total resistances to motion of the car are assumed to be constant and of magnitude 500 N.

a Calculate the power the car's engine must develop to maintain this speed.

b Comment on the assumption that the resistance to motion is constant.

6 A van is travelling along a horizontal road at a constant speed of $16\,\text{m}\,\text{s}^{-1}$. The van's engine is working at 8.8 kW. Calculate the magnitude of the resistance to motion.

(E) 7 A car of mass 850 kg is travelling along a straight horizontal road against resistances totalling 350 N. The car's engine is working at 9 kW. Calculate:

a the acceleration when the car is travelling at $7\,\text{m}\,\text{s}^{-1}$ **(3 marks)**

b the acceleration when the car is travelling at $15\,\text{m}\,\text{s}^{-1}$ **(2 marks)**

c the maximum speed of the car. **(2 marks)**

8 A car of mass 900 kg is travelling along a straight horizontal road at a speed of $20\,\text{m}\,\text{s}^{-1}$. The constant resistances to motion total 300 N. The car is accelerating at $0.3\,\text{m}\,\text{s}^{-2}$. Calculate the power developed by the engine.

(E) 9 A car of mass 1000 kg is travelling along a straight horizontal road. The car's engine is working at 12 kW. When its speed is $24\,\text{m}\,\text{s}^{-1}$ its acceleration is $0.2\,\text{m}\,\text{s}^{-2}$. The resistances to motion have a total magnitude of R newtons. Calculate the value of R. **(4 marks)**

10 A cyclist is travelling along a straight horizontal road. The resistance to his motion is constant and has magnitude 28 N. The maximum rate at which he can work is 280 W. Calculate his maximum speed.

(E) 11 A van of mass 1200 kg is travelling up a straight road inclined at 5° to the horizontal. The van moves at a constant speed of $20\,\text{m}\,\text{s}^{-1}$ and its engine is working at 24 kW. The resistance to motion from non-gravitational forces has magnitude R newtons.

a Calculate the value of R. **(3 marks)**

The road now becomes horizontal. The resistance to motion from non-gravitational forces is unchanged.

b Calculate the initial acceleration of the van. **(4 marks)**

(E) 12 A car of mass 800 kg is travelling at $18\,\text{m}\,\text{s}^{-1}$ along a straight horizontal road. The car's engine is working at a constant rate of 26 kW against a constant resistance of magnitude 750 N.

a Find the acceleration of the car. **(3 marks)**

The car now ascends a straight hill, inclined at 9° to the horizontal. The resistance to motion from non-gravitational forces is unchanged and the car's engine works at the same rate.

b Find the maximum speed at which the car can travel up the hill. **(4 marks)**

(E) 13 A van of mass 1500 kg is travelling at its maximum speed of $30\,\text{m}\,\text{s}^{-1}$ along a straight horizontal road against a constant resistance of magnitude 600 N.

a Find the power developed by the van's engine. **(3 marks)**

The van now travels up a hill along a straight road inclined at 8° to the horizontal. The van's engine works at the same rate and the resistance to motion from non-gravitational forces is unchanged.

b Find the maximum speed at which the van can ascend the hill. **(4 marks)**

(P) **14** A cyclist and her bicycle have a total mass of 80 kg. She travels at a constant speed of 7 m s^{-1}, first on flat ground, and then up a hill inclined at 2° to the horizontal. Find the increase in power required on the hill compared to the flat ground.

(E) **15** A train of mass 150 tonnes is moving up a straight track which is inclined at 2° to the horizontal. The resistance to the motion of the train from non-gravitational forces has magnitude 6 kN and the train's engine is working at a constant rate of 350 kW.

a Calculate the maximum speed of the train. **(5 marks)**

The track now becomes horizontal. The engine continues to work at 350 kW and the resistance to motion remains 6 kN.

b Find the initial acceleration of the train. **(3 marks)**

(E/P) **16** A car is moving along a straight horizontal road with speed v m s^{-1}. The magnitude of the resistance to motion of the car is given by the formula $(150 + 3v)$ N. The car's engine is working at 10 kW. Calculate the maximum value of v. **(6 marks)**

(E/P) **17** A van of mass 4000 kg is travelling in a straight line. At the instant when the speed of the car is v m s^{-1}, the total resistances to motion are modelled as a variable force of magnitude $(1200 + 8v)$ N. The engine of the van works at a constant rate of 28 kW.

Problem-solving

For part **b**, use $P = Fv$ to find the driving force in terms of w.

a Find the acceleration of the van at the instant when $v = 10$. **(4 marks)**

When the car is travelling at w m s^{-1}, it is decelerating at 0.2 m s^{-2}.

b Find the value of w. **(4 marks)**

Chapter review 4

Whenever a numerical value of g is required, take $g = 9.8$ m s^{-2}, and give your answers to either 2 significant figures or 3 significant figures.

1 A cyclist and her bicycle have a combined mass of 70 kg. She is cycling at a constant speed of 6 m s^{-1} on a straight road up a hill inclined at 5° to the horizontal. She is working at a constant rate of 480 W. Calculate the magnitude of the resistance to motion from non-gravitational forces.

2 A boy raises a bucket of water through a vertical distance of 25 m. The combined mass of the bucket and water is 12 kg. The bucket starts from rest and finishes at rest.

a Calculate the work done by the boy.

The boy takes 30 s to raise the bucket.

b Calculate the average rate of working of the boy.

(P) **3** A particle P of mass 0.5 kg is moving in a straight line from A to B on a rough horizontal plane. At A the speed of P is 12 m s^{-1}, and at B its speed is 8 m s^{-1}. The distance from A to B is 25 m. The only resistance to motion is the friction between the particle and the plane. Find:

a the work done by friction as P moves from A to B

b the coefficient of friction between the particle and the plane.

4 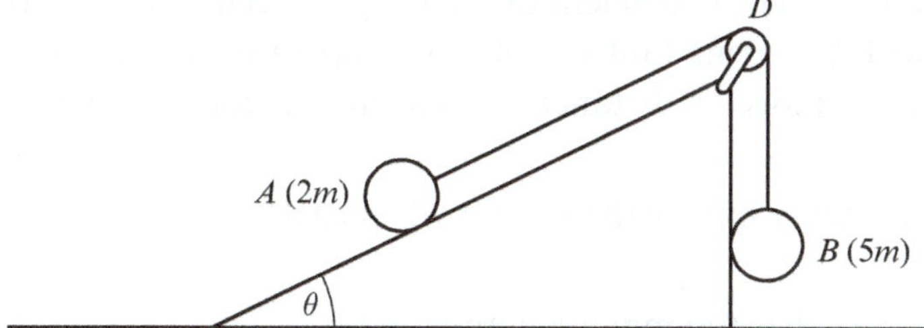

The diagram shows a particle A of mass $2m$ which can move on the rough surface of a plane inclined at an angle θ to the horizontal, where $\sin\theta = \frac{3}{5}$. A second particle B of mass $5m$ hangs freely attached to a light inextensible string which passes over a smooth light pulley fixed at D. The other end of the string is attached to A. The coefficient of friction between A and the plane is $\frac{3}{8}$. Particle B is initially hanging 2 m above the ground and A is 4 m from D. When the system is released from rest with the string taut, A moves up a line of greatest slope of the plane.

a Find the initial acceleration of A. **(7 marks)**

When B has descended 1 m the string breaks.

b By using the principle of conservation of energy, calculate the total distance moved by A before it first comes to rest. **(5 marks)**

E 5 A car of mass 800 kg is travelling along a straight horizontal road. The resistance to motion from non-gravitational forces has a constant magnitude of 500 N. The engine of the car is working at a rate of 16 kW.

a Calculate the acceleration of the car when its speed is 15 m s^{-1}. **(3 marks)**

The car comes to a hill at the moment when it is travelling at 15 m s^{-1}. The road is still straight but is now inclined at 5° to the horizontal. The resistance to motion from non-gravitational forces is unchanged. The rate of working of the engine is increased to 24 kW.

b Calculate the new acceleration of the car. **(4 marks)**

E/P 6 A car of mass 750 kg is moving at a constant speed of 18 m s^{-1} down a straight road inclined at an angle θ to the horizontal, where $\tan\theta = \frac{1}{20}$. The resistance to motion from non-gravitational forces has a constant magnitude of 1000 N.

a Find, in kW, the rate of working of the car's engine. **(3 marks)**

The engine of the car is now switched off and the car comes to rest t seconds later. The resistance to motion from non-gravitational forces is unchanged.

b Find the value of t. **(3 marks)**

7 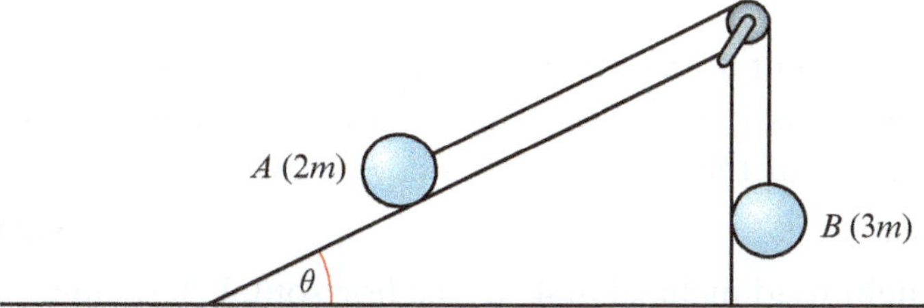

The diagram shows a particle A of mass $2m$ which can move on the rough surface of a plane inclined at an angle θ to the horizontal, where $\sin\theta = \frac{3}{5}$. A second particle B of mass $3m$ hangs freely attached to a light inextensible string which passes over a smooth pulley.

The other end of the string is attached to A. The coefficient of friction between A and the plane is $\frac{1}{4}$. The system is released from rest with the string taut and A moves up a line of greatest slope of the plane. When each particle has moved a distance s, A has not reached the pulley and B has not reached the ground.

a Find an expression for the potential energy lost by the system when each particle has moved a distance s.

When each particle has moved a distance s they are moving with speed v.

b Find an expression for v^2, in terms of s.

(E) **8** A parcel of mass 5 kg is resting on a platform inclined at 25° to the horizontal. The coefficient of friction between the parcel and the platform is 0.3. The parcel is released from rest and slides down a line of greatest slope of the platform. Calculate:

a the speed of the parcel after it has been moving for 2 s **(3 marks)**

b the potential energy lost by the parcel during this time. **(3 marks)**

9 A car of mass 2000 kg is travelling in a straight line at $10\,\text{m s}^{-1}$. The engine of the car produces a constant power of 4000 W. Find the acceleration of the car.

10 A lorry of mass 16 000 kg is travelling up a straight road inclined at 12° to the horizontal. The lorry is travelling at a constant speed of $14\,\text{m s}^{-1}$ and the resistance to motion from non-gravitational forces has a constant magnitude of 200 kN. Find the work done in 10 s by the engine of the lorry.

11 A particle P of mass 0.3 kg is moving in a straight line on a smooth horizontal surface under the action of a constant horizontal force. The particle passes point A with speed $6\,\text{m s}^{-1}$ and point B with speed $12\,\text{m s}^{-1}$.

a Find the kinetic energy gained by P while moving from A to B.

b Write down the work done by the constant force.

The distance from A to B is 4 m.

c Calculate the magnitude of the force.

(E) **12** A box of mass 5 kg slides in a straight line across a rough horizontal floor. The initial speed of the box is $10\,\text{m s}^{-1}$. The only resistance to the motion is the frictional force between the box and the floor. The box comes to rest after moving 8 m. Calculate:

a the kinetic energy lost by the box in coming to rest **(2 marks)**

b the coefficient of friction between the box and the floor. **(4 marks)**

(E) **13** A car of mass 900 kg is moving along a straight horizontal road. The resistance to motion has a constant magnitude. The engine of the car is working at a rate of 15 kW. When the car is moving with speed $20\,\text{m s}^{-1}$, the acceleration of the car is $0.3\,\text{m s}^{-2}$.

a Find the magnitude of the resistance. **(3 marks)**

The car now moves downhill on a straight road inclined at 4° to the horizontal. The engine of the car is now working at a rate of 8 kW. The resistance to motion from non-gravitational forces remains unchanged.

b Calculate the speed of the car when its acceleration is $0.5\,\text{m s}^{-2}$. **(3 marks)**

(E) **14** A bus of mass 7000 kg is travelling in a straight line on a hill inclined at 10° to the horizontal. The engine of the bus produces a constant power of 4000 W and the bus accelerates at $2\,\text{m s}^{-2}$. Find the speed of the bus. **(3 marks)**

(E) **15** A block of wood of mass 4 kg is pulled across a rough horizontal floor by a rope inclined at 15° to the horizontal. The tension in the rope is constant and has magnitude 75 N. The coefficient of friction between the block and the floor is $\frac{3}{8}$.

a Find the magnitude of the frictional force opposing the motion. **(4 marks)**

b Find the work done by the tension when the block moves 6 m. **(4 marks)**

The block is initially at rest.

c Find the speed of the block when it has moved 6 m. **(3 marks)**

16 The engine of a lorry works at a constant rate of 20 kW. The lorry has a mass of 1800 kg. When moving along a straight horizontal road there is a constant resistance to motion of magnitude 600 N. Calculate:

a the maximum speed of the lorry

b the acceleration of the lorry, in m s^{-2}, when its speed is $20\,\text{m s}^{-1}$.

(E) **17** A car of mass 1200 kg is travelling at a constant speed of $20\,\text{m s}^{-1}$ along a straight horizontal road. The constant resistance to motion has magnitude 600 N.

a Calculate the power, in kW, developed by the engine of the car. **(4 marks)**

The rate of working of the engine of the car is suddenly increased and the initial acceleration of the car is $0.5\,\text{m s}^{-2}$. The resistance to motion is unchanged.

b Find the new rate of working of the engine of the car. **(5 marks)**

The car now comes to a hill. The road is still straight but is now inclined at 20° to the horizontal. The rate of working of the engine of the car is increased further to 50 kW. The resistance to motion from non-gravitational forces still has magnitude 600 N. The car climbs the hill at a constant speed $v\,\text{m s}^{-1}$.

c Find the value of v. **(5 marks)**

Challenge

A car of mass 3000 kg drives on the inside of a cylindrical tube along a line of steepest slope. It maintains a constant speed of $20\,\text{m s}^{-1}$.

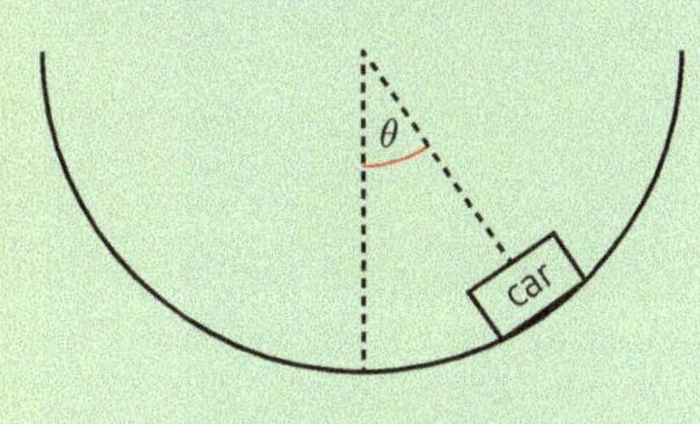

a Find the power generated by the engine when the car is in the position shown in the diagram, giving your answer in terms of θ.

b Explain what happens as $\theta \to 0$ and $\theta \to 90°$.

Summary of key points

1. You can calculate the work done by a force when its point of application moves along a straight line using the formula

   work done = $\dfrac{\text{component of force in}}{\text{direction of motion}} \times \dfrac{\text{distance moved}}{\text{in direction of force}}$

2. Work done against gravity = mgh, where m is the mass of the particle, g is the acceleration due to gravity and h is the vertical distance raised.

3. Kinetic energy (K.E.) = $\frac{1}{2}mv^2$, where m is the mass of the particle and v is its speed.

   Potential energy (P.E.) = mgh, where h is the height of the particle above an arbitrary fixed level.

4. For a particle moving horizontally, work done = change in kinetic energy.

5. You must choose a **zero level** of potential energy before calculating a particle's potential energy.

6. **Principle of conservation of mechanical energy**

   When no external forces (other than gravity) do work on a particle during its motion, the sum of the particle's kinetic and potential energy remains constant.

7. **Work-energy principle**

   The change in the total energy of a particle is equal to the work done on the particle.

8. Power is the rate of doing work.

   For a vehicle, power = Fv where F is the driving force produced by the engine and v is the speed of the vehicle.

5 IMPULSES AND COLLISIONS

4.1
4.2
4.3

Learning objectives

After completing this chapter you should be able to:

- Use the impulse–momentum principle and the principle of conservation of momentum in vector form → pages 123–126
- Solve problems involving the direct impact of two particles by using the principle of conservation of momentum and Newton's law of restitution → pages 126–132
- Apply Newton's law of restitution to problems involving the direct collision of a particle with a smooth plane surface → pages 132–135
- Find the change in energy due to an impact or the application of an impulse → pages 135–140
- Solve problems involving successive direct impacts → pages 141–147

Prior knowledge check

1. Two particles A and B of masses 0.4 kg and 0.5 kg respectively are moving towards each other on a straight line on a smooth horizontal surface. Just before the collision, both particles have speeds of $1\,\text{m s}^{-1}$. After the collision, the direction of motion of B is reversed and its speed is $0.8\,\text{m s}^{-1}$.
   a Calculate the speed and direction of A after the collision.
   b Calculate the magnitude of the impulse given by A to B during the collision. ← Mechanics 1 Section 6.2
2. A cricket ball has a mass of 0.16 kg and has kinetic energy of 50 J. Work out the speed of the cricket ball. ← Mechanics 2 Section 4.2
3. A rock of mass 2 kg falls vertically from the top of a cliff into the sea. Given that the rock is travelling at $25\,\text{m s}^{-1}$ when it hits the water, calculate the height of the cliff. ← Mechanics 1 Section 2.6

When a ball bounces, the speed with which it leaves the ground cannot be greater than the speed with which it approaches the ground. You can use Newton's **law of restitution** to model the ratio between these two speeds.

5.1 Momentum as a vector

You have used the **impulse–momentum principle** and the principle of conservation of linear **momentum** for motion in one dimension.

The **impulse–momentum principle** states that the impulse of a force is equal to the change in momentum:

impulse = force × time

$I = mv - mu$

Impulse is measured in newton seconds (N s).

For two-particle collisions, the **principle of conservation of momentum** states that the total momentum before **impact** equals the total momentum after impact:

momentum = mass × velocity

$m_1u_1 + m_2u_2 = m_1v_1 + m_2v_2$

Momentum is measured in newton seconds (N s) or kg m s^{-1}.

Impulse and momentum are both vector quantities. You can write the impulse–momentum principle and the principle of conservation of momentum as vector equations, and use them to solve problems involving collisions where the velocities and any impulse are given in vector form.

- **$\mathbf{I} = m\mathbf{v} - m\mathbf{u}$**
  where m is the mass of the body, u the initial velocity and v the final velocity.
- **$m_1\mathbf{u}_1 + m_2\mathbf{u}_2 = m_1\mathbf{v}_1 + m_2\mathbf{v}_2$**
  where a body of mass m_1 moving with velocity $\mathbf{u}_1$ collides with a body of mass m_2 moving with a velocity of $\mathbf{u}_2$, $\mathbf{v}_1$ and $\mathbf{v}_2$ are the velocities of the bodies after the collision.

Example 1

A particle of mass 0.2 kg is moving with velocity (10**i** – 5**j**) m s^{-1} when it receives an impulse (3**i** – 2**j**) N s. Find the new velocity of the particle.

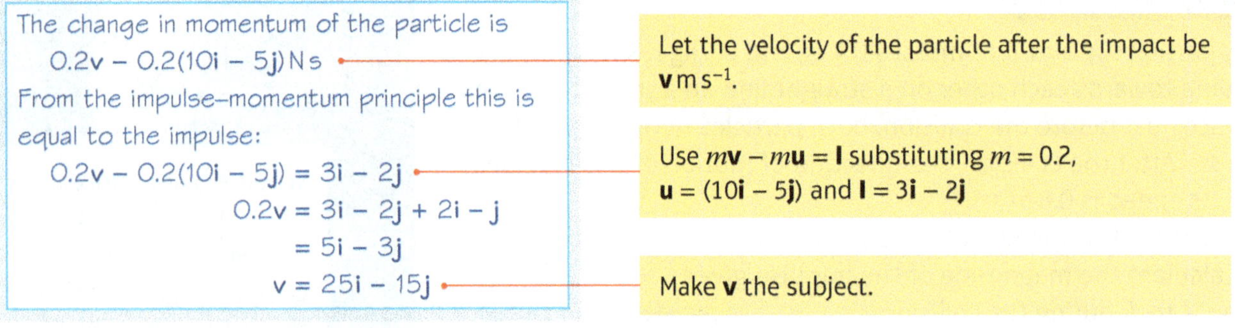
The change in momentum of the particle is

$0.2\mathbf{v} - 0.2(10\mathbf{i} - 5\mathbf{j})$ N s

From the impulse–momentum principle this is equal to the impulse:

$0.2\mathbf{v} - 0.2(10\mathbf{i} - 5\mathbf{j}) = 3\mathbf{i} - 2\mathbf{j}$

$0.2\mathbf{v} = 3\mathbf{i} - 2\mathbf{j} + 2\mathbf{i} - \mathbf{j}$

$= 5\mathbf{i} - 3\mathbf{j}$

$\mathbf{v} = 25\mathbf{i} - 15\mathbf{j}$

Let the velocity of the particle after the impact be **v** m s^{-1}.

Use $m\mathbf{v} - m\mathbf{u} = \mathbf{I}$ substituting $m = 0.2$, $\mathbf{u} = (10\mathbf{i} - 5\mathbf{j})$ and $\mathbf{I} = 3\mathbf{i} - 2\mathbf{j}$

Make **v** the subject.

Example 2 SKILLS PROBLEM-SOLVING

An ice hockey puck of mass 0.17 kg receives an impulse **Q** N s. Immediately before the impulse the velocity of the puck is $(10\mathbf{i} + 5\mathbf{j})\,\text{m s}^{-1}$ and immediately afterwards its velocity is $(15\mathbf{i} - 7\mathbf{j})\,\text{m s}^{-1}$. Find the magnitude of **Q** and the angle between **Q** and **i**.

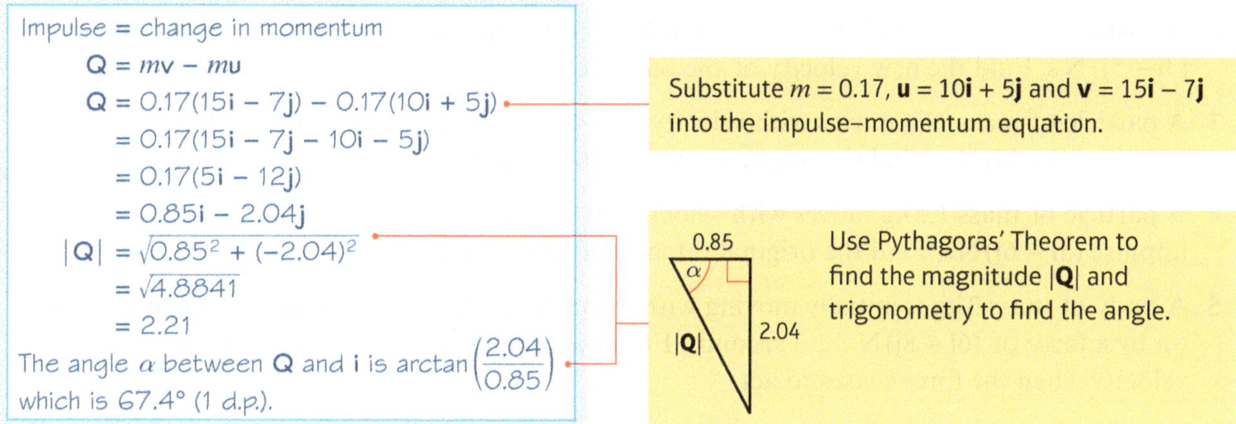

Example 3

A tennis ball of mass 0.025 kg is moving with velocity $(22\mathbf{i} + 37\mathbf{j})\,\text{m s}^{-1}$ when it hits a wall. It **rebounds** with velocity $(10\mathbf{i} - 11\mathbf{j})\,\text{m s}^{-1}$. Find the impulse exerted by the wall on the tennis ball.

Impulse $= m\mathbf{v} - m\mathbf{u}$

Impulse $= 0.025((10\mathbf{i} - 11\mathbf{j}) - (22\mathbf{i} + 37\mathbf{j}))$

$= 0.025(-12\mathbf{i} - 48\mathbf{j})$

$= (-0.3\mathbf{i} - 1.2\mathbf{j})\,\text{N s}$

The impulse exerted by the wall on the tennis ball is equal to the change in momentum of the ball.

Example 4

A particle of mass 0.15 kg is moving with velocity $(20\mathbf{i} - 10\mathbf{j})\,\text{m s}^{-1}$ when it collides with a particle of mass 0.25 kg moving with velocity $(16\mathbf{i} - 8\mathbf{j})\,\text{m s}^{-1}$. The two particles **coalesce** and form one particle of mass 0.4 kg. Find the velocity of the combined particle.

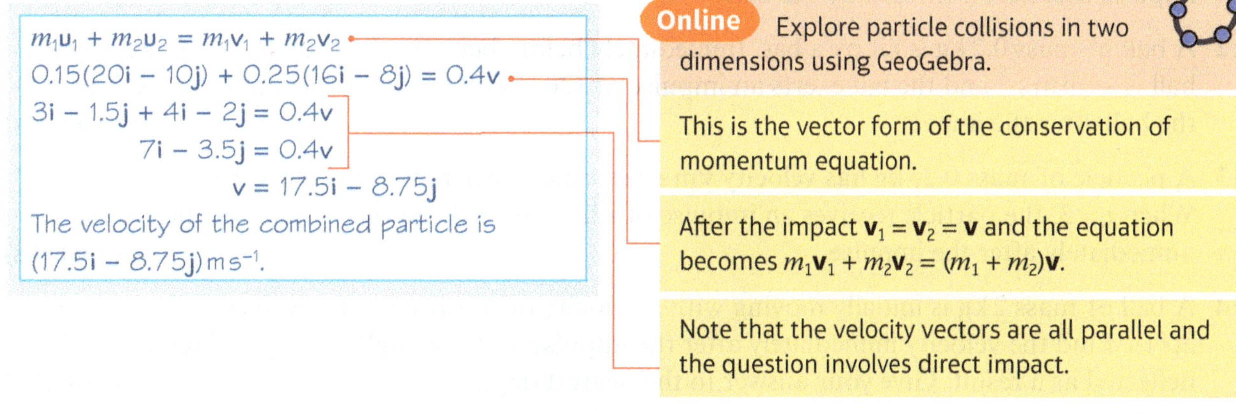

Exercise 5A SKILLS PROBLEM-SOLVING

In this exercise, **i** and **j** are perpendicular unit vectors.

1 A particle of mass 0.25 kg is moving with velocity (12**i** + 4**j**) m s^{-1} when it receives an impulse (8**i** – 7**j**) N s. Find the new velocity of the particle.

2 A particle of mass 0.5 kg is moving with velocity (2**i** – 2**j**) m s^{-1} when it receives an impulse (3**i** + 5**j**) N s. Find the new velocity of the particle.

3 A particle of mass 2 kg moves with velocity (3**i** + 2**j**) m s^{-1} immediately after it has received an impulse (4**i** + 8**j**) N s. Find the original velocity of the particle.

4 A particle of mass 1.5 kg moves with velocity (5**i** – 8**j**) m s^{-1} immediately after it has received an impulse (3**i** – 6**j**) N s. Find the original velocity of the particle.

5 A body of mass 3 kg is initially moving with a constant velocity of (**i** + **j**) m s^{-1} when it is acted on by a force of (6**i** – 8**j**) N for 3 seconds. Find the impulse exerted on the body and find its velocity when the force ceases to act.

6 A body of mass 0.5 kg is initially moving with a constant velocity of (5**i** + 12**j**) m s^{-1} when it is acted on by a force of (2**i** – **j**) N for 5 seconds. Find the impulse exerted on the body and find its velocity when the force ceases to act.

7 A particle of mass 2 kg is moving with velocity (5**i** + 3**j**) m s^{-1} when it hits a wall. It rebounds with velocity (–**i** – 3**j**) m s^{-1}. Find the impulse exerted by the wall on the particle.

8 A particle of mass 0.5 kg is moving with velocity (11**i** – 2**j**) m s^{-1} when it hits a wall. It rebounds with velocity (–**i** + 7**j**) m s^{-1}. Find the impulse exerted by the wall on the particle.

9 A particle P of mass 3 kg receives an impulse **Q** N s. Immediately before the impulse the velocity of P is 5**i** m s^{-1} and immediately afterwards it is (13**i** – 6**j**) m s^{-1}. Find the magnitude of **Q** and the angle between **Q** and **i**.

10 A particle P of mass 0.5 kg receives an impulse **Q** N s. Immediately before the impulse the velocity of P is (–**i** – 2**j**) m s^{-1} and immediately afterwards it is (3**i** – 4**j**) m s^{-1}. Find the magnitude of **Q** and the angle between **Q** and **i**.

E 11 A cricket ball of mass 0.5 kg is hit by a bat. Immediately before being hit the velocity of the ball is (20**i** – 4**j**) m s^{-1} and immediately afterwards it is (–16**i** + 8**j**) m s^{-1}. Find the magnitude of the impulse exerted on the ball by the bat. **(3 marks)**

E 12 A ball of mass 0.2 kg is hit by a bat. Immediately before being hit by the bat the velocity of the ball is –15**i** m s^{-1} and the bat exerts an impulse of (2**i** + 6**j**) N s on the ball. Find the velocity of the ball after the impact. **(3 marks)**

E 13 A particle of mass 0.25 kg has velocity **v** m s^{-1} at time t s where $\mathbf{v} = (t^2 - 3)\mathbf{i} + 4t\mathbf{j}$, $t \leqslant 3$. When $t = 3$, the particle receives an impulse of (2**i** + 2**j**) N s. Find the velocity of the particle immediately after the impulse. **(3 marks)**

E/P 14 A ball of mass 2 kg is initially moving with a velocity of (**i** + **j**) m s^{-1}. It receives an impulse of 2**j** N s. Find the velocity immediately after the impulse and the angle through which the ball is deflected as a result. Give your answer to the nearest degree. **(5 marks)**

(E) **15** A particle of mass 0.5 kg moving with velocity $3\mathbf{i}\,\text{m s}^{-1}$ collides with a particle of mass 0.25 kg moving with velocity $12\mathbf{i}\,\text{m s}^{-1}$. The two particles coalesce and move as one particle of mass 0.75 kg. Find the velocity of the combined particle. **(3 marks)**

(E) **16** A particle of mass 5 kg moving with velocity $(\mathbf{i} - \mathbf{j})\,\text{m s}^{-1}$ collides with a particle of mass 2 kg moving with velocity $(-\mathbf{i} + \mathbf{j})\,\text{m s}^{-1}$. The two particles coalesce and move as one particle of mass 7 kg. Find the magnitude of the velocity $\mathbf{v}\,\text{m s}^{-1}$ of the combined particle. **(3 marks)**

Challenge

A particle of mass m kg moves with a constant velocity of $(a\mathbf{i} + b\mathbf{j})\,\text{m s}^{-1}$. After being given an impulse, the particle then moves with a constant velocity of $(c\mathbf{i} + d\mathbf{j})\,\text{m s}^{-1}$. Given that the direction of the impulse makes an angle of 45° above the direction of **i**, show that $b + c = a + d$.

5.2 Direct impact and Newton's law of restitution

You can solve problems involving the direct impact of two particles by using the principle of conservation of momentum and **Newton's law of restitution**.

A direct impact is a collision between particles which are moving along the same straight line. When two particles collide, their speeds after the collision depend upon the materials from which they are made.

Newton's law of restitution (sometimes called Newton's experimental law) defines how the speeds of the particles after the collision depend on the nature of the particles as well as their speeds before the collision. This law holds only when the collision takes place in free space or on a smooth surface.

- **Newton's law of restitution states that**

$$\frac{\text{speed of separation of particles}}{\text{speed of approach of particles}} = e$$

The constant e is the coefficient of restitution between the particles; $0 \leqslant e \leqslant 1$.

The value of the coefficient of restitution e depends on the materials from which the particles are made. In a perfectly **elastic collision**, $e = 1$ so the speed of separation is the same as the speed of approach. In a totally **inelastic collision**, $e = 0$ so the particles coalesce on impact.

Notation **Coalesce** means join together. Two balls of sticky dough would produce a totally inelastic collision, with $e = 0$.

	Perfectly elastic collision ($e = 1$)	Totally inelastic collision ($e = 0$)
Before collision:	→ $v\,\text{m s}^{-1}$ ←	→ $v\,\text{m s}^{-1}$ ←
After collision:	← $v\,\text{m s}^{-1}$ →	$0\,\text{m s}^{-1}$

Example 5

In each part of this question, the two diagrams show the speeds and directions of motion of two particles A and B just before and just after a collision. The particles move on a smooth horizontal plane. Find the coefficient of restitution e in each case.

a

Before impact		After impact	
$8\,\text{m s}^{-1}$ →	At rest	At rest	$2\,\text{m s}^{-1}$ →
A	B	A	B

b

Before impact		After impact	
$6\,\text{m s}^{-1}$ →	$3\,\text{m s}^{-1}$ →	$4\,\text{m s}^{-1}$ →	$5\,\text{m s}^{-1}$ →
A	B	A	B

c

Before impact		After impact	
$11\,\text{m s}^{-1}$ →	← $7\,\text{m s}^{-1}$	← $6\,\text{m s}^{-1}$	$3\,\text{m s}^{-1}$ →
A	B	A	B

a The speed of approach is $8 - 0 = 8\,\text{m s}^{-1}$

The speed of separation is $2 - 0 = 2\,\text{m s}^{-1}$

$e = \frac{2}{8} = \frac{1}{4}$

b The speed of approach is $6 - 3 = 3\,\text{m s}^{-1}$

The speed of separation is $5 - 4 = 1\,\text{m s}^{-1}$

$e = \frac{1}{3}$

c The speed of approach is $11 + 7 = 18\,\text{m s}^{-1}$

The speed of separation is $6 + 3 = 9\,\text{m s}^{-1}$

$e = \frac{9}{18} = \frac{1}{2}$

Find the difference in the velocities before impact, called the speed of approach.

Find the difference in the velocities after impact, called the speed of separation.

Find e using

$$e = \frac{\text{speed of separation of particles}}{\text{speed of approach of particles}}$$

Watch out In part **c** the particles are moving in **opposite directions** so the speed of approach/separation will be the sum of the speeds of each particle.

Example 6

Two particles A and B are travelling in the same direction on a smooth surface with speeds $4\,\text{m s}^{-1}$ and $3\,\text{m s}^{-1}$ respectively. They collide directly, and immediately after the collision continue to travel in the same direction with speeds $2\,\text{m s}^{-1}$ and $v\,\text{m s}^{-1}$ respectively.

Given that the coefficient of restitution between A and B is $\frac{1}{3}$, find v.

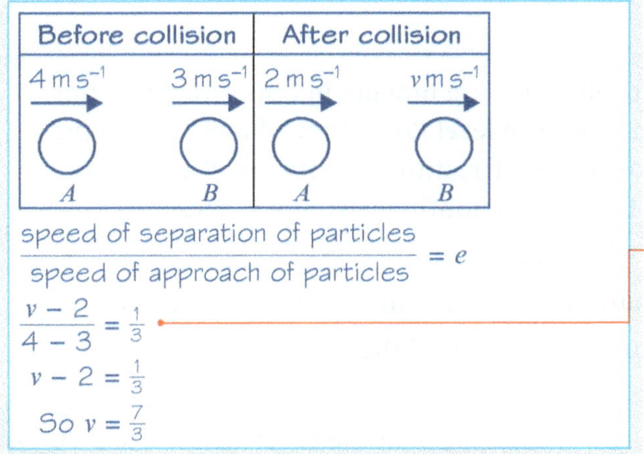

$$\frac{\text{speed of separation of particles}}{\text{speed of approach of particles}} = e$$

$$\frac{v-2}{4-3} = \frac{1}{3}$$

$$v - 2 = \frac{1}{3}$$

$$\text{So } v = \frac{7}{3}$$

Substitute the speed of approach, 4 – 3, and the speed of separation, v – 2, then make v the subject of the formula.

You can use the **principle of conservation of linear momentum** together with Newton's law of restitution to solve problems involving two unknown velocities.

Links For particles of masses m_1 and m_2 colliding in a straight line, with initial velocities u_1 and u_2 respectively, and final velocities v_1 and v_2 respectively:
$m_1u_1 + m_2u_2 = m_1v_1 + m_2v_2$
← **Mechanics 1 Section 6.2**

Example 7 SKILLS PROBLEM-SOLVING

Two particles A and B of masses 0.2 kg and 0.4 kg respectively are travelling in opposite directions towards each other on a smooth surface with speeds $5\,m\,s^{-1}$ and $4\,m\,s^{-1}$ respectively. They collide directly, and immediately after the collision have velocities $v_1\,m\,s^{-1}$ and $v_2\,m\,s^{-1}$ respectively, measured in the direction of motion of A before the collision.

Given that the coefficient of restitution between A and B is $\frac{1}{2}$, find v_1 and v_2.

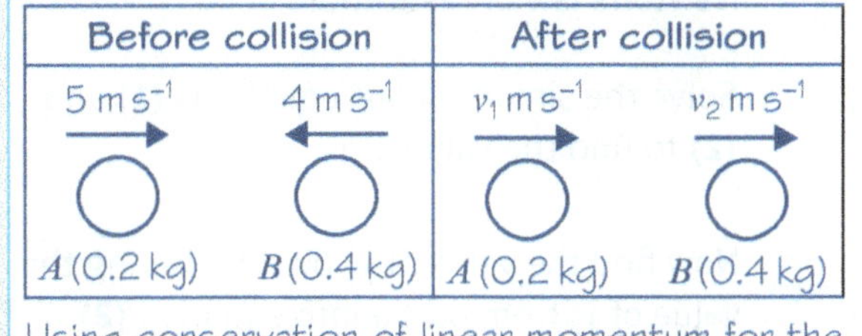

Online Explore direct impact using GeoGebra.

Draw a diagram showing the situation before and after the collision.

Using conservation of linear momentum for the system (→):

$$0.2 \times 5 + 0.4 \times (-4) = 0.2v_1 + 0.4v_2$$
$$1 - 1.6 = 0.2v_1 + 0.4v_2$$
$$-0.6 = 0.2v_1 + 0.4v_2$$
$$-3 = v_1 + 2v_2 \quad \textbf{(1)}$$

Problem-solving

The final velocities are measured in the direction of motion of A before the collision, so choose this as the positive direction. Initially, B is moving in the opposite direction, so its velocity will be negative.

$$\frac{\text{speed of separation of particles}}{\text{speed of approach of particles}} = e$$

$$\frac{v_2 - v_1}{5+4} = \frac{1}{2}$$

$$v_2 - v_1 = \frac{9}{2} \quad \textbf{(2)}$$

Calculate the speed of approach and the speed of separation and substitute into Newton's law of restitution.

Eliminating v_1 between equations **(1)** and **(2)** gives

$v_2 = \frac{1}{2}$

Substituting this value into equation **(1)** gives

$v_1 = -4$

Solve the simult aneous equations **(1)** and **(2)** to find the values of v_1 and v_2.

Example 8

Two balls P and Q have masses $3m$ and $4m$ respectively. They are moving in opposite directions towards each other along the same straight line on a smooth level floor. Immediately before they collide, P has speed $3u$ and Q has speed $2u$. The coefficient of restitution between P and Q is e. By modelling the balls as smooth spheres and the floor as a smooth horizontal plane,

a show that the speed of Q after the collision is $\frac{u}{7}(15e + 1)$.

b Given that the direction of motion of P is unchanged, find the range of possible values of e.

c Given that the magnitude of the impulse of P on Q is $\frac{80mu}{9}$, find the value of e.

a

Before impact		After impact	
$3u$ →	← $2u$	v_1 →	v_2 →
$P\ (3m)$	$Q\ (4m)$	$P\ (3m)$	$Q\ (4m)$

Choose a positive direction and draw a diagram showing masses and velocities before and after the impact. Use v_1 and v_2 for the unknown velocities after impact.

Using conservation of linear momentum for the system (→):

$9mu - 8mu = 3mv_1 + 4mv_2$

$\Rightarrow 3v_1 + 4v_2 = u \quad$ **(1)**

Use $m_1u_1 + m_2u_2 = m_1v_1 + m_2v_2$, noting that u_2 is negative as Q is moving in the opposite direction.

Newton's law of restitution gives

$\frac{v_2 - v_1}{3u + 2u} = e$

$\Rightarrow v_2 - v_1 = 5eu \quad$ **(2)**

Calculate the speed of approach and the speed of separation then substitute into Newton's law of restitution.

Eliminating v_1 between equations **(1)** and **(2)** gives

$7v_2 = 15ue + u$

$v_2 = \frac{u}{7}(15e + 1)$

Solve the simultaneous equations **(1)** and **(2)** to find the value of v_2.

b Substituting this value into equation **(2)** gives

$v_1 = \frac{u}{7}(15e + 1) - 5eu$

Now find the value of v_1 by substituting the value of v_2 from part **a** into equation **(2)**.

$v_1 = \frac{u}{7}(1 - 20e)$

As $v_1 > 0$, $\frac{u}{7}(1 - 20e) > 0$

So $0 \leqslant e < \frac{1}{20}$

Problem-solving

As the direction of motion of P is unchanged by the impact, v_1 must be positive.

c Impulse of P on Q = change in momentum of Q

$= 4mv_2 - 4m(-2u)$

This is $m_2v_2 - m_2u_2$

$= \frac{4mu}{7}(15e + 1) + 8mu$

$= \frac{60mu}{7}(1 + e)$

However the impulse is given as $\frac{80mu}{9}$

So $\frac{60mu}{7}(1 + e) = \frac{80mu}{9}$

$(1 + e) = \frac{28}{27}$

$e = \frac{1}{27}$

Online Explore direct impact with a known impulse using GeoGebra.

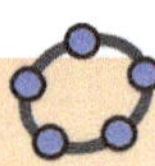

Exercise 5B SKILLS PROBLEM-SOLVING

1 In each part of this question, the two diagrams show the speeds and directions of motion of two particles *A* and *B* just before and just after a collision. The particles move on a smooth horizontal plane. Find the coefficient of restitution *e* in each case.

	Before collision		After collision	
a	$6\,\mathrm{m\,s^{-1}}$ → *A*	At rest *B*	At rest *A*	$4\,\mathrm{m\,s^{-1}}$ → *B*
b	$4\,\mathrm{m\,s^{-1}}$ → *A*	$2\,\mathrm{m\,s^{-1}}$ → *B*	$2\,\mathrm{m\,s^{-1}}$ → *A*	$3\,\mathrm{m\,s^{-1}}$ → *B*
c	$9\,\mathrm{m\,s^{-1}}$ → *A*	$6\,\mathrm{m\,s^{-1}}$ ← *B*	$3\,\mathrm{m\,s^{-1}}$ ← *A*	$2\,\mathrm{m\,s^{-1}}$ → *B*

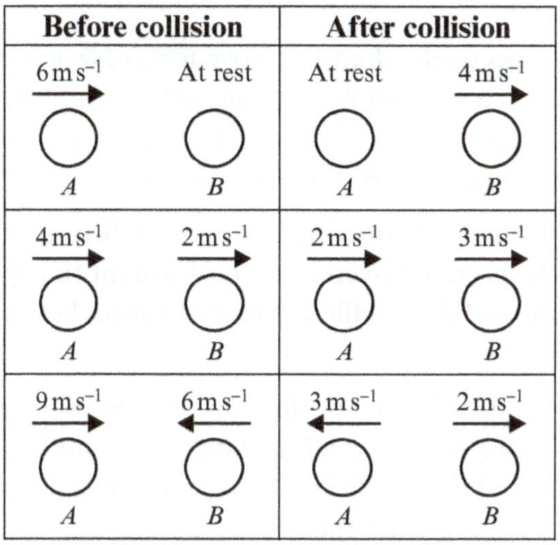

2 In each part of this question, the two diagrams show the speeds and directions of motion of two particles *A* and *B* just before a collision, and their velocities relative to the initial directon of *A* just after a collision. The particles move on a smooth horizontal plane. The masses of *A* and *B* and the coefficients of restitution *e* are also given. Find the values of v_1 and v_2 in each case.

	Before collision		After collision	
a $e = \frac{1}{2}$	$6\,\mathrm{m\,s^{-1}}$ → *A* (0.25 kg)	At rest *B* (0.5 kg)	$v_1\,\mathrm{m\,s^{-1}}$ → *A* (0.25 kg)	$v_2\,\mathrm{m\,s^{-1}}$ → *B* (0.5 kg)
b $e = 0.25$	$4\,\mathrm{m\,s^{-1}}$ → *A* (2 kg)	$2\,\mathrm{m\,s^{-1}}$ → *B* (3 kg)	$v_1\,\mathrm{m\,s^{-1}}$ → *A* (2 kg)	$v_2\,\mathrm{m\,s^{-1}}$ → *B* (3 kg)
c $e = \frac{1}{7}$	$8\,\mathrm{m\,s^{-1}}$ → *A* (3 kg)	$6\,\mathrm{m\,s^{-1}}$ ← *B* (1 kg)	$v_1\,\mathrm{m\,s^{-1}}$ → *A* (3 kg)	$v_2\,\mathrm{m\,s^{-1}}$ → *B* (1 kg)
d $e = \frac{2}{3}$	$6\,\mathrm{m\,s^{-1}}$ → *A* (400 g)	$6\,\mathrm{m\,s^{-1}}$ ← *B* (400 g)	$v_1\,\mathrm{m\,s^{-1}}$ → *A* (400 g)	$v_2\,\mathrm{m\,s^{-1}}$ → *B* (400 g)
e $e = \frac{1}{5}$	$3\,\mathrm{m\,s^{-1}}$ → *A* (5 kg)	$12\,\mathrm{m\,s^{-1}}$ ← *B* (4 kg)	$v_1\,\mathrm{m\,s^{-1}}$ → *A* (5 kg)	$v_2\,\mathrm{m\,s^{-1}}$ → *B* (4 kg)

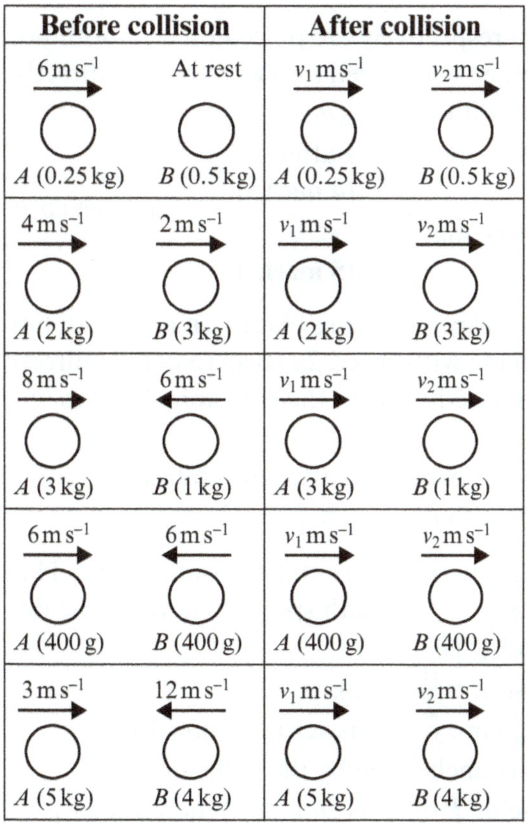

3 A small smooth sphere A of mass 1 kg is travelling along a straight line on a smooth horizontal plane with speed $4\,\mathrm{m\,s^{-1}}$ when it collides with a second smooth sphere B of the same radius, with mass 2 kg and travelling in the same direction as A with speed $2.5\,\mathrm{m\,s^{-1}}$. After the collision, A continues in the same direction with speed $2\,\mathrm{m\,s^{-1}}$. Find:

a the speed of B after the collision

b the coefficient of restitution for the spheres.

4 Two spheres A and B have masses 2 kg and 6 kg respectively. A and B move towards each other in opposite directions along the same straight line on a smooth horizontal surface with speeds $4\,\mathrm{m\,s^{-1}}$ and $6\,\mathrm{m\,s^{-1}}$ respectively. If the coefficient of restitution is $\frac{1}{5}$, find the velocities of the spheres after the collision and the magnitude of the impulse given to each sphere.

(P) **5** Two particles P and Q of masses $2m$ and $3m$ respectively are moving in opposite directions towards each other. Each particle is travelling with speed u. Given that Q is brought to rest by the collision, find the speed of P after the collision, and the coefficient of restitution between the particles.

(P) **6** Two particles A and B are travelling along the same straight line in the same direction on a smooth horizontal surface with speeds $3u$ and u respectively. Particle A catches up and collides with particle B. If the mass of B is twice that of A and the coefficient of restitution is e, find, in terms of e and u, expressions for the speeds of A and B after the collision.

(P) **7** Two identical particles of mass m are projected towards each other along the same straight line on a smooth horizontal surface with speeds of $2u$ and $3u$. After the collision, the directions of motion of both particles are reversed. Show that this implies that the coefficient of restitution e satisfies the inequality $e > \frac{1}{5}$.

(E/P) **8** Two particles A and B of masses m and km respectively are placed on a smooth horizontal plane. Particle A is made to move on the plane with speed u so as to collide directly with B, which is at rest. After the collision B moves with speed $\frac{3}{10}u$.

a Find, in terms of u and the constant k, the speed of A after the collision. **(4 marks)**

b By using Newton's law of restitution, show that $\frac{7}{3} \leqslant k \leqslant \frac{17}{3}$. **(5 marks)**

Problem-solving

In part **b** use the limits of e to set up an inequality.

(E/P) **9** Two particles A and B of masses m and $3m$ respectively are placed on a smooth horizontal plane. Particle A is made to move on the plane with speed $2u$ so as to collide directly with B, which is moving in the same direction with speed u. After the collision B moves with speed ku, where k is a constant.

a Find, in terms of u and the constant k, the speed of A after the collision. **(4 marks)**

b By using Newton's law of restitution, show that $\frac{5}{4} \leqslant k \leqslant \frac{3}{2}$. **(5 marks)**

(E/P) **10** A particle P of mass m is moving with speed $4u$ on a smooth horizontal plane. The particle collides directly with a particle Q of mass $3m$ moving with speed $2u$ in the same direction as P. The coefficient of restitution between P and Q is e.

a Show that the speed of Q after the collision is $\frac{u}{2}(5 + e)$. **(6 marks)**

b Find the speed of P after the collision, giving your answer in terms of e. **(4 marks)**

c Show that the direction of motion of P is unchanged by the collision. **(2 marks)**

d Given that the magnitude of the impulse of P on Q is $2mu$, find the value of e. **(4 marks)**

Challenge

Two particles P and Q of masses $3m$ kg and m kg respectively move towards each other in a straight line in opposite directions on a smooth horizontal surface. P has initial speed 2 m s^{-1} and Q has initial speed $u\text{ m s}^{-1}$. After P and Q collide, both particles move in the same direction as P's original motion. Given that, after the impact, Q moves with twice the speed of P and that the coefficient of restitution between P and Q is $\frac{1}{4}$, show that $u = \frac{14}{9}$.

5.3 Direct collision with a smooth plane

You can also apply Newton's law of restitution to problems involving the **direct collision** of a particle with a smooth plane surface perpendicular to the direction of motion of the particle.

In the figure below, a particle is shown moving horizontally with speed u before impact with a vertical plane surface. After impact the particle moves in the opposite direction with speed v.

The speed of the particle after the impact depends on the speed of the particle before the impact and the coefficient of restitution e between the particle and the plane.

- **For the direct collision of a particle with a smooth plane, Newton's law of restitution can be written as:**

$$\frac{\text{speed of rebound}}{\text{speed of approach}} = e$$

Example 9

A particle collides normally with a fixed vertical plane. The diagram shows the speeds of the particle before and after the collision. Find the value of the coefficient of restitution e.

Notation If a particle collides **normally** then its direction of motion immediately before the instant of impact is perpendicular to the plane.

Before impact

8 m s^{-1} → Vertical plane

After impact

2 m s^{-1} ← Vertical plane

Using Newton's law of restitution,

$$e = \frac{\text{speed of rebound}}{\text{speed of approach}}$$

$$= \frac{2}{8} = \frac{1}{4}$$

This is the same as $e = \frac{2-0}{8-0}$.

The coefficient of restitution is $\frac{1}{4}$.

Example 10

A small sphere collides normally with a fixed vertical wall. Before the impact the sphere is moving with a speed of 4 m s^{-1} on a smooth horizontal floor. The coefficient of restitution between the sphere and the wall is 0.2. Find the speed of the sphere after the collision.

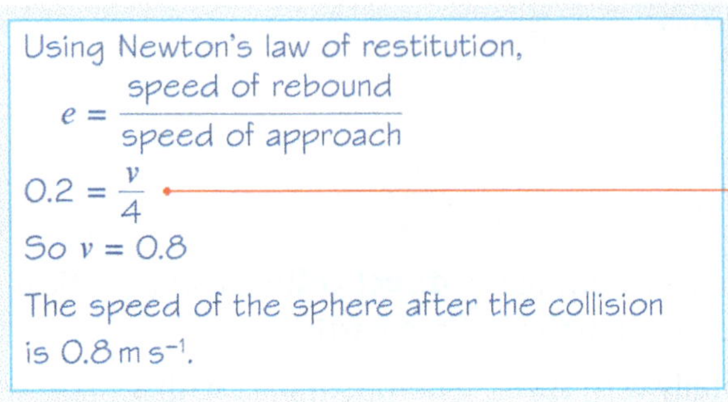

Using Newton's law of restitution,

$e = \dfrac{\text{speed of rebound}}{\text{speed of approach}}$

$0.2 = \dfrac{v}{4}$

So $v = 0.8$

The speed of the sphere after the collision is 0.8 m s^{-1}.

Let the speed of the sphere after the collision be v m s^{-1}.

Example 11

SKILLS PROBLEM-SOLVING

A particle falls 22.5 cm from rest onto a smooth horizontal plane. It then rebounds to a height of 10 cm. Find the coefficient of restitution between the particle and the plane. Give your answer to 2 significant figures.

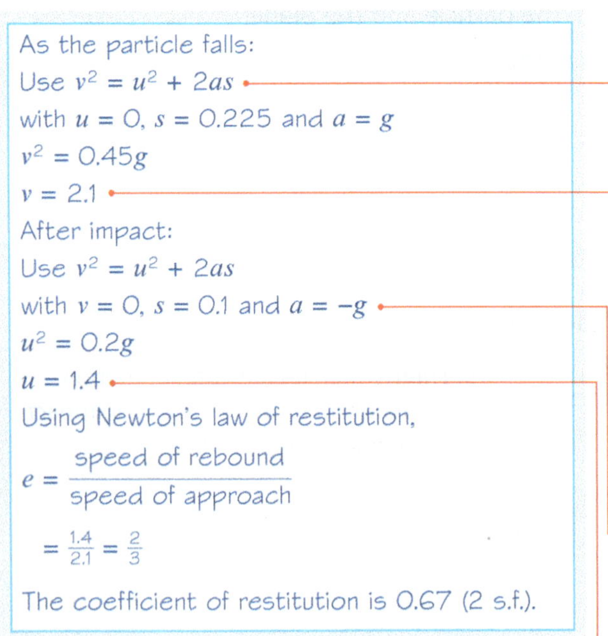

As the particle falls:

Use $v^2 = u^2 + 2as$

with $u = 0$, $s = 0.225$ and $a = g$

$v^2 = 0.45g$

$v = 2.1$

After impact:

Use $v^2 = u^2 + 2as$

with $v = 0$, $s = 0.1$ and $a = -g$

$u^2 = 0.2g$

$u = 1.4$

Using Newton's law of restitution,

$e = \dfrac{\text{speed of rebound}}{\text{speed of approach}}$

$= \frac{1.4}{2.1} = \frac{2}{3}$

The coefficient of restitution is 0.67 (2 s.f.).

Online Explore the direct collision of a falling particle with a smooth horizontal plane using GeoGebra.

The particle is falling under gravity so use the appropriate constant acceleration formula to find its speed when it hits the plane.

← Mechanics 1 Section 2.6

Using $g = 9.8$ m s^{-2}, calculate v m s^{-1}, the speed of the particle when it hits the plane.

After it rebounds it initially moves upward under gravity. As the upward direction is taken as positive here, the acceleration is negative.

u m s^{-1} is the rebound speed of the particle.

Whenever a numerical value of g is required, take $g = 9.8\,\text{m s}^{-2}$.

1 A smooth sphere collides normally with a fixed vertical wall. The two diagrams show the speed and direction of motion of the sphere before and after the collision. In each case, find the value of the coefficient of restitution e.

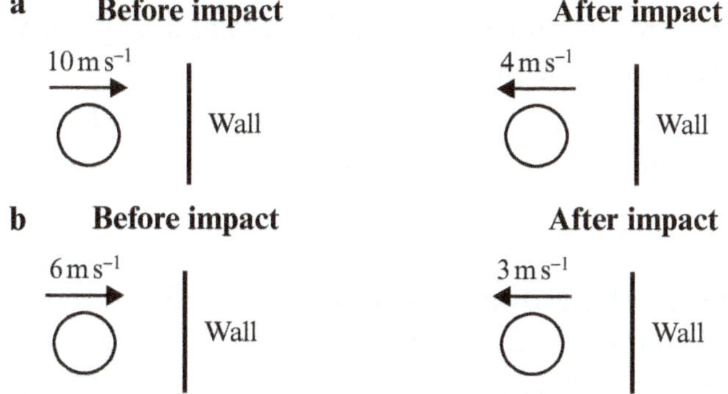

2 A smooth sphere collides normally with a fixed vertical wall. The two diagrams show the speed and direction of motion of the sphere before and after the collision. The value of e is given in each case. Find the speed of the sphere after the collision in each case.

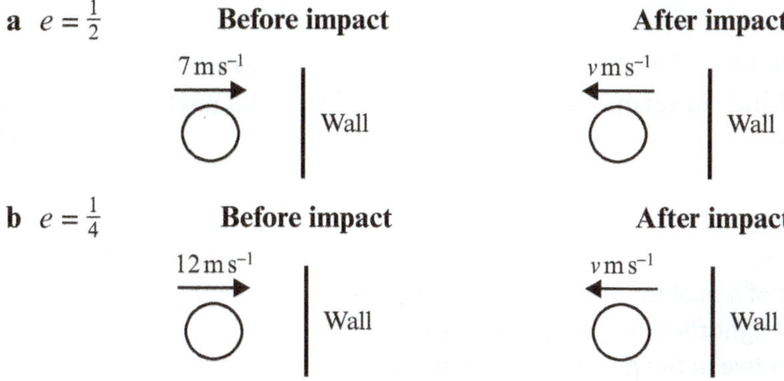

3 A smooth sphere collides normally with a fixed vertical wall. The two diagrams show the speed and direction of motion of the sphere before and after the collision. The value of e is also given. Find the speed of the sphere before the collision in each case.

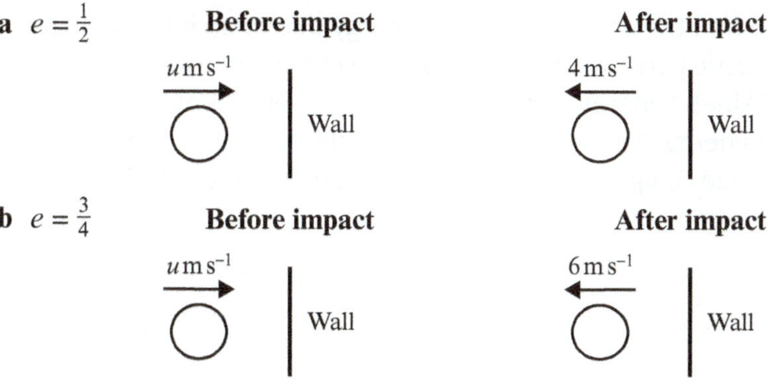

4 A small smooth sphere of mass 0.3 kg is moving on a smooth horizontal table with a speed of $10\,\text{m}\,\text{s}^{-1}$ when it collides normally with a fixed smooth wall. It rebounds with a speed of $7.5\,\text{m}\,\text{s}^{-1}$. Find the coefficient of restitution between the sphere and the wall.

5 A particle falls 2.5 m from rest onto a smooth horizontal plane. It then rebounds to a height of 1.5 m. Find the coefficient of restitution between the particle and the plane. Give your answer to 2 significant figures.

(E) 6 A particle falls 3 m from rest onto a smooth horizontal plane. It then rebounds to a height h m. The coefficient of restitution between the particle and the plane is 0.25.

a Find the value of h. **(4 marks)**

b Without further calculation, state how your answer to part **a** would change if $e > 0.25$. **(1 mark)**

(E/P) 7 A small smooth sphere falls from rest onto a smooth horizontal plane and rebounds from the plane. It takes 2 seconds to reach the plane, then another 2 seconds to reach the plane a second time. Find the coefficient of restitution between the sphere and the plane. **(7 marks)**

(E/P) 8 A small smooth sphere falls from rest onto a smooth horizontal plane. It takes 3 seconds to reach the plane. The coefficient of restitution between the sphere and the plane is 0.49. Find the time it takes after rebound for the sphere to reach the plane a second time. **(7 marks)**

(E/P) 9 A particle falls from rest from a height of h m above level ground. After rebounding, the particle reaches a maximum height of $\frac{1}{2}h$. Find, in terms of g and h, the height of the particle 1 second after it hits the ground. **(8 marks)**

Challenge

A particle P falls from rest from a height of h m above level ground. Show that after hitting the ground, the maximum height that the particle reaches is he^2, where e is the coefficient of restitution between the particle and the ground.

5.4 Loss of kinetic energy

You can solve problems that ask you to find the change in kinetic energy due to an impact or the application of an impulse. When two elastic particles collide there is no change in total linear momentum but there is a loss in **total** kinetic energy ($e \neq 1$). This energy is transformed into other forms of energy such as heat or sound energy. If a collision is perfectly elastic ($e = 1$) then no kinetic energy is lost. An individual particle can gain or lose kinetic energy due to a collision or the application of an impulse.

Example 12 SKILLS PROBLEM-SOLVING

Two spheres A and B have masses 3 kg and 5 kg respectively. A and B move towards each other in opposite directions along the same straight line on a smooth horizontal surface with speeds $3\,\text{m s}^{-1}$ and $2\,\text{m s}^{-1}$ respectively.

a Given the coefficient of restitution is $\frac{3}{5}$, find the velocities of the spheres after the collision.

b Find the loss of kinetic energy due to the impact.

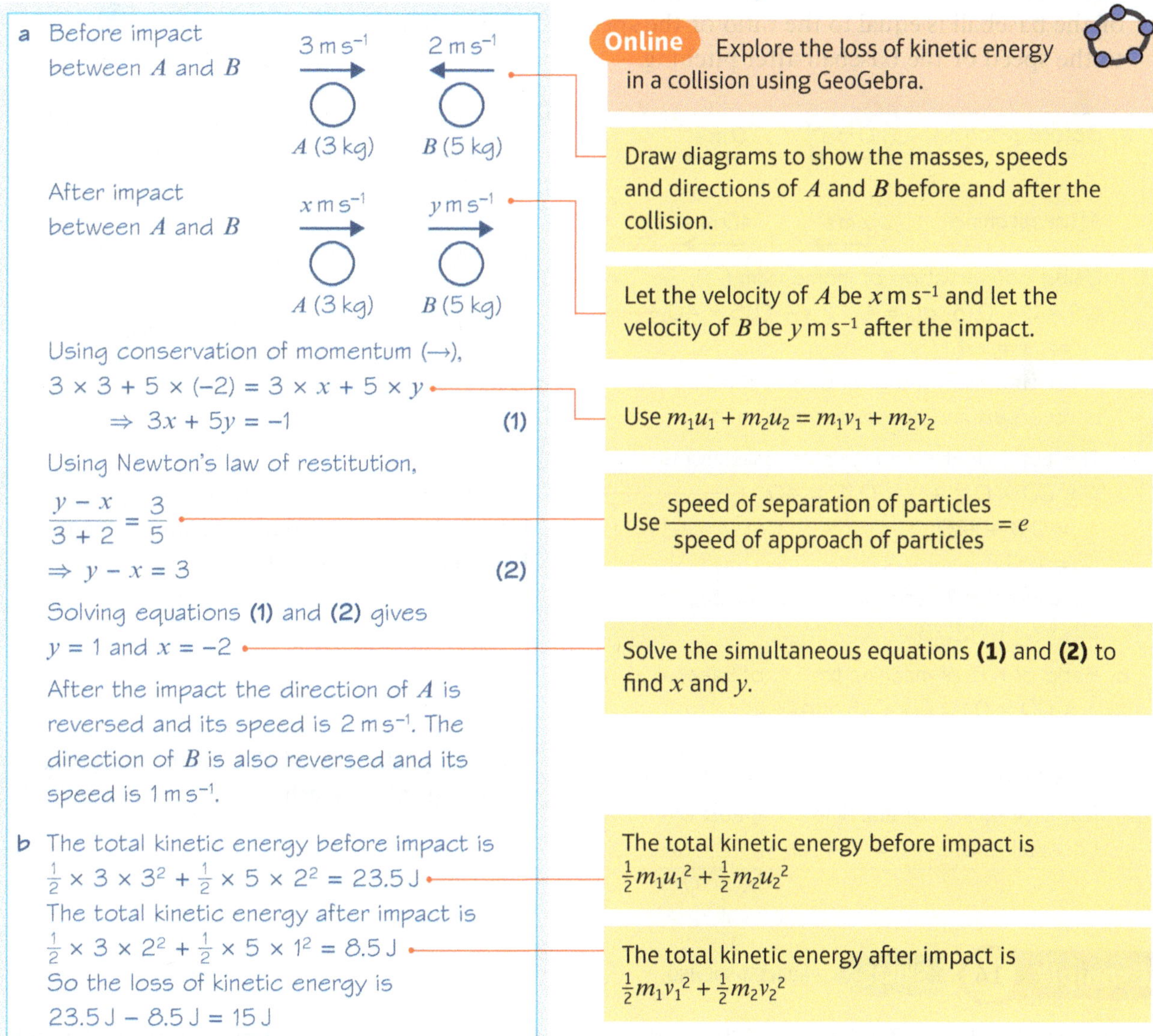

a Before impact between A and B

After impact between A and B

Using conservation of momentum (⟶),

$3 \times 3 + 5 \times (-2) = 3 \times x + 5 \times y$

$\Rightarrow 3x + 5y = -1$ **(1)**

Using Newton's law of restitution,

$\frac{y - x}{3 + 2} = \frac{3}{5}$

$\Rightarrow y - x = 3$ **(2)**

Solving equations **(1)** and **(2)** gives

$y = 1$ and $x = -2$

After the impact the direction of A is reversed and its speed is $2\,\text{m s}^{-1}$. The direction of B is also reversed and its speed is $1\,\text{m s}^{-1}$.

b The total kinetic energy before impact is

$\frac{1}{2} \times 3 \times 3^2 + \frac{1}{2} \times 5 \times 2^2 = 23.5\,\text{J}$

The total kinetic energy after impact is

$\frac{1}{2} \times 3 \times 2^2 + \frac{1}{2} \times 5 \times 1^2 = 8.5\,\text{J}$

So the loss of kinetic energy is

$23.5\,\text{J} - 8.5\,\text{J} = 15\,\text{J}$

Online Explore the loss of kinetic energy in a collision using GeoGebra.

Draw diagrams to show the masses, speeds and directions of A and B before and after the collision.

Let the velocity of A be $x\,\text{m s}^{-1}$ and let the velocity of B be $y\,\text{m s}^{-1}$ after the impact.

Use $m_1u_1 + m_2u_2 = m_1v_1 + m_2v_2$

Use $\frac{\text{speed of separation of particles}}{\text{speed of approach of particles}} = e$

Solve the simultaneous equations **(1)** and **(2)** to find x and y.

The total kinetic energy before impact is $\frac{1}{2}m_1u_1^2 + \frac{1}{2}m_2u_2^2$

The total kinetic energy after impact is $\frac{1}{2}m_1v_1^2 + \frac{1}{2}m_2v_2^2$

- **The loss of kinetic energy due to impact is**

$$\left(\tfrac{1}{2}m_1u_1^2 + \tfrac{1}{2}m_2u_2^2\right) - \left(\tfrac{1}{2}m_1v_1^2 + \tfrac{1}{2}m_2v_2^2\right)$$

Watch out When the particles are perfectly elastic ($e = 1$) you will find that there is no loss of kinetic energy due to impact. In all practical situations $e < 1$ and some kinetic energy is converted into heat or sound energy at impact.

Example 13

A baseball-pitching machine of mass 60 kg fires a baseball of mass 0.15 kg horizontally with speed 30 m s^{-1}.

a Find the velocity of the machine after the baseball has been pitched.

b Find the total kinetic energy generated on pitching.

c Show that the ratio of the energy of the machine to the energy of the baseball is equal to the ratio of the speed of the machine to the speed of the baseball after pitching.

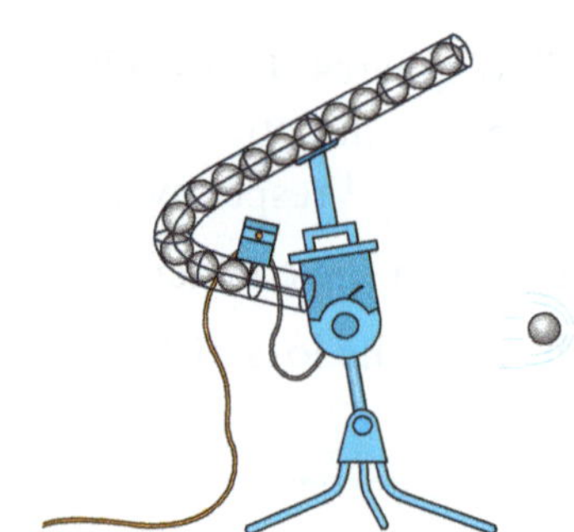

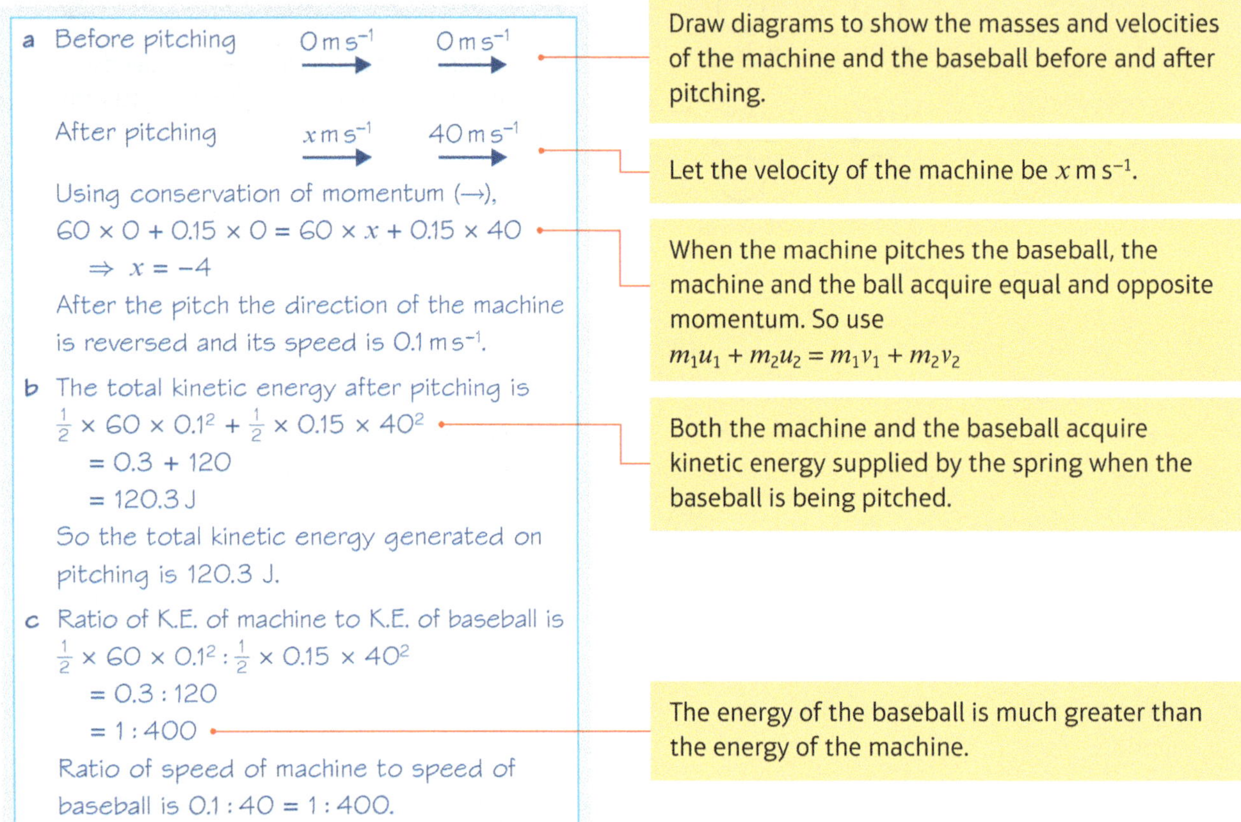

a Before pitching $\quad 0\,\text{m s}^{-1} \rightarrow \quad 0\,\text{m s}^{-1} \rightarrow$

After pitching $\quad x\,\text{m s}^{-1} \rightarrow \quad 40\,\text{m s}^{-1} \rightarrow$

Using conservation of momentum ($\rightarrow$),

$60 \times 0 + 0.15 \times 0 = 60 \times x + 0.15 \times 40$

$\Rightarrow x = -4$

After the pitch the direction of the machine is reversed and its speed is 0.1 m s^{-1}.

b The total kinetic energy after pitching is

$\frac{1}{2} \times 60 \times 0.1^2 + \frac{1}{2} \times 0.15 \times 40^2$

$= 0.3 + 120$

$= 120.3\,\text{J}$

So the total kinetic energy generated on pitching is 120.3 J.

c Ratio of K.E. of machine to K.E. of baseball is

$\frac{1}{2} \times 60 \times 0.1^2 : \frac{1}{2} \times 0.15 \times 40^2$

$= 0.3 : 120$

$= 1 : 400$

Ratio of speed of machine to speed of baseball is $0.1 : 40 = 1 : 400$.

Draw diagrams to show the masses and velocities of the machine and the baseball before and after pitching.

Let the velocity of the machine be x m s^{-1}.

When the machine pitches the baseball, the machine and the ball acquire equal and opposite momentum. So use $m_1u_1 + m_2u_2 = m_1v_1 + m_2v_2$

Both the machine and the baseball acquire kinetic energy supplied by the spring when the baseball is being pitched.

The energy of the baseball is much greater than the energy of the machine.

Example 14 SKILLS PROBLEM-SOLVING

Two particles A and B, of masses 200 g and 300 g respectively, are connected by a light inextensible string. The particles are side by side at rest on a smooth floor and A is projected with speed 6 m s^{-1} directly away from B. When the string becomes taut, particle B is pulled suddenly into motion and A and B then move with a common speed in the direction of projection of A. Find:

a the common speed of the particles after the string becomes taut

b the loss of total kinetic energy due to the pull.

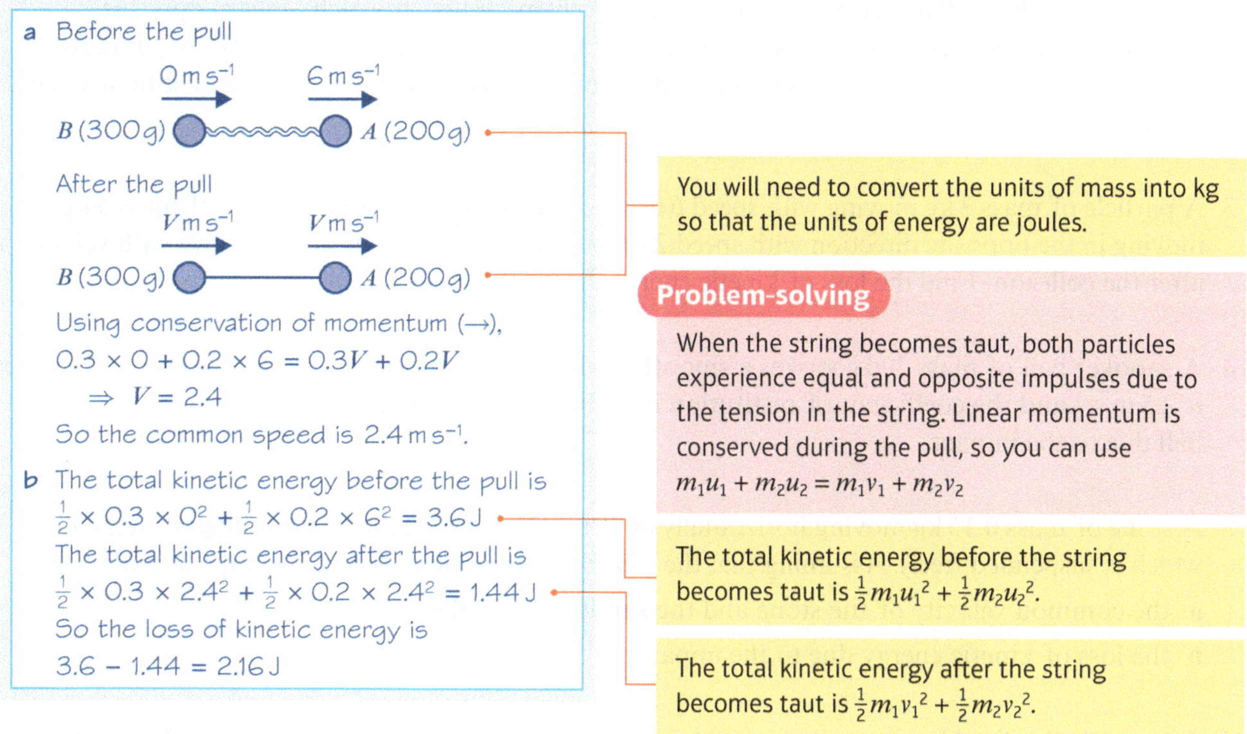

Example 15

A particle of mass 0.1 kg is moving with velocity $(2\mathbf{i} + 5\mathbf{j})$ m s^{-1} when it receives an impulse of $(-\mathbf{i} + 3\mathbf{j})$ N s. Find the change in the kinetic energy of the particle due to the impulse.

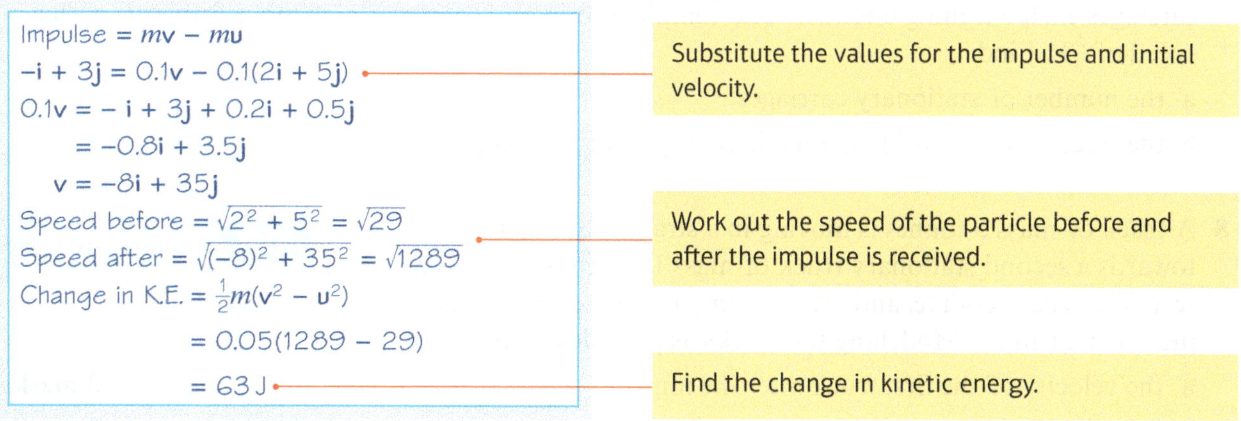

Exercise 5D

SKILLS PROBLEM-SOLVING

1 A particle A of mass 500 g lies at rest on a smooth horizontal table. A second particle B of mass 600 g is projected along the table with speed 6 m s^{-1} and collides directly with A. If the collision reduces the speed of B to 1 m s^{-1}, without changing its direction, find:

a the speed of A after the collision

b the loss of kinetic energy due to the collision.

2 Two particles A and B of masses m and $2m$ respectively move towards each other in opposite directions with speeds u and $2u$. If the coefficient of restitution between the particles is $\frac{2}{3}$, find the velocities of A and of B after the collision. Find also, in terms of m and u, the loss of kinetic energy due to the collision.

3 A particle of mass 3 kg moving with speed $6\,\text{m}\,\text{s}^{-1}$ collides directly with a particle of mass 5 kg moving in the opposite direction with speed $2\,\text{m}\,\text{s}^{-1}$. The particles coalesce and move with velocity v after the collision. Find the loss of kinetic energy due to the impact.

4 A snooker ball of mass 200 g strikes a smooth cushion at right angles. Its speed before the impact is $2.5\,\text{m}\,\text{s}^{-1}$ and the coefficient of restitution is $\frac{4}{5}$. Find the loss in kinetic energy of the snooker ball due to the impact.

5 A stone of mass 0.15 kg moving horizontally at $40\,\text{m}\,\text{s}^{-1}$ embeds itself in a sandbag of mass 30 kg, which is suspended freely. Assuming that the sandbag is **stationary** before the impact, find:

a the common velocity of the stone and the sandbag after the impact

b the loss of kinetic energy due to the impact.

6 A tennis ball is fired horizontally from a 'serving' machine. The machine has mass 8 kg and the tennis ball has mass 100 g. The initial speed of the tennis ball is $40\,\text{m}\,\text{s}^{-1}$. Find:

a the initial speed with which the machine recoils

b the total kinetic energy generated as a result of serving the tennis ball.

(P) **7** A train of mass 30 tonnes moving with a small speed V impacts upon a number of stationary carriages each weighing 6 tonnes. The complete train and carriages now move forwards with a speed of $\frac{5}{8}V$. Find:

a the number of stationary carriages

b the fraction of the original kinetic energy lost in the impact.

(E) **8** A truck of mass 5 tonnes is moving in a straight line at $1.5\,\text{m}\,\text{s}^{-1}$ towards a second stationary truck of mass 10 tonnes which is at rest. The trucks collide, and after the impact the second truck moves at $0.6\,\text{m}\,\text{s}^{-1}$. Modelling the trucks as particles, find:

Hint 1 tonne = 1000 kg

a the velocity of the first truck after the impact **(3 marks)**

b the coefficient of restitution between the two trucks **(2 marks)**

c the loss of kinetic energy due to the impact. **(3 marks)**

(P) **9** A particle of mass moves in a straight line with speed v when it explodes into two parts, one of mass $\frac{1}{3}m$ and the other of mass $\frac{2}{3}m$, both moving in the same direction as before. If the explosion increases the energy of the system by $\frac{1}{4}mu^2$, where u is a positive constant, find the speeds of the particles immediately after the explosion. Give your answers in terms of u and v.

E/P **10** A small smooth sphere A of mass 2 kg moves at $4\,m\,s^{-1}$ on a smooth horizontal table. It collides directly with a second small smooth sphere B of mass 3 kg, which is moving in the same direction at a speed of $1\,m\,s^{-1}$. The loss of kinetic energy due to the collision is 3 J. After the collision A and B continue to move in the same direction with speeds u and v respectively.

a Show that $5v^2 - 22v + 21 = 0$. **(6 marks)**

b Hence find u and v, carefully justifying your choice of solutions. **(5 marks)**

E **11** Two particles A and B, of masses 2 kg and 5 kg respectively, are connected by a light inextensible string. The particles are side by side on a smooth floor and A is projected with speed $7\,m\,s^{-1}$ directly away from B. When the string becomes taut, particle B is pulled suddenly into motion and A and B then move with a common speed in the direction of the original velocity of A. Find:

a the common speed of the particles after the string becomes taut **(4 marks)**

b the loss of total kinetic energy due to the pull. **(3 marks)**

E/P **12** Two particles A and B, of masses m and M respectively, are connected by a light inextensible string. The particles are side by side on a smooth floor and A is projected with speed u directly away from B. When the string becomes taut, particle B is pulled suddenly into motion and A and B then move with a common speed in the direction of the original projection of A.

Show that the loss of total kinetic energy due to the pull is $\frac{mMu^2}{2(m + M)}$. **(8 marks)**

Problem-solving

Start by finding an expression for the common speed, v, after the string becomes taut.

E **13** Two particles of masses 3 kg and 5 kg lie on a smooth table and are connected by a slack inextensible string. The first particle is projected along the table with a velocity of $20\,m\,s^{-1}$ directly away from the second particle. Find:

a the velocity of each particle after the string has become taut **(6 marks)**

b the difference between the kinetic energies of the system when the string is slack and when it is taut. **(3 marks)**

E/P **14** Three small spheres of masses 20 g, 40 g and 60 g respectively lie in order in a straight line on a large smooth table. The distance between adjacent spheres is 10 cm. Two slack strings, each 70 cm in length, connect the first sphere with the second, and the second sphere with the third. The 60 g sphere is projected with a speed of $5\,m\,s^{-1}$, directly away from the other two. Find:

a the time which elapses before the 20 g sphere begins to move and the speed with which it starts **(8 marks)**

b the loss in kinetic energy resulting from the two pulls. **(5 marks)**

Challenge

Two small spheres A and B with masses 4 kg and 1 kg respectively lie on a large smooth surface. A and B are connected by a light inextensible string and are projected directly towards each other with speeds $2\,m\,s^{-1}$ and $3\,m\,s^{-1}$ respectively. The coefficient of restitution for the collision of A and B is 0.8. After the collision both particles move in the direction in which A was originally moving. Find the kinetic energy of the system when the string becomes taut.

5.5 Successive direct impacts

You can solve problems involving **successive** direct impacts of particles with each other, or with a smooth plane surface. When you are solving such problems, you should draw a clear diagram showing the 'before' and 'after' information for each collision.

Example  SKILLS PROBLEM-SOLVING

Three spheres A, B and C have masses m, $2m$ and $3m$ respectively. The spheres move along the same straight line on a horizontal plane with A following B, which is following C. Initially the speeds of A, B and C are 7 m s^{-1}, 3 m s^{-1} and 1 m s^{-1} respectively, in the direction ABC. Sphere A collides with sphere B and then sphere B collides with sphere C. The coefficient of restitution between A and B is $\frac{1}{2}$ and the coefficient of restitution between B and C is $\frac{1}{4}$.

a Find the velocities of the three spheres after the second collision.

b Explain how you can predict that there will be a further collision between A and B.

a First collision:

Before impact between A and B

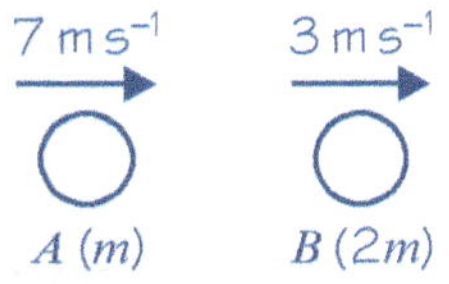

7 m s^{-1} → A (m) 3 m s^{-1} → B $(2m)$

After impact between A and B

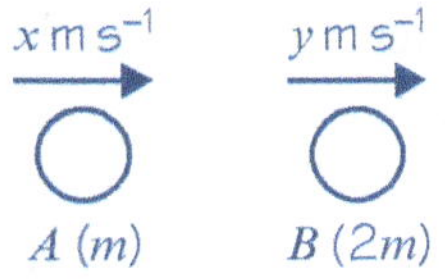

$x\text{ m s}^{-1}$ → A (m) $y\text{ m s}^{-1}$ → B $(2m)$

Using conservation of momentum (→),

$m \times 7 + 2m \times 3 = m \times x + 2m \times y$

$\Rightarrow x + 2y = 13$ **(1)**

Using Newton's law of restitution,

$\frac{y - x}{7 - 3} = \frac{1}{2}$

$\Rightarrow y - x = 2$ **(2)**

Solving equations **(1)** and **(2)** gives

$y = 5$ and $x = 3$

After the first impact, the velocity of A is 3 m s^{-1} and the velocity of B is 5 m s^{-1}.

Second collision:

Before impact between B and C

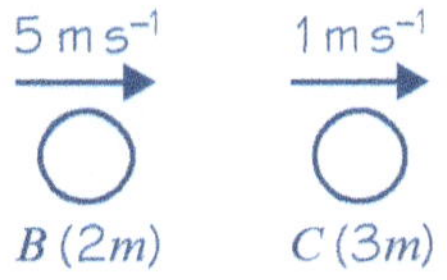

5 m s^{-1} → B $(2m)$ 1 m s^{-1} → C $(3m)$

Online Explore successive collisions using GeoGebra.

Draw diagrams to show the masses, speeds and directions of A and B before and after the first collision.

Let the velocity of A be $x\text{ m s}^{-1}$ and let the velocity of B be $y\text{ m s}^{-1}$ after the impact. Make it clear in your diagram which velocity corresponds to which particle and which direction is positive.

Use $m_1u_1 + m_2u_2 = m_1v_1 + m_2v_2$

Use $\frac{\text{speed of separation of particles}}{\text{speed of approach of particles}} = e$

Add equations **(1)** and **(2)** to eliminate x and to give $3y = 15$, then substitute $y = 5$ into equation **(2)** to give $x = 3$.

Draw diagrams to show the masses, speeds and directions of B and C before and after the collision.

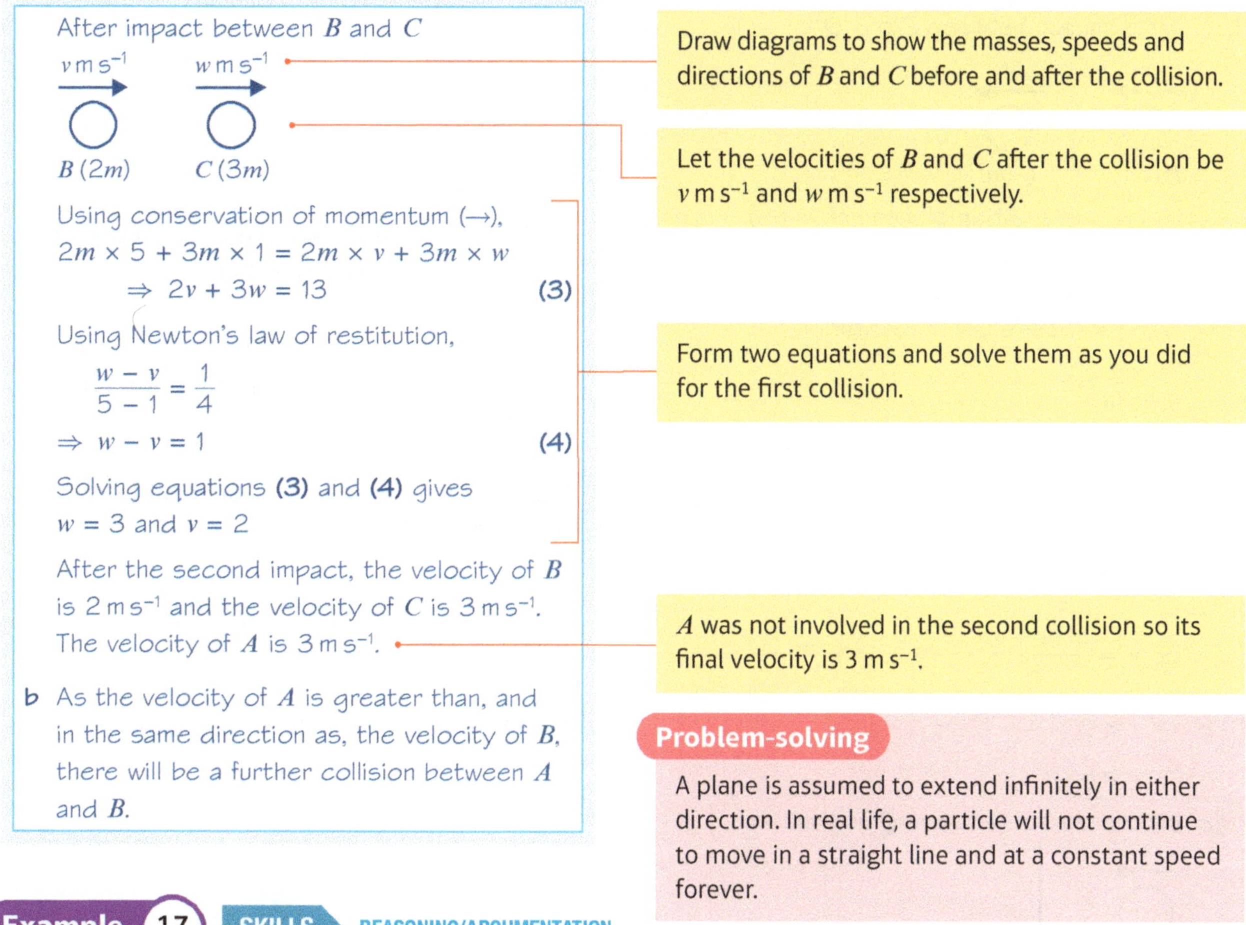

Using conservation of momentum (→),
$2m \times 5 + 3m \times 1 = 2m \times v + 3m \times w$
$\Rightarrow 2v + 3w = 13$ **(3)**

Using Newton's law of restitution,
$\frac{w - v}{5 - 1} = \frac{1}{4}$
$\Rightarrow w - v = 1$ **(4)**

Solving equations **(3)** and **(4)** gives
$w = 3$ and $v = 2$

After the second impact, the velocity of B is 2 m s^{-1} and the velocity of C is 3 m s^{-1}.
The velocity of A is 3 m s^{-1}.

b As the velocity of A is greater than, and in the same direction as, the velocity of B, there will be a further collision between A and B.

Draw diagrams to show the masses, speeds and directions of B and C before and after the collision.

Let the velocities of B and C after the collision be v m s^{-1} and w m s^{-1} respectively.

Form two equations and solve them as you did for the first collision.

A was not involved in the second collision so its final velocity is 3 m s^{-1}.

Problem-solving

A plane is assumed to extend infinitely in either direction. In real life, a particle will not continue to move in a straight line and at a constant speed forever.

Example 17 SKILLS REASONING/ARGUMENTATION

A uniform smooth sphere P of mass $3m$ is moving in a straight line with speed u on a smooth horizontal table. Another uniform smooth sphere Q of mass m is moving with speed $2u$ in the same straight line as P, but in the opposite direction. The sphere P collides with the sphere Q directly. The velocities of P and Q after the collision are v and w respectively, measured in the direction of motion of P before the collision. The coefficient of restitution between P and Q is e.

a Find expressions for v and w in terms of u and e.

b Show that, if the direction of motion of P is changed by the collision, then $e > \frac{1}{3}$.

Following the collision with P, the sphere Q then collides with and rebounds from a vertical wall, which is perpendicular to the direction of motion of Q. The coefficient of restitution between Q and the wall is e'.

c Given that $e = \frac{5}{9}$ and that P and Q collide again in the subsequent motion, show that $e' > \frac{1}{9}$.

a First collision:

Before impact between P and Q

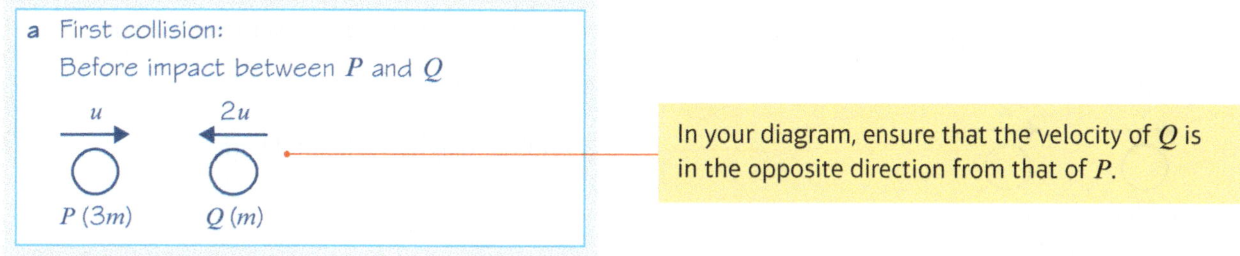

In your diagram, ensure that the velocity of Q is in the opposite direction from that of P.

After impact between P and Q

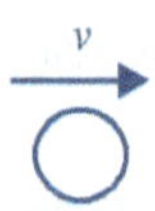

Using conservation of momentum (→),

$3m \times u - m \times 2u = 3m \times v + m \times w$

$\Rightarrow 3v + w = u$ **(1)**

Each velocity must be given the correct sign in relation to the positive direction.

Using Newton's law of restitution,

$\frac{w - v}{u + 2u} = e$

$\Rightarrow w - v = 3eu$ **(2)**

Note that the spheres are initially moving in opposite directions, so the speed of approach is $u + 2u$.

Subtract equation **(2)** from equation **(1)** to give

$4v = u(1 - 3e)$

$v = \frac{u(1 - 3e)}{4}$

Solve the two simultaneous equations to obtain an expression for v.

So $w = \frac{u(9e + 1)}{4}$

Substitute v into equation **(2)** to find the expression for w.

b As $v < 0$, $\frac{u(1 - 3e)}{4} < 0$

So $e > \frac{1}{3}$

If P changes direction, then v will be negative.

c Before impact between Q and the wall

After impact between Q and the wall

Draw diagrams for the new collision.

Using Newton's law of restitution,

$\frac{V}{w} = e'$

Use $\frac{\text{speed of rebound}}{\text{speed of approach}} = e$, with $e = e'$.

Using $e = \frac{5}{9}$, $w = \frac{u(5 + 1)}{4} = \frac{3u}{2}$

So $V = \frac{3e'u}{2}$

Use your answer to part **a** to find w and use Newton's law of restitution to find V.

Also when $e = \frac{5}{9}$, $v = -\frac{u}{6}$

Evaluate v using your answer to part **a** and $e = \frac{5}{9}$.

After impact between Q and t he wall

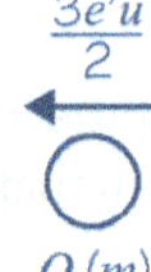

Another diagram is helpful here.

Since Q and P collide again

$\frac{3e'u}{2} > \frac{u}{6}$

So $e' > \frac{1}{9}$

Problem-solving

If the two particles collide again, then the speed of Q must be greater than the speed of P.

Example 18

A tennis ball, which may be modelled as a particle, is dropped from rest at a height of 90 cm onto a smooth horizontal plane. The coefficient of restitution between the ball and the plane is 0.5. Assume that there is no air resistance and that the ball falls under gravity and hits the plane at right angles.

a Find the height to which the ball rebounds after the first bounce.

b Find the height to which the ball rebounds after the second bounce.

c Find the total distance travelled by the ball before it comes to rest, according to this model.

d Criticise this model with respect to the motion of the ball as it continues to bounce.

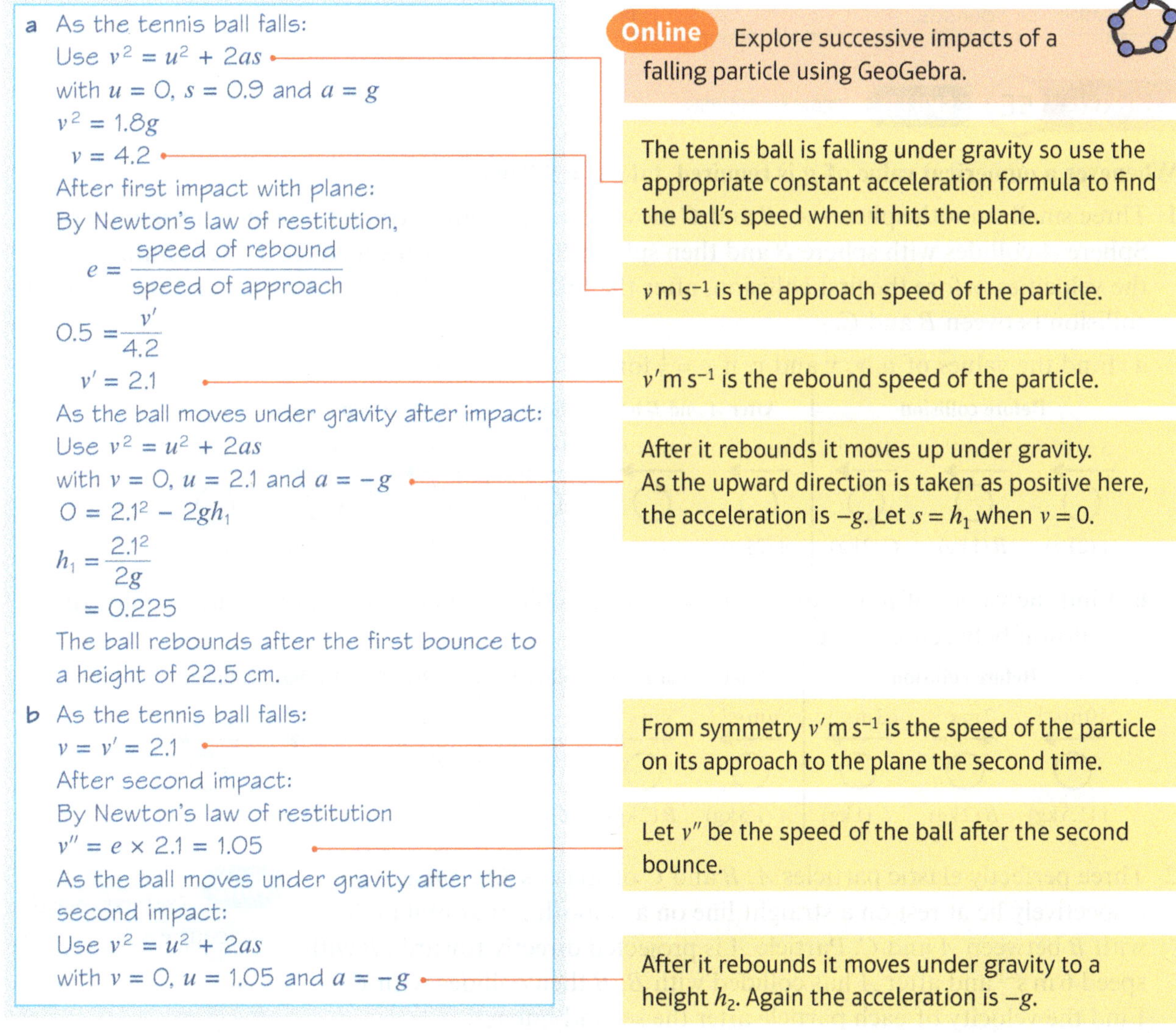

a As the tennis ball falls:

Use $v^2 = u^2 + 2as$

with $u = 0$, $s = 0.9$ and $a = g$

$v^2 = 1.8g$

$v = 4.2$

After first impact with plane:

By Newton's law of restitution,

$e = \frac{\text{speed of rebound}}{\text{speed of approach}}$

$0.5 = \frac{v'}{4.2}$

$v' = 2.1$

As the ball moves under gravity after impact:

Use $v^2 = u^2 + 2as$

with $v = 0$, $u = 2.1$ and $a = -g$

$0 = 2.1^2 - 2gh_1$

$h_1 = \frac{2.1^2}{2g}$

$= 0.225$

The ball rebounds after the first bounce to a height of 22.5 cm.

b As the tennis ball falls:

$v = v' = 2.1$

After second impact:

By Newton's law of restitution

$v'' = e \times 2.1 = 1.05$

As the ball moves under gravity after the second impact:

Use $v^2 = u^2 + 2as$

with $v = 0$, $u = 1.05$ and $a = -g$

Online Explore successive impacts of a falling particle using GeoGebra.

The tennis ball is falling under gravity so use the appropriate constant acceleration formula to find the ball's speed when it hits the plane.

v m s^{-1} is the approach speed of the particle.

v' m s^{-1} is the rebound speed of the particle.

After it rebounds it moves up under gravity. As the upward direction is taken as positive here, the acceleration is $-g$. Let $s = h_1$ when $v = 0$.

From symmetry v' m s^{-1} is the speed of the particle on its approach to the plane the second time.

Let v'' be the speed of the ball after the second bounce.

After it rebounds it moves under gravity to a height h_2. Again the acceleration is $-g$.

$0 = 1.05^2 - 2gh_2$

$h_2 = \frac{1.05^2}{2g} = 0.056\,25$

The ball rebounds after the second bounce to a height of 5.625 cm.

c The total distance travelled is

$0.9 + 0.225 + 0.225 + 0.056\,25 + 0.056\,25 + \ldots$

$= 0.9 + 2(0.225 + 0.225 \times \frac{1}{4} + 0.225 \times (\frac{1}{4})^2 + 0.225 \times (\frac{1}{4})^3 + \ldots)$

$= 0.9 + 2 \times \frac{0.225}{(1 - \frac{1}{4})} = 1.5$

The total distance travelled by the ball before it comes to rest is 1.5 m.

d According to this model the ball will bounce an infinite number of times. In real life, the ball will stop bouncing after a finite number of bounces.

Problem-solving

The ratio of the heights on each bounce is constant.
Initial height = 0.9 m, $h_1 = 0.9 \times \frac{1}{4} = 0.225$ m, $h_2 = 0.9 \times (\frac{1}{4})^2 = 0.056\,25$ m, and so on.

The ball moves 0.9 m before first impact, then moves up 0.225 m and down 0.225 m before second impact then moves up 0.056 25 m and down 0.056 25 m before third impact.

Use the formula for the sum of an infinite geometric series:
$S = \frac{a}{1 - r}$ with $a = 0.225$ and $r = \frac{1}{4}$

← Pure 2 Section 5.5

Exercise  5E SKILLS PROBLEM-SOLVING

Whenever a numerical value of g is required, take $g = 9.8\,\text{m s}^{-2}$.

1 Three small smooth spheres A, B and C move along the same straight line on a horizontal plane. Sphere A collides with sphere B and then sphere B collides with sphere C. The diagrams show the velocities before the first collision, after the first collision between A and B and then after the collision between B and C.

a Find the values of u, v, x and y, if $e = \frac{1}{2}$ for both collisions.

Before collision			After A and B have collided			After B and C have collided		
$5\,\text{m s}^{-1}$ →	$1\,\text{m s}^{-1}$ →	$4\,\text{m s}^{-1}$ →	$u\,\text{m s}^{-1}$ →	$v\,\text{m s}^{-1}$ →	$4\,\text{m s}^{-1}$ →	$u\,\text{m s}^{-1}$ →	$x\,\text{m s}^{-1}$ →	$y\,\text{m s}^{-1}$ →
A (2 kg)	B (1 kg)	C (2 kg)	A (2 kg)	B (1 kg)	C (2 kg)	A (2 kg)	B (1 kg)	C (2 kg)

b Find the values of u, v, x and y, if $e = \frac{1}{6}$ for the collision between A and B and $e = \frac{1}{2}$ for the collision between B and C.

Before collision			After A and B have collided			After B and C have collided		
$10\,\text{m s}^{-1}$ →	$2\,\text{m s}^{-1}$ ←	$3\,\text{m s}^{-1}$ →	$u\,\text{m s}^{-1}$ →	$v\,\text{m s}^{-1}$ →	$3\,\text{m s}^{-1}$ →	$u\,\text{m s}^{-1}$ →	$x\,\text{m s}^{-1}$ →	$y\,\text{m s}^{-1}$ →
A (1.5 kg)	B (2 kg)	C (1 kg)	A (1.5 kg)	B (2 kg)	C (1 kg)	A (1.5 kg)	B (2 kg)	C (1 kg)

(P) 2 Three perfectly elastic particles A, B and C of masses $3m$, $5m$ and $4m$ respectively lie at rest on a straight line on a smooth horizontal table with B between A and C. Particle A is projected directly towards B with speed $6\,\text{m s}^{-1}$ and after A has collided with B, B then collides with C. Find the velocity of each particle after the second impact.

Hint Perfectly elastic means $e = 1$.

(P) **3** Three identical smooth spheres A, B and C, each of mass m, lie at rest on a straight line on a smooth horizontal table. Sphere A is projected with speed u to strike sphere B directly. Sphere B then strikes sphere C directly. The coefficient of restitution between any two spheres is e, $e \neq 1$.

a Find the speeds in terms of u and e of the spheres after these two collisions.

b Show that A will catch up with B and there will be a further collision.

(E/P) **4** Three identical spheres A, B and C of equal mass m move along the same straight line on a horizontal plane. A and B are moving in opposite directions towards each other with speeds $4u$ and $2u$ respectively. C is moving in the same direction as A with speed $3u$.

a If the coefficient of restitution between any two of the spheres is e, show that B will only collide with C if $e > \frac{2}{3}$. **(8 marks)**

b Find the direction of motion of A after collision, if $e > \frac{2}{3}$. **(2 marks)**

(E/P) **5** Two particles P and Q of masses $2m$ and $3m$ respectively are moving in opposite directions on a smooth plane with speeds $4u$ and $2u$ respectively. The particles collide directly. The direction of motion of Q is reversed by the impact and its speed after impact is u. This particle then hits a smooth vertical wall perpendicular to its direction of motion. The coefficient of restitution between Q and the wall is $\frac{2}{3}$. In the subsequent motion, there is a further collision between Q and P. Find the velocities of P and Q after this collision. **(8 marks)**

(E/P) **6** Two small smooth spheres P and Q have masses m and $3m$ respectively. Sphere P is moving with speed $12u$ on a smooth horizontal table when it collides directly with Q which is at rest on the table. The coefficient of restitution between P and Q is $\frac{2}{3}$.

a Find the velocities of P and Q immediately after the collision. **(6 marks)**

After the collision Q hits a smooth vertical wall perpendicular to the direction of its motion. The coefficient of restitution between Q and the wall is $\frac{4}{5}$. Q then collides with P a second time.

b Find the velocities of P and Q after the second collision between P and Q. **(8 marks)**

(E/P) **7** A small table tennis ball, which may be modelled as a particle, falls from rest at a height 40 cm onto a smooth horizontal plane. The coefficient of restitution between the ball and the plane is 0.7.

a Find the height to which the ball rebounds after:

i the first bounce

ii the second bounce. **(8 marks)**

b Describe the subsequent motion of the ball. **(1 mark)**

c Find the total distance travelled by the ball before it comes to rest. **(5 marks)**

Problem-solving

Find the ratio between the heights of successive bounces according to the model.

d Give one reason why your answer to part **c** is unrealistic. **(1 mark)**

E/P **8** A small smooth ball, which may be modelled as a particle, falls from rest at a height H onto a smooth horizontal plane. The coefficient of restitution between the ball and the plane is e.

a Find, in terms of H and e, the height to which the ball rebounds after the first bounce. **(9 marks)**

b Find, in terms of H and e, the height to which the ball rebounds after the second bounce. **(6 marks)**

c Find an expression for the total distance travelled by the ball before it comes to rest. **(6 marks)**

Problem-solving

Draw clear diagrams for each stage of the motion of the ball.

E/P **9** A ball B lies on a smooth horizontal plane between two smooth, parallel, vertical walls W_1 and W_2 that are d m apart. B is initially halfway between the walls and is projected with velocity $2\,\text{m s}^{-1}$ towards W_2. The coefficient of restitution between B and W_2 is e_2.

The ball then travels towards and strikes W_1. The coefficient of restitution between B and W_1 is e_1.

The ball then travels towards W_2 again.

Find, in terms of d, e_1 and e_2, the total time elapsed from the moment the ball is projected to the time when it strikes W_2 for the second time. **(9 marks)**

Challenge

Two particles P and Q, of equal mass, lie on a smooth horizontal plane between two smooth parallel, vertical walls W_1 and W_2 that are 4 m apart. P and Q are projected towards each other with speeds $2\,\text{m s}^{-1}$ and $1\,\text{m s}^{-1}$ respectively in such a way that they collide directly at a point equidistant (i.e. at an equal distance) from W_1 and W_2. The coefficient of restitution between the particles is 0.5. P then travels towards W_1, and Q travels towards W_2 and strikes W_2 perpendicularly. The coefficient of restitution between Q and W_2 is 0.4. Show that in the subsequent motion, P strikes W_1 before Q collides with P for a second time.

Chapter review 

E **1** A cricket ball of mass 0.5 kg is struck by a bat. Immediately before being struck the velocity of the ball is $-25\mathbf{i}\,\text{m s}^{-1}$. Immediately after being struck the velocity of the ball is $(23\mathbf{i} + 20\mathbf{j})\,\text{m s}^{-1}$. Find the magnitude of the impulse exerted on the ball by the bat and the angle between the impulse and the direction of $\mathbf{i}$. **(5 marks)**

E **2** A ball of mass 0.2 kg is hit by a bat which gives it an impulse of $(2.4\mathbf{i} + 3.6\mathbf{j})\,\text{N s}$. The velocity of the ball immediately after being hit is $(12\mathbf{i} + 5\mathbf{j})\,\text{m s}^{-1}$. Find the velocity of the ball immediately before it is hit. **(3 marks)**

E **3** A body P of mass 4 kg is moving with velocity $(2\mathbf{i} + 16\mathbf{j})\,\text{m s}^{-1}$ when it collides with a body Q of mass 3 kg moving with velocity $(-\mathbf{i} - 8\mathbf{j})\,\text{m s}^{-1}$. Immediately after the collision the velocity of P is $(-4\mathbf{i} - 32\mathbf{j})\,\text{m s}^{-1}$. Find the velocity of Q immediately after the collision. **(3 marks)**

E/P **4** A particle P of mass 0.3 kg is moving so that its position vector $\mathbf{r}$ metres at time t seconds is given by

$$\mathbf{r} = (t^3 + t^2 + 4t)\mathbf{i} + (11t)\mathbf{j},\ t \leqslant 4.$$

a Calculate the speed of P when $t = 4$. **(4 marks)**

When $t = 4$, the particle is given an impulse $(2.4\mathbf{i} + 3.6\mathbf{j})$ N s.

b Find the velocity of P immediately after the impulse. **(3 marks)**

5 Two identical spheres, moving in opposite directions, collide directly. As a result of the impact one of the spheres is brought to rest. The coefficient of restitution between the spheres is $\frac{1}{3}$. Show that the ratio of the speeds of the spheres before the impact is 2 : 1.

6 A particle P of mass m is moving in a straight line with speed $\frac{1}{4}u$ at the instant when it collides directly with a particle Q, of mass λm, which is at rest. The coefficient of restitution between P and Q is $\frac{1}{4}$. Given that P comes to rest immediately after hitting Q, find the value of λ.

E/P **7** A boy of mass m dives off a boat of mass M which was previously at rest. Immediately after diving off, the boy has horizontal speed v and the boat has horizontal speed V.

a Modelling the boy and the boat as particles in free space, find an expression for V in terms of v. **(4 marks)**

b Prove that the total kinetic energy of the boy and the boat is $\dfrac{m(m + M)v^2}{2M}$. **(4 marks)**

c Criticise the model in respect of the boat. **(1 mark)**

8 Two spheres P and Q of masses 4 kg and 2 kg respectively are travelling towards each other in opposite directions along a straight line on a smooth horizontal surface. Initially, P has a speed of $5\,\text{m}\,\text{s}^{-1}$ and Q has a speed of $3\,\text{m}\,\text{s}^{-1}$. After the collision the direction of Q is reversed and it is travelling at a speed of $2\,\text{m}\,\text{s}^{-1}$. Find the velocity of P after the collision and the loss of kinetic energy due to the collision.

E/P **9** A particle P of mass $3m$ is moving in a straight line with speed u at the instant when it collides directly with a particle Q of mass m which is at rest. The coefficient of restitution between P and Q is e.

a Show that after the collision P is moving with speed $\dfrac{u(3 - e)}{4}$. **(6 marks)**

b Show that the loss of kinetic energy due to the collision is $\dfrac{3mu^2(1 - e^2)}{8}$. **(10 marks)**

c Find, in terms of m, u and e, the impulse exerted on Q by P in the collision. **(4 marks)**

E/P **10** Two spheres of masses 70 g and 100 g respectively are moving in opposite directions with speeds $4\,\text{m}\,\text{s}^{-1}$ and $8\,\text{m}\,\text{s}^{-1}$ respectively. The spheres collide directly. The coefficient of restitution between the spheres is $\frac{5}{12}$. Find:

a the velocities of the spheres after impact **(4 marks)**

b the amount of kinetic energy lost in the collision. **(3 marks)**

E/P **11** A sphere of mass 2 kg moving at $35\,\text{m s}^{-1}$ catches up and collides directly with a sphere of mass 10 kg moving in the same direction at $20\,\text{m s}^{-1}$. Five seconds after the impact the 10 kg sphere encounters a fixed barrier which reduces it to rest. Assuming the coefficient of restitution between the spheres is $\frac{3}{5}$, find the time that will elapse before the 2 kg sphere strikes the 10 kg sphere again. You may assume that the spheres are moving on a smooth surface and have constant speed between collisions. **(10 marks)**

E/P **12** Three balls A, B and C of masses $4m$, $3m$ and $3m$ respectively lie at rest on a smooth horizontal table with their centres in a straight line. The coefficient of restitution between any pair of balls is $\frac{3}{4}$. Show that if A is projected towards B with speed V there are three impacts and the final velocities are $\frac{5}{32}V$, $\frac{1}{4}V$ and $\frac{7}{8}V$ respectively. **(12 marks)**

E **13** A cricket ball of mass 200 g is thrown horizontally at a fixed vertical metal barrier. The cricket ball hits the barrier when it is travelling at $30\,\text{m s}^{-1}$ and then rebounds.

a Find the kinetic energy lost at the impact if $e = 0.4$. **(8 marks)**

b Give one possible form of energy into which the lost kinetic energy has been transformed. **(1 mark)**

E **14** A particle A of mass $4m$ moving with speed u on a horizontal plane strikes directly a particle B of mass $3m$ which is at rest on the plane. The coefficient of restitution between A and B is e.

a Find, in terms of e and u, the speeds of A and B immediately after the collision. **(8 marks)**

b Given that the magnitude of the impulse exerted by A on B is $2mu$, show that $e = \frac{1}{6}$. **(6 marks)**

E/P **15** A ball of mass m moving with speed kV on a smooth table catches up and collides with another ball of mass λm moving with speed V travelling in the same direction on the table. The impact reduces the first ball to rest.

a Show that the coefficient of restitution is $\dfrac{\lambda + k}{\lambda(k - 1)}$. **(6 marks)**

b Show further that $\lambda > \dfrac{k}{k - 2}$ and $k > 2$. **(6 marks)**

E **16** A ball is dropped from zero velocity and after falling for 1 s under gravity meets another identical ball which is moving upwards at $7\,\text{m s}^{-1}$.

a Taking the value of g as $9.8\,\text{m s}^{-2}$, calculate the velocity of each ball immediately after the impact, given that the coefficient of restitution is $\frac{1}{4}$. **(8 marks)**

b Find the percentage loss in kinetic energy due to the impact, giving your answer to 2 significant figures. **(4 marks)**

E **17** A particle falls from a height 8 m onto a fixed horizontal plane. The coefficient of restitution between the particle and the plane is $\frac{1}{4}$.

a Find the height to which the particle rises after impact. **(6 marks)**

b Find the time the particle takes from leaving the plane after impact to reach the plane again. **(2 marks)**

c Find the speed of the particle after the second rebound. **(2 marks)**

You may leave your answers in terms of g.

E/P **18** A particle falls from a height h onto a fixed horizontal plane. If e is the coefficient of restitution between the particle and the plane, show that the total time taken before the particle finishes bouncing is $\frac{1+e}{1-e}\sqrt{\frac{2h}{g}}$. **(10 marks)**

E/P **19** A sphere P of mass m lies on a smooth table between a sphere Q of mass $8m$ and a fixed vertical plane. Sphere Q is initially at rest. Sphere P is projected towards sphere Q so they impact directly. The coefficient of restitution between the two spheres is $\frac{7}{8}$. Given that sphere P is reduced to rest by a second impact with sphere Q, find the coefficient of restitution between sphere P and the fixed vertical plane. **(12 marks)**

E/P **20** A machine that bowls cricket balls, of mass M kg, is free to move horizontally. The machine bowls a cricket ball of mass m kg in a horizontal direction. The energy released by the spring, which occurred on bowling the cricket ball, is E J. Find the velocity of the cricket ball in terms of m, M and E if all of this energy is given to the cricket ball and the machine. **(8 marks)**

E/P **21** A snooker ball of mass m kg is dropped from a point H m above a horizontal floor. The ball falls freely under gravity, strikes the floor and bounces to a height of h m. Given that the coefficient of restitution between the ball and the floor is e,

a show that $e = \sqrt{\frac{h}{H}}$ **(6 marks)**

b find the height the ball reaches after it bounces a second time **(3 marks)**

c describe the subsequent motion of the ball according to this model. **(2 marks)**

E/P **22** A small sphere B of mass 2 kg is held at rest 2 m up a smooth slope that is angled at 30° to the horizontal. When sphere B is released from rest it moves down the slope onto a smooth, horizontal plane where it collides directly with a stationary small sphere C of mass 1 kg. Given that the coefficient of restitution between the balls is 0.75, calculate:

a the speed and direction of motion of B and C after the collision **(12 marks)**

b the kinetic energy lost in the collision. **(5 marks)**

c Without doing any further calculations, state how the amount of kinetic energy lost in the collision would change if $e < 0.75$. **(1 mark)**

Problem-solving

Resolve to find the acceleration of B.

E/P **23** Two spheres A and B sit on a smooth horizontal plane at a point P between two parallel, vertical walls W_1 and W_2 such that the ratio of distance $W_1P : PW_2$ is 2 : 1. Sphere A is projected towards W_1 with speed 2 m s^{-1} and B is projected towards W_2 with speed 3 m s^{-1}. The spheres rebound off their respective walls before colliding at a point Q. Given that the coefficient of restitution between both walls and spheres is 0.6, calculate the ratio of the distances $W_1Q : W_2Q$. **(8 marks)**

Challenge

1 A particle P of mass m kg moves at a speed of $u\,\text{m s}^{-1}$ when it collides head on with a particle Q of mass km kg also travelling at a speed of $u\,\text{m s}^{-1}$, travelling along the same line.

After collision, the two particles move off together with a common speed of $v\,\text{m s}^{-1}$.

Show that k can be written as $\frac{u-v}{u+v}$ or $\frac{u+v}{u-v}$ and explain how these expressions relate to:

a the relative size of u and v

b the subsequent motion of P and Q after collision.

2 Three small spheres A, B and C of masses m_1, m_2 and m_3 respectively lie in order in a straight line on a large smooth table. Two slack strings connect A to B, and B to C. Sphere C is projected with speed $u\,\text{m s}^{-1}$ away from the other two spheres. Show that the total kinetic energy of A, B and C when both strings are taut is $\frac{m_3{}^2u^2}{2(m_1+m_2+m_3)}$.

Summary of key points

1 You can write the impulse–momentum principle and the principle of conservation of momentum as vector equations:

- $\mathbf{I} = m\mathbf{v} - m\mathbf{u}$

  where m is the mass of the body, $\mathbf{u}$ the initial velocity and $\mathbf{v}$ the final velocity.

- $m_1\mathbf{u_1} + m_2\mathbf{u_2} = m_1\mathbf{v_1} + m_2\mathbf{v_2}$

  where a body of mass m_1 moving with velocity $\mathbf{u_1}$ collides with a body of mass m_2 moving with velocity $\mathbf{u_2}$, $\mathbf{v_1}$ and $\mathbf{v_2}$ are the velocities of the bodies after the collision.

2 **Newton's law of restitution** states that

$$\frac{\text{speed of separation of particles}}{\text{speed of approach of particles}} = e$$

The constant e is the **coefficient of restitution** between the particles; $0 \leqslant e \leqslant 1$.

3 For the direct collision of a particle with a smooth plane, Newton's law of restitution can be written as

$$\frac{\text{speed of rebound}}{\text{speed of approach}} = e$$

4 The loss of kinetic energy due to impact is

$$\left(\tfrac{1}{2}m_1u_1^2 + \tfrac{1}{2}m_2u_2^2\right) - \left(\tfrac{1}{2}m_1v_1^2 + \tfrac{1}{2}m_2v_2^2\right)$$

6 STATICS OF RIGID BODIES

6.1

Learning objectives

- Understand the moment of a force → pages 153–157
- Solve problems involving ladders resting against smooth or rough vertical walls and on smooth or rough ground → pages 153–157

Prior knowledge check

1 Calculate the moment about P of each force acting on a lamina.

a

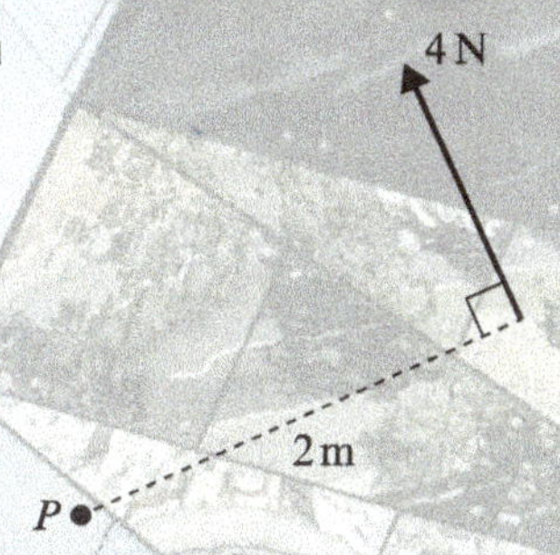

b

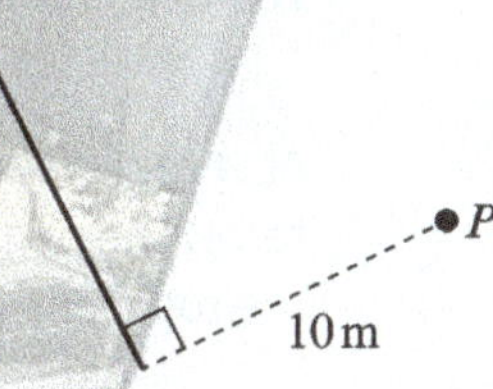

← Mechanics 1 Section 8.1

2 A uniform rod PQ of mass m kg and length 6 m rests on two pivots A and B as shown below, with $PA = 1$ m and $BQ = 2$ m. Work out the normal reaction acting on each pivot.

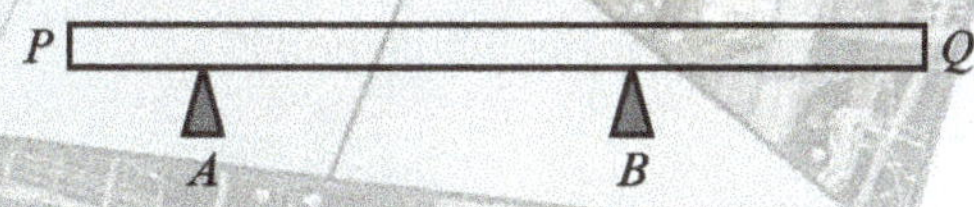

← Mechanics 1 Section 8.2

3 In the triangle ABC, work out the length AD.

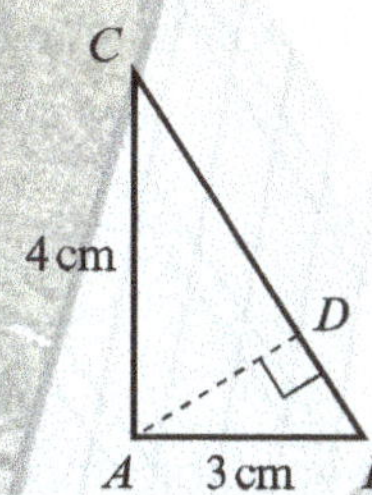

← International GCSE Mathematics

When designing and building the Torre Mare Nostrum building in Barcelona, the engineers would have needed to use moments to calculate the forces acting on the building.

6.1 Static rigid bodies

If you need to consider the rotational forces acting on an object you can model it as a **rigid body**.

- For a rigid body in **static equilibrium**:
  - the body is stationary
  - the resultant force in any direction is zero
  - the resultant moment about any given point is zero.

You will sometimes need to consider the moments acting on a body and the resultant force acting on the body separately.

Links The moment of a force of magnitude F N about a point P is Fd, where d is the **perpendicular distance** from the line of action of the force to P.
← Mechanics 1 Section 8.1

Example 1 SKILLS PROBLEM-SOLVING

A uniform rod AB of mass 40 kg and length 10 m rests with the end A on rough horizontal ground. The rod rests against a smooth peg C where $AC = 8$ m. The rod is in limiting equilibrium at an angle of 15° to the horizontal. Find:

a the magnitude of the reaction at C

b the coefficient of friction between the rod and the ground.

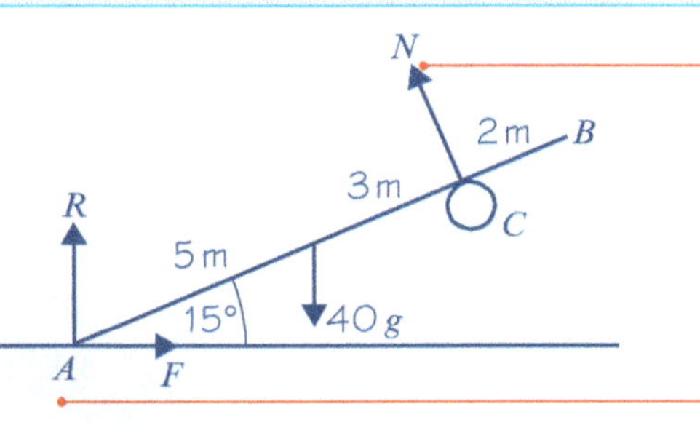

Start with a diagram showing all the forces.

N, the reaction at C, is perpendicular to the rod. The peg is smooth, so there is no friction here.

At A there is a normal reaction and a frictional force. The peg is smooth so the only force stopping the rod from sliding is the frictional force at A.

a Taking moments about A:

$40g \times 5\cos 15° = N \times 8$

$N = \dfrac{200g\cos 15°}{8}$

$= 25g\cos 15°$

$= 236.65\ldots$ N

The reaction at C has magnitude 240 N (2 s.f.).

Problem-solving Solve part **a** by taking moments. If you take moments about A you can ignore the frictional force. For part **b** you can resolve forces horizontally and vertically for the whole body.

b R(→), $F = N\cos 75° = 61.25$ N

R(↑), $R + N\cos 15° = 40g$

$R = 40g - N\cos 15° = 163.41\ldots$ N

The rod is in limiting equilibrium, so

$F = \mu R,\ \mu = \dfrac{F}{R} = 0.37$ (2 s.f.)

Resolve horizontally and vertically.

It is important to consider whether the rigid body is in limiting equilibrium. Only if this is the case does the frictional force take its maximum value μR where μ is the coefficient of friction. Otherwise you can only assume $F \leqslant \mu R$.

Example 2 SKILLS PROBLEM-SOLVING

A ladder AB, of mass m and length $3a$, has one end A resting on rough horizontal ground. The other end B rests against a smooth vertical wall. A load of mass $2m$ is fixed on the ladder at the point C, where $AC = a$. The ladder is modelled as a uniform rod in a vertical plane perpendicular to the wall and the load is modelled as a particle. The ladder rests in limiting equilibrium at an angle of 60° with the ground.

a Find the coefficient of friction between the ladder and the ground.

b State how you have used the assumption that the ladder is uniform in your calculations.

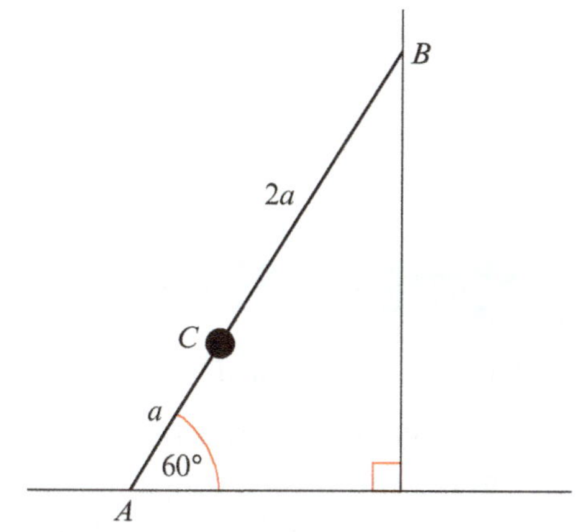

a

The reaction at B is perpendicular to the wall. The wall is smooth, so there is no friction at B.

Watch out The reactions at the wall and the floor are different and so must be labelled with different letters.

Online Explore the forces in this question in a more detailed diagram using GeoGebra.

R(→), $F = P$

R(↑), $R = 2mg + mg = 3mg$

Resolve vertically and horizontally for the whole system.

Taking moments about B:

You want to find $\mu = \frac{F}{R}$ so take moments at B. Remember to use the **perpendicular distance** from each force to B.

$2mg \times 2a\cos 60° + mg \times 1.5a\cos 60°$
$+ F \times 3a\sin 60° = R \times 3a\cos 60°$

$5.5mg\cos 60° + 3F\sin 60°$
$= 3R\cos 60°$

You can divide both sides by a.

$$2.75mg + \frac{3\sqrt{3}}{2}F = 1.5R$$

Since $R = 3mg$, and $F = \mu R$ (as the ladder is in limiting equilibrium)

$$2.75mg + \frac{3\sqrt{3}}{2}\mu \times 3mg = 1.5 \times 3mg$$

Divide through by mg and solve to find μ. The answer is independent of g so round to 3 s.f.

$$\mu = \frac{4.5 - 2.75}{\left(\frac{9\sqrt{3}}{2}\right)} = \frac{7}{18\sqrt{3}} = 0.225 \text{ (3 s.f.)}$$

b The assumption that the ladder is uniform allows you to assume that its weight acts at its midpoint.

Problem-solving

There are other options for points to take moments about. For example, if you were to take moments about the point where the lines of action of R and P meet you would eliminate R and P from your working and simplify your calculation.

Exercise 6A  SKILLS PROBLEM-SOLVING

Whenever a numerical value of g is required, take $g = 9.8\text{ m s}^{-2}$.

1 A uniform rod AB of weight 80 N rests with its lower end A on a rough horizontal floor. A string attached to end B keeps the rod in equilibrium. The string is held at 90° to the rod. The tension in the string is T. The coefficient of friction between the rod and the ground is μ. R is the normal reaction at A and F is the frictional force at A. Find the magnitudes of T, R and F, and the least possible value of μ.

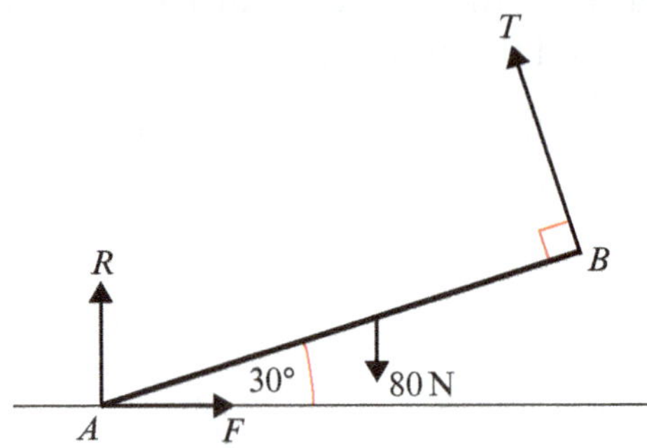

2 A uniform ladder of mass 10 kg and length 5 m rests against a smooth vertical wall with its lower end on rough horizontal ground. The ladder rests in equilibrium at an angle of 65° to the horizontal. Find:

a the magnitude of the normal reaction S at the wall

b the magnitude of the normal reaction R at the ground and the frictional force at the ground

c the least possible value of the coefficient of friction between the ladder and the ground.

d State how you have used the assumption that the ladder is uniform in your calculations.

(P) **3** A uniform ladder AB of mass 20 kg rests with its top A against a smooth vertical wall and its base B on rough horizontal ground. The coefficient of friction between the ladder and the ground is $\frac{3}{4}$. A mass of 10 kg is attached to the ladder. Given that the ladder is about to slip, find the inclination of the ladder to the horizontal,

a if the 10 kg mass is attached at A

b if the 10 kg mass is attached at B.

c State how you have used the assumption that the wall is smooth in your calculations.

(E) **4** A uniform ladder of mass 20 kg and length 8 m rests against a smooth vertical wall with its lower end on rough horizontal ground. The coefficient of friction between the ground and the ladder is 0.3. The ladder is inclined at an angle θ to the horizontal, where $\tan\theta = 2$. A boy of mass 30 kg climbs up the ladder. By modelling the ladder as a uniform rod, the boy as a particle and the wall as smooth and vertical,

a find how far up the ladder the boy can climb before the ladder slips. **(8 marks)**

b Criticise this model with respect to:

i the ladder **ii** the wall. **(2 marks)**

5 A smooth horizontal rail is fixed at a height of 3 m above a rough horizontal surface. A uniform pole (a long stick used to hold something up) AB of weight 4 N and length 6 m is resting with end A on the rough ground and touching the rail at point C. The vertical plane containing the pole is perpendicular to the rail. The distance AC is 4.5 m and the pole is in limiting equilibrium. Calculate:

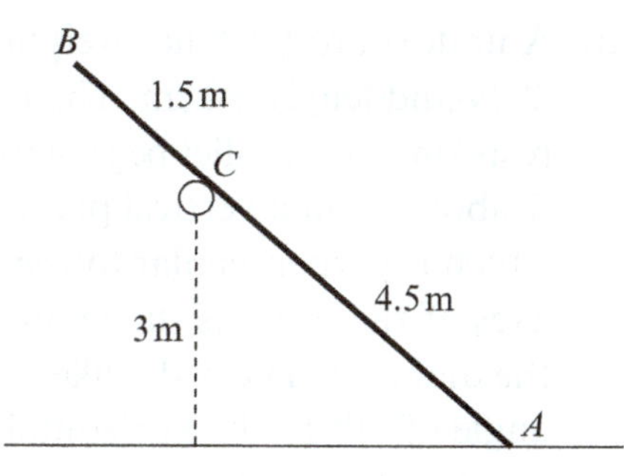

a the magnitude of the force exerted by the rail on the pole

b the coefficient of friction between the pole and the ground.

c State how you have used the assumption that the rail is smooth in your calculations.

(P) 6 A uniform ladder rests in limiting equilibrium with its top against a smooth vertical wall and its base on a rough horizontal floor. The coefficient of friction between the ladder and the floor is μ. Given that the ladder makes an angle θ with the floor, show that $2\mu \tan \theta = 1$.

7 A uniform ladder AB has length 7 m and mass 20 kg. The ladder is resting against a smooth cylindrical drum at P, where AP is 5 m, with end A in contact with rough horizontal ground. The ladder is inclined at 35° to the horizontal.

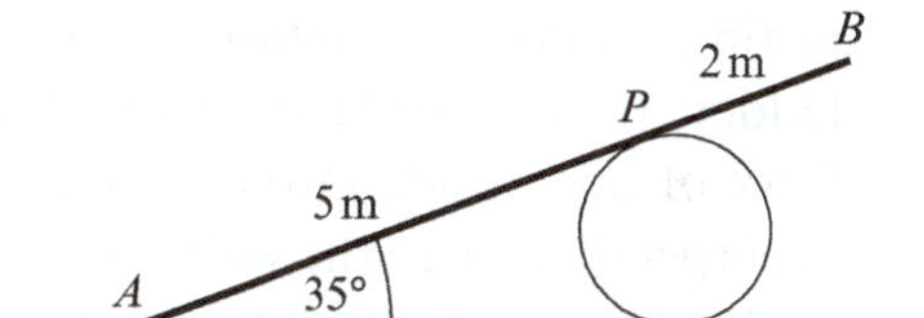

Find the normal and frictional components of the contact force at A, and hence find the least possible value of the coefficient of friction between the ladder and the ground.

(E/P) 8 A uniform ladder rests in limiting equilibrium with one end on rough horizontal ground and the other end against a rough vertical wall. The coefficient of friction between the ladder and the ground is μ_1 and the coefficient of friction between the ladder and the wall is μ_2. Given that the ladder makes an angle θ with the horizontal, show that $\tan \theta = \frac{1 - \mu_1\mu_2}{2\mu_1}$. **(8 marks)**

(E/P) 9 A uniform ladder of weight W rests in equilibrium with one end on rough horizontal ground and the other resting against a smooth vertical wall. The vertical plane containing the ladder is at right angles to the wall and the ladder is inclined at 60° to the horizontal. The coefficient of friction between the ladder and the ground is μ.

a Find, in terms of W, the magnitude of the force exerted by the wall on the ladder. **(6 marks)**

b Show that $\mu \geqslant \frac{\sqrt{3}}{6}$. **(3 marks)**

A load of weight w is attached to the ladder at its upper end (resting against the wall).

c Given that $\mu = \frac{\sqrt{3}}{5}$ and that the equilibrium is limiting, find w in terms of W. **(8 marks)**

10 A uniform rod XY has weight 20 N and length 90 cm. The rod rests on two parallel pegs, with X above Y, in a vertical plane which is perpendicular to the axes of the pegs, as shown in the diagram. The rod makes an angle of 30° to the horizontal and touches the two pegs at P and Q, where $XP = 20$ cm and $XQ = 60$ cm.

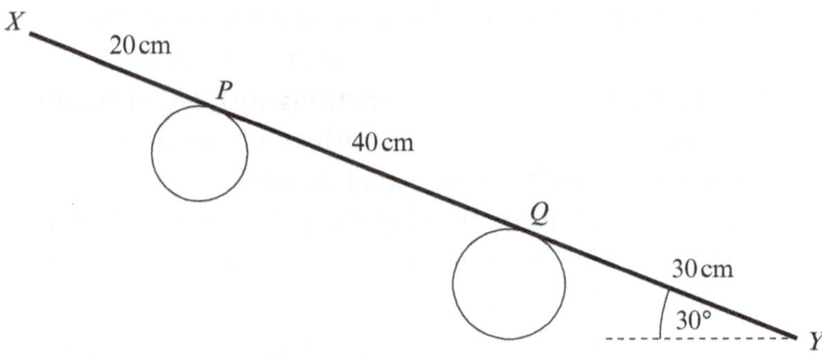

a Calculate the normal components of the forces on the rod at P and at Q. **(8 marks)**

The coefficient of friction between the rod and each peg is μ.

b Given that the rod is about to slip, find μ. **(2 marks)**

E **11** A ladder XY, of length l and weight W, has its end X on rough horizontal ground. The coefficient of friction between the ladder and the ground is $\frac{1}{5}$. The end Y of the ladder is resting against a smooth vertical wall. A window cleaner of weight $9W$ stands at the top of the ladder. To stop the ladder from slipping, the window cleaner's assistant applies a horizontal force of magnitude P to the ladder at X, towards the wall. The force acts in a direction which is perpendicular to the wall. The ladder rests in equilibrium in a vertical plane perpendicular to the wall and makes an angle θ with the horizontal ground, where $\tan\theta = \sqrt{3}$. The window cleaner is modelled as a particle and the ladder is modelled as a uniform rod.

a Find, in terms of W, the reaction of the wall on the ladder at Y. **(5 marks)**

b Find, in terms of W, the range of possible values of P for which the ladder remains in equilibrium. **(5 marks)**

c State how you have used the modelling assumption that the ladder is uniform in your calculations. **(1 mark)**

In practice, the ladder is wider and heavier at the bottom. The model is adjusted so the ladder is modelled as a **non-uniform** rod with its centre of mass closer to the base.

d State, with a reason, the effect this will have on

i the magnitude of the reaction of the wall on the ladder at Y

ii the range of possible values of P for which the ladder remains in equilibrium. **(4 marks)**

12 A uniform rod AB, of mass m and length $4a$, is freely hinged to a fixed point A.

A particle of mass km is fixed to the rod at B. The rod is held in equilibrium, at an angle θ to the horizontal, by a force of magnitude F acting at the point C on the rod, where $AC = \left(\frac{3}{5}\right)AB$. The force at C acts at right angles to AB and in the vertical plane containing AB.

Given that $\theta = 30°$,

a show that $F = \frac{5\sqrt{3}}{12}mg(1 + 2k)$ **(4 marks)**

Find, in terms of m, g and k,

b **i** the horizontal component of the force exerted by the hinge on the rod at A

ii the vertical component of the force exerted by the hinge on the rod at A. **(5 marks)**

Given also that the force acting on the rod at A acts at 75° above the horizontal,

c find the value of k. **(3 marks)**

Chapter review 6

(E/P) **1** A uniform ladder AB has one end A on smooth horizontal ground. The other end B rests against a smooth vertical wall. The ladder is modelled as a uniform rod of mass m and length $5a$. The ladder is kept in equilibrium by a horizontal force F acting at a point C on the ladder where $AC = a$. The force F and the ladder lie in a vertical plane perpendicular to the wall. The ladder is inclined to the horizontal at an angle θ, where $\tan\theta = \frac{9}{5}$, as shown in the diagram.

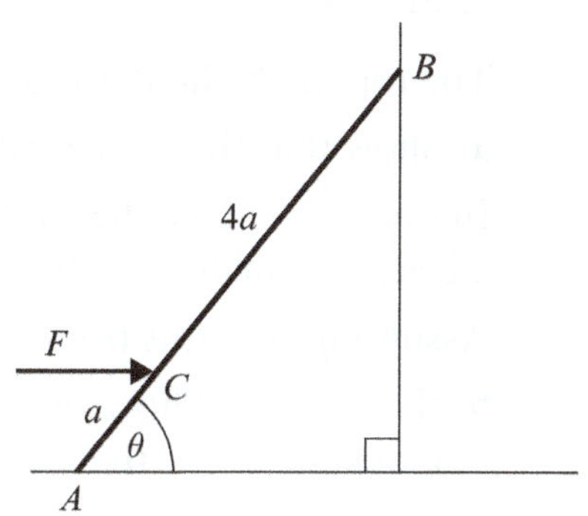

Show that $F = \dfrac{25mg}{72}$. **(8 marks)**

(E/P) **2** A uniform ladder AB, of mass m and length $2a$, has one end A on rough horizontal ground. The other end B rests against a smooth vertical wall. The ladder is in a vertical plane perpendicular to the wall. The ladder makes an angle α with the vertical, where $\tan\alpha = \frac{3}{4}$. A child of mass $2m$ stands on the ladder at C where $AC = \frac{2}{3}a$, as shown in the diagram. The ladder and the child are in equilibrium.

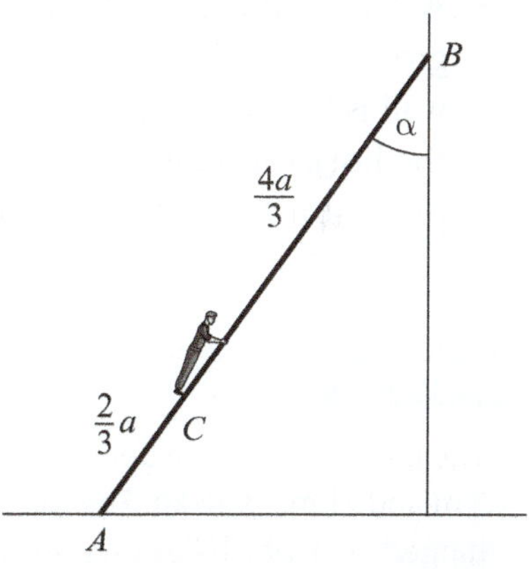

By modelling the ladder as a rod and the child as a particle, calculate the least possible value of the coefficient of friction between the ladder and the ground. **(8 marks)**

(E/P) **3** A uniform ladder, of weight W and length $2a$, rests in equilibrium with one end A on a smooth horizontal floor and the other end B against a rough vertical wall. The ladder is in a vertical plane perpendicular to the wall. The coefficient of friction between the wall and the ladder is μ. The ladder makes an angle θ with the floor, where $\tan\theta = \frac{4}{3}$. A horizontal light inextensible string CD is attached to the ladder at the point C, where $AC = \frac{1}{4}a$. The string is attached to the wall at the point D, with BD vertical, as shown in the diagram.

The tension in the string is $\frac{1}{3}W$. By modelling the ladder as a rod,

a find the magnitude of the force of the floor on the ladder **(5 marks)**

b show that $\mu \geqslant \frac{1}{3}$. **(3 marks)**

c State how you have used the modelling assumption that the ladder is a rod. **(1 mark)**

4 A uniform ladder, of weight W and length 5 m, has one end on rough horizontal ground and the other touching a smooth vertical wall. The coefficient of friction between the ladder and the ground is 0.3.

The top of the ladder touches the wall at a point 4 m vertically above the level of the ground.

a Show that the ladder cannot rest in equilibrium in this position. **(6 marks)**

In order to enable the ladder to rest in equilibrium in the position described above, a brick is attached to the bottom of the ladder.

Assuming that this brick is at the lowest point of the ladder, but not touching the ground,

b show that the horizontal frictional force exerted by the ladder on the ground is independent of the mass of the brick **(4 marks)**

c find, in terms of W and g, the smallest mass of the brick for which the ladder will rest in equilibrium. **(3 marks)**

E/P 5 A non-uniform ladder PQ of mass 20 kg and length 4 metres, rests with P on smooth horizontal ground and Q against a rough vertical wall. The coefficient of friction between the ladder and the wall is 0.2. The centre of mass of the ladder is 1 m from P. The ladder is inclined at an angle α to the horizontal, where $\tan \alpha = \frac{5}{2}$. A horizontal force F applied to the base of the ladder can just prevent it from slipping. By modelling the ladder as a rod, determine the value of F. **(10 marks)**

Challenge

The diagram shows a uniform rod AB of length 3 m and of mass 10 kg. The rod is smoothly hinged at A which lies on a vertical wall. A particle of mass 5 kg is suspended 1 m from B. The rod is kept in a horizontal position by a light inextensible string BC, where C lies on the wall directly above A. The plane ABC is perpendicular to the wall and $\angle ABC$ is 60°.

a Calculate the tension in the string.

b Work out the magnitude and direction of the reaction at the hinge.

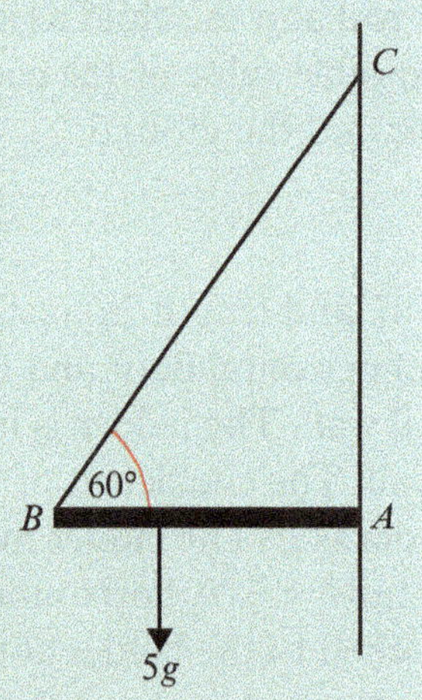

Watch out The reaction at the hinge does not have to be normal (perpendicular) to the wall.

Summary of key points

1. The moment of a force F acting at a perpendicular distance d from a point is give by $F \times d$.
2. A particle is in static equilibrium if it is at rest and the resultant force acting on the particle is zero.
3. The maximum value of the frictional force $F_{MAX} = \mu R$ is reached when the body you are considering is on the point of moving. The body is then said to be in limiting equilibrium.
4. In general, the force of friction F is such that $F \leqslant \mu R$, and the direction of the frictional force is opposite to the direction in which the body would move if the frictional force were absent.
5. For a rigid body in static equilibrium:
   - the body is stationary
   - the resultant force in any direction is zero
   - the resultant moment about any given point is zero.

Review exercise 2

(E) **1** In this question, **i** and **j** are perpendicular unit vectors in a horizontal plane.

A ball has a mass 0.2 kg. It is moving with velocity $30\mathbf{i}\,\text{m s}^{-1}$ when it is struck by a bat. The bat exerts an impulse of $(-4\mathbf{i} + 4\mathbf{j})$ N s on the ball. Find:

a the velocity of the ball immediately after the impact **(3)**

b the angle through which the ball is deflected as a result of the impact **(2)**

c the kinetic energy lost by the ball in the impact. **(4)**

← Mechanics 2 Sections 4.2, 5.1

(E/P) **2** A particle P of mass 0.75 kg is moving under the action of a single force **F** newtons. At time t seconds, the velocity $\mathbf{v}\,\text{m s}^{-1}$ of P is given by

$$\mathbf{v} = (t^2 + 2)\mathbf{i} - 6t\mathbf{j}$$

a Find the magnitude of **F** when $t = 4$. **(5)**

When $t = 5$, the particle P receives an impulse of magnitude $9\sqrt{2}$ N s in the direction of the vector $\mathbf{i} - \mathbf{j}$.

b Find the velocity of P immediately after the impulse. **(4)**

← Mechanics 2 Section 5.1

(E) **3** A tennis ball of mass 0.2 kg is moving with velocity $-10\mathbf{i}\,\text{m s}^{-1}$ when it is struck by a tennis racket. Immediately after being struck, the ball has velocity $(15\mathbf{i} + 15\mathbf{j})\,\text{m s}^{-1}$. Find:

a the magnitude of the impulse exerted by the racket on the ball **(4)**

b the angle, to the nearest degree, between the vector **i** and the impulse exerted by the racket **(2)**

c the kinetic energy gained by the ball as a result of being struck. **(2)**

← Mechanics 2 Sections 4.2, 5.1

(E) **4** At time t seconds the acceleration, $\mathbf{a}\,\text{m s}^{-2}$, of a particle P relative to a fixed origin O, is given by $\mathbf{a} = 2\mathbf{i} + t\mathbf{j}$. Initially the velocity of P is $(2\mathbf{i} - 4\mathbf{j})\,\text{m s}^{-1}$.

a Find the velocity of P at time t seconds. **(3)**

At time $t = 2$ seconds the particle P is given an impulse $(3\mathbf{i} - 1.5\mathbf{j})$ N s.

Given that the particle P has mass 0.5 kg,

b find the speed of P immediately after the impulse has been applied. **(6)**

← Mechanics 2 Section 5.1

(E) **5** A rough plane is inclined at an angle of 5° to the horizontal. A sled of mass 1250 kg is pushed up a line of greatest slope of the plane. The coefficient of friction between the sled and the plane is 0.05.

a Find the magnitude of the frictional force acting on the sled. **(5)**

The sled moves a total distance of 750 m, as measured along the line of greatest slope of the plane. Find:

b the work done against friction **(3)**

c the work done against gravity. **(4)**

← Mechanics 2 Section 4.1

(E) **6** A rock of mass 4 kg is dropped from rest at the top of a cliff. It falls 40 m vertically down before hitting the surface of the sea.

a Modelling the rock as a particle falling freely under gravity, find its kinetic energy when it hits the surface of the sea. **(2)**

In reality the falling rock will be subject to air resistance which will oppose its motion.

b Without further calculation, state how this will affect the kinetic energy of the rock when it hits the surface of the water. **(1)**

← Mechanics 2 Section 4.2

E/P 7 A winch pulls a sled of mass 1000 kg, a distance of 25 m up a line of greatest slope of a ramp. Given that the work done against gravity by the winch is 19.6 kJ, show that the slope is inclined at an angle of $\arcsin\left(\frac{2}{25}\right)$ to the horizontal. **(5)**

← Mechanics 2 Section 4.1

E 8 A cable car of mass 200 kg moves along a section of cable which can be modelled as a straight line inclined at 30° above the horizontal. As the cable car moves up the cable for 200 m its speed reduces from $2\,\text{m}\,\text{s}^{-1}$ to $1.5\,\text{m}\,\text{s}^{-1}$. Calculate:

a the loss in kinetic energy of the cable car **(4)**

b the gain in potential energy of the cable car. **(4)**

← Mechanics 2 Section 4.2

E 9

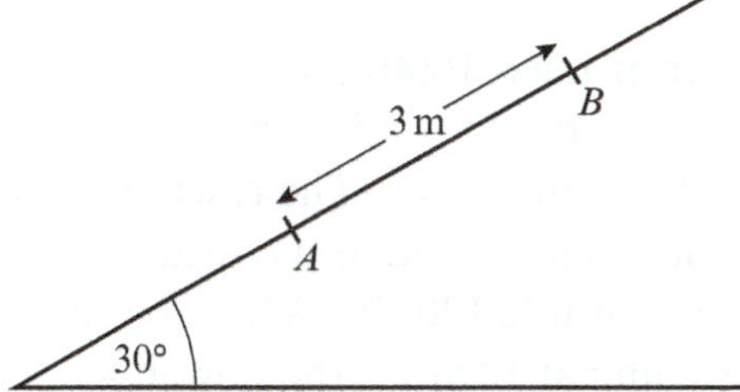

A particle P of mass 2 kg is projected from a point A up a line of greatest slope AB of a fixed plane. The plane is inclined at an angle of 30° to the horizontal and $AB = 3\,\text{m}$ with B above A, as shown in the figure. The speed of P at A is $10\,\text{m}\,\text{s}^{-1}$.

a Assuming the plane is smooth, find the speed of P at B. **(4)**

The plane is now assumed to be rough. At A the speed of P is $10\,\text{m}\,\text{s}^{-1}$ and at B the speed of P is $7\,\text{m}\,\text{s}^{-1}$.

b By using the work–energy principle, or otherwise, find the coefficient of friction between P and the plane. **(5)**

← Mechanics 2 Section 4.3

E 10

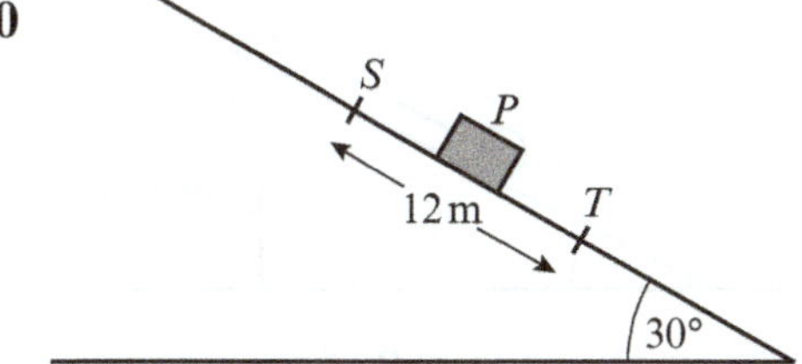

A small package is modelled as a particle P of mass 0.6 kg. The package slides down a rough plane from a point S to a point T, where $ST = 12\,\text{m}$. The plane is inclined at 30° to the horizontal and ST is a line of greatest slope of the plane, as shown in the figure. The speed of P at S is $10\,\text{m}\,\text{s}^{-1}$ and the speed of P at T is $9\,\text{m}\,\text{s}^{-1}$. Calculate:

a the total loss of energy of P in moving from S to T **(4)**

b the coefficient of friction between P and the plane. **(5)**

← Mechanics 2 Section 4.3

E/P 11 A particle P has mass 4 kg. It is projected from a point A up a line of greatest slope of a rough plane inclined at an angle α to the horizontal, where $\tan\alpha = \frac{3}{4}$. The coefficient of friction between P and the plane is $\frac{2}{7}$. The particle comes to rest instantaneously at the point B on the plane, where $AB = 2.5\,\text{m}$. It then moves back down the plane to A.

a Find the work done by friction as P moves from A to B. **(4)**

b Using the work–energy principle, find the speed with which P is projected from A. **(4)**

c Find the speed of P when it returns to A. **(4)**

← Mechanics 2 Section 4.3

E/P **12**

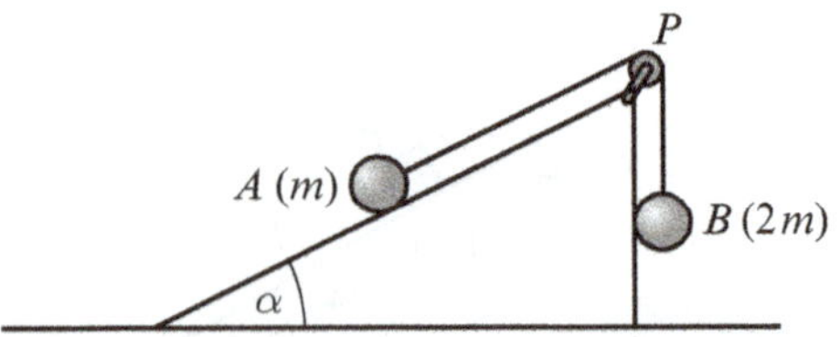

Two particles A and B of masses m and $2m$ respectively are attached to the ends of a light inextensible string. The particle A lies on a rough plane inclined at an angle α to the horizontal, where $\tan \alpha = \frac{3}{4}$. The string passes over a small light pulley P fixed at the top of the plane. The particle B hangs freely below P, as shown in the figure. The particles are released from rest with the string taut and the section of the string from A to P parallel to a line of greatest slope of the plane. The coefficient of friction between A and the plane is $\frac{5}{8}$. When each particle has moved a distance h, B has not reached the ground and A has not reached P.

a Find an expression for the potential energy lost by the system when each particle has moved a distance h. **(2)**

When each particle has moved a distance h, they are moving with speed v.

b Using the work–energy principle, find an expression for v^2, giving your answer in the form kgh where k is a number. **(5)**

← Mechanics 2 Section 4.3

E **13** A car of mass 1000 kg is towing a trailer of mass 1500 kg along a straight horizontal road. The tow-bar joining the car to the trailer is modelled as a light rod parallel to the road. The total resistance to motion of the car is modelled as having constant magnitude 750 N. The total resistance to motion of the trailer is modelled as a force of magnitude R newtons, where R is a constant. When the engine is working at a rate of 50 kW, the car and the trailer travel at a constant speed of 25 m s^{-1}.

a Show that $R = 1250$. **(3)**

When travelling at 25 m s^{-1} the driver of the car disengages the engine and applies the brakes. The brakes provide a constant braking force of magnitude 1500 N to the car. The resisting forces of magnitude 750 N and 1250 N are assumed to remain unchanged. Calculate:

b the deceleration of the car while braking **(3)**

c the thrust in the tow-bar while braking **(2)**

d the work done, in kJ, by the braking force in bringing the car and the trailer to rest. **(4)**

e Suggest how the modelling assumption that the resistances to motion are constant could be refined to be more realistic. **(1)**

← Mechanics 2 Section 4.4

E **14** A car of mass 1000 kg is moving along a straight road with constant acceleration a m s^{-2}. The resistance to motion is modelled as a constant force of magnitude 1200 N. When the car is travelling at 12 m s^{-1}, the power generated by the engine of the car is 24 kW.

a Calculate the value of a. **(4)**

When the car is travelling at 14 m s^{-1}, the engine is switched off and the car comes to rest without braking in a distance d metres.

b Assuming the same model for resistance, use the work–energy principle to calculate the value of d. **(3)**

c Give a reason why the model used for resistance may not be realistic. **(1)**

← Mechanics 2 Sections 4.3, 4.4

E **15**

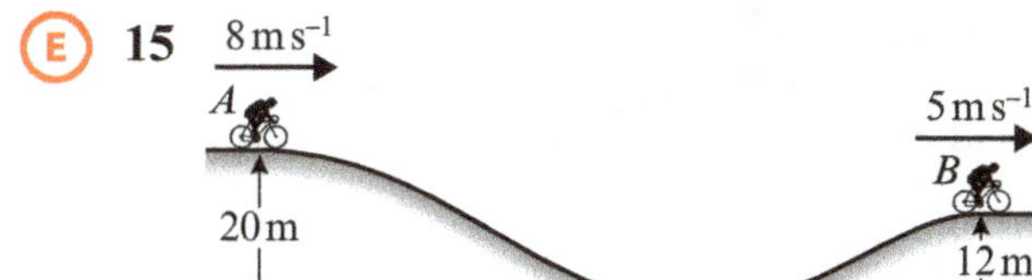

The figure shows the path taken by a cyclist in travelling on a section of a road. When the cyclist comes to the point A on the top of the hill she is travelling at $8\,\text{m}\,\text{s}^{-1}$. She descends a vertical distance of 20 m to the bottom of the hill. The road then rises to the point B through a vertical distance of 12 m. When she reaches B her speed is $5\,\text{m}\,\text{s}^{-1}$. The total mass of the cyclist and the bicycle is 80 kg and the total distance along the road from A to B is 500 m. By modelling the resistance to the motion of the cyclist as of constant magnitude 20 N,

a find the work done by the cyclist in moving from A to B. **(5)**

At B the road is horizontal.

b Given that at B the cyclist is accelerating at $0.5\,\text{m}\,\text{s}^{-2}$, find the power generated by the cyclist at B. **(4)**

← Mechanics 2 Sections 4.3, 4.4

E **16** A car of mass 400 kg is moving up a straight road inclined at an angle θ to the horizontal where $\sin\theta = \frac{1}{14}$. The resistance to motion of the car from non-gravitational forces is modelled as a constant force of magnitude R newtons. When the car is moving at a constant speed of $20\,\text{m}\,\text{s}^{-1}$, the power developed by the car's engine is 10 kW.

Find the value of R. **(5)**

← Mechanics 2 Section 4.4

E **17** A lorry of mass 1500 kg moves along a straight horizontal road. The resistance to motion of the lorry has magnitude 750 N and the lorry's engine is working at a rate of 36 kW.

a Find the acceleration of the lorry when its speed is $20\,\text{m}\,\text{s}^{-1}$. **(4)**

The lorry comes to a hill inclined at an angle α to the horizontal, where $\sin\alpha = \frac{1}{10}$. The magnitude of the resistance to motion from non-gravitational forces remains 750 N.

The lorry moves up the hill at a constant speed of $20\,\text{m}\,\text{s}^{-1}$.

b Find the rate at which the lorry is now working. **(3)**

← Mechanics 2 Section 4.4

E **18** A car of mass 1200 kg moves along a straight horizontal road. The resistance to motion of the car from non-gravitational forces is of constant magnitude 600 N. The car moves with constant speed and the engine of the car is working at a rate of 21 kW.

a Find the speed of the car. **(2)**

The car moves up a hill inclined at an angle α to the horizontal, where $\sin\alpha = \frac{1}{14}$. The car's engine continues to work at 21 kW and the resistance to motion from non-gravitational forces remains of magnitude 600 N.

b Find the constant speed at which the car moves up the hill. **(4)**

← Mechanics 2 Section 4.4

E **19** A car of mass 1000 kg is moving along a straight horizontal road. The resistance to motion is modelled as a constant force of magnitude R newtons. The engine of the car is working at a constant rate of 12 kW. When the car is moving with speed $15\,\text{m}\,\text{s}^{-1}$, the acceleration of the car is $0.2\,\text{m}\,\text{s}^{-2}$.

a Show that $R = 600$. **(4)**

The car now moves with constant speed U m s^{-1} downhill on a straight road inclined at θ to the horizontal, where $\sin\theta = \frac{1}{40}$. The engine of the car is now working at a rate of 7 kW. The resistance to motion from non-gravitational forces remains of magnitude R newtons.

b Calculate the value of U. **(5)**

← Mechanics 2 Section 4.4

E/P **20** A motorcycle of mass 600 kg moves along a straight road at a speed of v m s^{-1}. The total resistances to motion of the motorcycle are modelled as a variable force of magnitude $(500 + 2v^2)$ N.

Calculate the power that must be generated by the motorcycle engine to maintain a constant speed of 15 m s^{-1}

a when the road is horizontal **(4)**

b when the road slopes downhill at an angle of 5° to the horizontal. **(5)**

← Mechanics 2 Section 4.4

E/P **21** A van of mass 1500 kg is driving up a straight road inclined at angle α to the horizontal, where $\sin\alpha = \frac{1}{12}$. The resistance to motion due to non-gravitational forces is modelled as a variable force of magnitude $(700 + 10v)$ N, where v m s^{-1} is the speed of the van.

a Given that initially the speed of the van is 30 m s^{-1} and that the van's engine is working at a rate of 60 kW, calculate the magnitude of the initial deceleration of the van. **(4)**

When travelling up the same hill, the rate of working of the van's engine is increased to 80 kW.

b Using the same model for the resistance due to non-gravitational forces, calculate in m s^{-1} the maximum constant speed which can be sustained by the van at this rate of working. **(5)**

← Mechanics 2 Section 4.4

E **22**

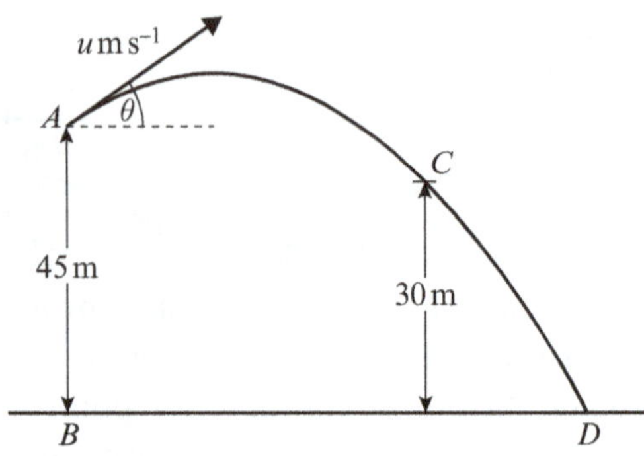

A particle P is projected from a point A with speed u m s^{-1} at an angle of elevation θ, where $\cos\theta = \frac{4}{5}$. The point B, on horizontal ground, is vertically below A and $AB = 45$ m. After projection, P moves freely under gravity, passing through a point C, 30 m above the horizontal ground, before striking the ground at the point D, as shown in the figure above.

Given that P passes through C with speed 24.5 m s^{-1},

a using conservation of energy, or otherwise, show that $u = 17.5$ **(4)**

b find the size of the angle which the velocity of P makes with the horizontal as P passes through C **(7)**

c find the distance BD. **(3)**

← Mechanics 2 Section 4.3

E **23**

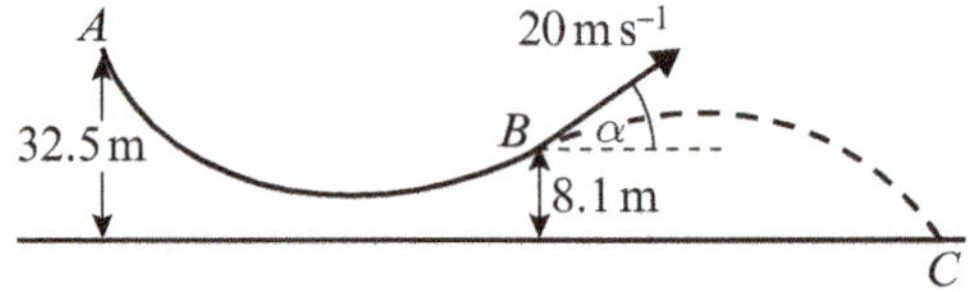

In a ski-jumping competition, a skier of mass 80 kg moves from rest at a point A on a ski slope. The skier's path is an arc AB. The starting point A of the slope is 32.5 m above horizontal ground. The end B of the slope is 8.1 m above the ground. When the skier reaches B she is travelling at 20 m s^{-1} and moving upwards at an angle α to the horizontal, where $\tan\alpha = \frac{3}{4}$, as shown in the figure. The distance along the slope from A to B is 60 m.

The resistance to motion while she is on the slope is modelled as a force of constant magnitude R newtons.

a By using the work–energy principle, find the value of R. **(5)**

On reaching B, the skier then moves through the air and reaches the ground at the point C. The motion of the skier in moving from B to C is modelled as that of a particle moving freely under gravity.

b Find the time the skier takes to move from B to C. **(5)**

c Find the horizontal distance from B to C. **(2)**

d Find the speed of the skier immediately before she reaches C. **(4)**

← Mechanics 2 Section 4.3

E/P **24** A ladder PQ of mass 25 kg and length 6 metres, rests with its base P on rough horizontal ground and its top Q leaning against a smooth vertical wall. The coefficient of friction between the ladder and the ground is 0.25. The ladder lies in a vertical plane perpendicular to the wall and the ground, and is inclined at an angle 60° to the horizontal.

A builder of mass 75 kg climbs up the ladder. Modelling the builder as a particle and the ladder as a uniform rod, find the maximum distance up the ladder the builder can climb before the ladder begins to slip. **(10)**

← Mechanics 2 Section 6.2

E/P **25** A uniform ladder PQ of mass m kg and length l metres, rests with one end P on rough horizontal ground and the other end Q against a smooth vertical wall. The coefficient of friction between the ladder and the ground is μ. The ladder lies in a vertical plane perpendicular to the wall and the ground, and is inclined at an angle α to the horizontal. Given that the ladder is on the point of slipping, find an expression for μ in terms of α. **(10)**

← Mechanics 2 Section 6.2

E/P **26** A non-uniform ladder AB of weight 240 N and length 6 m rests with its end A on smooth horizontal ground and its end B against a rough vertical wall. The coefficient of friction between the ladder and the wall is 0.3.The centre of mass of the ladder is 2 m from A. The ladder lies in a vertical plane perpendicular to the wall and the ground, and is inclined at an angle α to the horizontal, where $\tan \alpha = \frac{3}{2}$.

The ladder can be prevented from sliding down the wall by applying a horizontal force of magnitude P N to the bottom of the ladder. By modelling the ladder as a non-uniform rod, determine the minimum value of P. **(10)**

← Mechanics 2 Section 6.2

E/P **27** A smooth sphere S of mass m is moving on a smooth horizontal plane with speed u. It collides directly with another smooth sphere T, of mass $3m$, whose radius is the same as S. The sphere T is moving in the same direction as S with speed $\frac{1}{6}u$. The sphere S is brought to rest by the impact. Find the coefficient of restitution between the spheres. **(7)**

← Mechanics 2 Section 5.2

E/P **28** A smooth sphere S of mass m is moving with speed u on a smooth horizontal plane. The sphere S collides with another smooth sphere T, of equal radius to S but of mass km, moving in the same straight line and in the same direction with speed λu, $0 < \lambda < \frac{1}{2}$. The coefficient of restitution between S and T is e.

Given that S is brought to rest by the impact,

a show that $e = \dfrac{1 + k\lambda}{k(1 - \lambda)}$ **(6)**

b deduce that $k > 1$. **(3)**

← Mechanics 2 Section 5.2

E/P **29** A smooth uniform sphere S of mass m is moving on a smooth horizontal plane with speed u. The sphere collides directly with another smooth uniform sphere T, of the same radius as S and of mass $2m$, which is at rest on the plane.
The coefficient of restitution between the spheres is e.

a Show that the speed of T after the collision is $\frac{1}{3}u(1 + e)$.

Given that $e > \frac{1}{2}$,

b i find the speed of S after the collision **(4)**

ii determine whether the direction of motion of S is reversed by the collision. **(4)**

← Mechanics 2 Section 5.2

E/P **30** A particle P of mass $3m$ is moving with speed $2u$ in a straight line on a smooth horizontal table. The particle P collides with a particle Q of mass $2m$ moving with speed u in the opposite direction to P. The coefficient of restitution between P and Q is e.

a Show that the speed of Q after the collision is $\frac{1}{5}u(9e + 4)$. **(5)**

As a result of the collision, the direction of motion of P is reversed.

b Find the range of possible values of e. **(5)**

Given that the magnitude of the impulse of P on Q is $\frac{32}{5}mu$,

c find the value of e. **(4)**

← Mechanics 2 Section 5.2

E **31** A ball falls from rest and hits a smooth horizontal plane 2 seconds later. Given that the coefficient of restitution between the ball and the plane is $\frac{6}{7}$, find the maximum height reached by the ball on its first bounce. **(4)**

← Mechanics 2 Section 5.3

E/P **32** A small sphere S is projected from a point P along a smooth horizontal plane towards a fixed vertical wall, which is perpendicular to the direction of motion of the sphere. S strikes the wall 2 seconds after it is projected, and passes through P again 3 seconds after that.

a Find the value of the coefficient of restitution between the sphere and the wall. **(3)**

b Without further calculation, state with justification how your answer to part **a** would change if the plane was rough. **(2)**

← Mechanics 2 Section 5.3

E **33** A ball B falls from rest from a cliff of height 50 m onto horizontal ground. After rebounding, the ball reaches a maximum height of 35 m.

a Find the value of the coefficient of restitution between the ball and the ground. **(3)**

b Show that the time taken from the instant when the ball was dropped until the instant when it hits the ground for a second time is 8.54 s (3 s.f.). **(4)**

← Mechanics 2 Section 5.3

E/P **34** Two particles, A and B, of mass m and $3m$ respectively, lie at rest on a smooth horizontal table. The coefficient of restitution between the particles is 0.25. A and B are moving towards each other at speeds of $7u$ and u respectively and they collide directly. Find:

a the speeds of A and B after the collision **(7)**

b the loss in kinetic energy due to the collision. **(2)**

← Mechanics 2 Sections 5.2, 5.4

E/P **35** Two uniform smooth spheres A and B are of equal size and have masses $3m$ and $2m$ respectively. They are both moving in the same straight line with speed u, but in opposite directions, when they are in direct collision with each other. Given that A is brought to rest by the collision, find:

a the coefficient of restitution between the spheres **(6)**

b the kinetic energy lost in the impact. **(3)**

← Mechanics 2 Sections 5.2, 5.4

E/P **36** A smooth sphere A of mass m is moving with speed u on a smooth horizontal table when it collides directly with another smooth sphere B of mass $3m$, which is at rest on the table. The coefficient of restitution between A and B is e. The spheres have equal radius and are modelled as particles.

a Show that the speed of B immediately after the collision is $\frac{1}{4}(1 + e)u$. **(5)**

b Find the speed of A immediately after the collision. **(2)**

Immediately after the collision the total kinetic energy of the spheres is $\frac{1}{6}mu^2$.

c Find the value of e. **(6)**

d Hence show that A is at rest after the collision. **(1)**

← Mechanics 2 Sections 5.2, 5.4

E **37** A train engine of mass 8 tonnes is moving at $4\,\text{m}\,\text{s}^{-1}$ when it hits a carriage of mass 12 tonnes which is at rest. After the impact, the engine and the carriage move off together with speed $1.5\,\text{m}\,\text{s}^{-1}$ in the direction in which the engine was originally moving. Calculate the total loss of kinetic energy due to the impact. **(4)**

← Mechanics 2 Section 5.4

E/P **38** Two particles, P and Q, of masses 0.05 kg and 0.25 kg respectively, are connected by a light inextensible string. P is projected with speed $2\,\text{m}\,\text{s}^{-1}$ directly away from Q. When the string becomes taut, particle Q is pulled suddenly into motion, and P and Q then move with a common speed in the direction in which P was originally moving. Find the loss of total kinetic energy due to the pull. **(10)**

← Mechanics 2 Section 5.4

E/P **39** Two small spheres A and B have masses $3m$ and $2m$ respectively. They are moving towards each other in opposite directions on a smooth horizontal plane, both with speed $2u$, when they collide directly. As a result of the collision, the direction of motion of B is reversed and its speed is unchanged.

a Find the coefficient of restitution between the spheres. **(7)**

Subsequently, B collides directly with another small sphere C of mass $5m$ which is at rest. The coefficient of restitution between B and C is $\frac{3}{5}$.

b Show that, after B collides with C, there will be no further collisions between the spheres. **(7)**

← Mechanics 2 Sections 5.2, 5.5

E/P **40** A smooth sphere P of mass $2m$ is moving in a straight line with speed u on a smooth horizontal table. Another smooth sphere Q of mass m is at rest on the table. P collides directly with Q. The coefficient of restitution between P and Q is $\frac{1}{3}$. The spheres are modelled as particles.

a Show that, immediately after the collision, the speeds of P and Q are $\frac{5}{9}u$ and $\frac{8}{9}u$ respectively. **(7)**

After the collision, Q strikes a fixed vertical wall which is perpendicular to the direction of motion of P and Q. The coefficient of restitution between Q and the wall is e. When P and Q collide again, P is brought to rest.

b Find the value of e. **(7)**

c Explain why there must be a third collision between P and Q. **(1)**

← Mechanics 2 Sections 5.2, 5.3, 5.5

E **41** Two small smooth spheres, P and Q, of equal radius, have masses $2m$ and $3m$ respectively. The sphere P is moving with speed $5u$ on a smooth horizontal table when it collides directly with Q, which is at rest on the table. The coefficient of restitution between P and Q is e.

a Show that the speed of Q immediately after the collision is $2(1 + e)u$. **(5)**

After the collision, Q hits a smooth vertical wall which is at the edge of the table and perpendicular to the direction of motion of Q. The coefficient of restitution between Q and the wall is f, $0 < f \leqslant 1$.

b Show that, when $e = 0.4$, there is a second collision between P and Q. **(3)**

Given that $e = 0.8$ and there is a second collision between P and Q,

c find the range of possible values of f. **(3)**

← Mechanics 2 Sections 5.2, 5.3, 5.5

E/P **42** A particle A of mass $2m$, moving with speed $2u$ in a straight line on a smooth horizontal table, collides with a particle B of mass $3m$, moving with speed u in the same direction as A. The coefficient of restitution between A and B is e.

a Show that the speed of B after the collision is $\frac{1}{5}u(7 + 2e)$. **(5)**

b Find the speed of A after the collision, in terms of u and e. **(2)**

The speed of A after the collision is $\frac{11}{10}u$.

c Show that $e = \frac{1}{2}$. **(2)**

At the instant of collision, A and B are at a distance d from a vertical barrier fixed to the surface at right-angles to their direction of motion. Given that B hits the barrier, and that the coefficient of restitution between B and the barrier is $\frac{11}{16}$,

d find the distance of A from the barrier at the instant that B hits the barrier **(4)**

e show that, after B rebounds from the barrier, it collides with A again at a distance $\frac{5}{32}d$ from the barrier. **(4)**

← Mechanics 2 Sections 5.2, 5.3, 5.5

E/P **43** A ball is dropped from a height of 2 m onto a smooth horizontal plane. The coefficient of restitution between the ball and the plane is 0.8.

a Modelling the ball as a particle, find the total distance travelled by the ball before it comes to rest. **(10)**

b Criticise this model with respect to the number of times the ball bounces. **(1)**

← Mechanics 2 Section 5.5

E/P **44** Three small smooth spheres A, B and C, of masses m, $2m$ and $3m$ respectively, lie in order in a straight line on a smooth horizontal surface. The coefficient of restitution between A and B is 0.7 and the coefficient of restitution between B and C is 0.4. Sphere A is projected towards sphere B with speed $4\,\text{m}\,\text{s}^{-1}$. Show that exactly two collisions will occur. **(12)**

← Mechanics 2 Section 5.5

45 A uniform ladder AB of length 10 m and mass 20 kg rests in equilibrium with one end A on rough horizontal ground and the other end B against a smooth vertical wall such that the ladder is inclined at $\theta°$ to the horizontal.

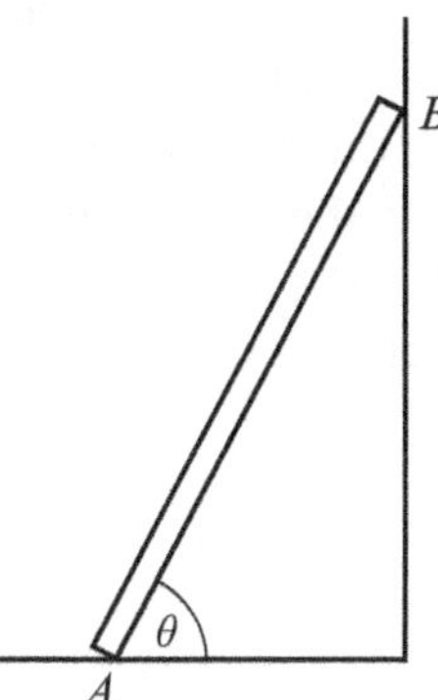

Given that the ladder is at the point of slipping and that the coefficient of friction between the ladder and the ground is 0.25,

a work out the magnitude of the frictional force of the ground on the ladder

b show that $\theta = \tan^{-1}(2)$.

46 A uniform ladder PQ has length l m and mass m kg. P rests on smooth horizontal ground and Q rests against a smooth vertical wall. The ladder is kept in equilibrium by a horizontal force of magnitude $\frac{1}{3}mg$ that acts at a point R on the ladder, where $PR = \frac{1}{4}PQ$. The angle between the ladder and the wall is θ.

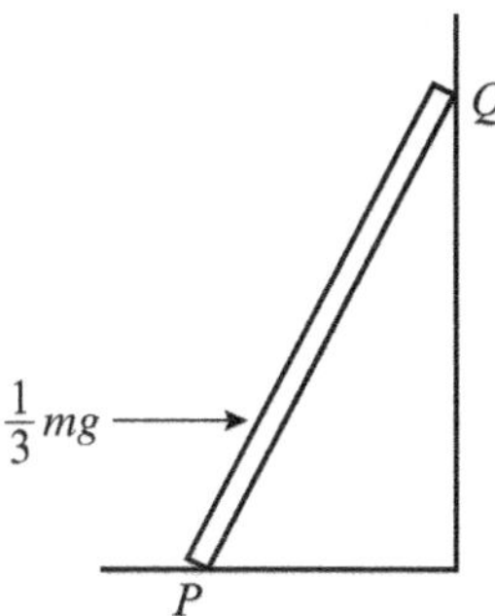

By modelling the ladder as a uniform rod, show that $\tan\theta = \frac{1}{2}$.

47 A uniform ladder PQ of length l and of mass m rests with P on a rough horizontal surface and Q against a smooth vertical wall. The ladder is inclined at an angle θ to the ground.

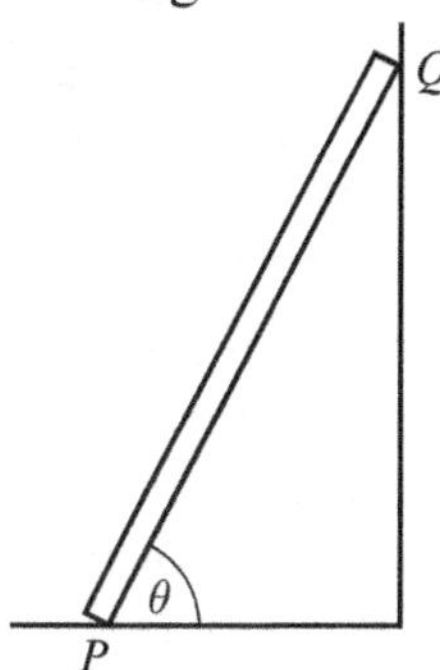

When a builder of mass $3m$ stands on the ladder at B, the ladder is on the point of slipping. Given that $PB = \frac{1}{5}l$ and that the coefficient of friction between the ladder and the ground is $\frac{3}{5}$, work out the value of θ.

48 A uniform rod AB of length 6 m and mass 10 kg rests in equilibrium with A on rough horizontal ground and the point C, where $AC = 4$ metres, on a smooth peg. The rod is inclined at 45° to the ground. Work out:

a the reaction on the rod at A

b the reaction on the rod at C

c the friction acting on the rod.

49 A uniform plank PQ of length l m and mass m kg rests on a smooth peg at the point R, where PR is $0.75PQ$. The end P rests in limiting equilibrium on rough ground. The coefficient of friction between the plank and the ground is μ. The plank is inclined at angle θ to the horizontal. Show that the magnitude of the force between the plank and the peg is $\frac{2}{3}mg\cos\theta$.

50 A uniform plank AB of mass 5 kg and length 5 m is at rest on a smooth cylinder which is fixed with its axis horizontal to rough horizontal ground.

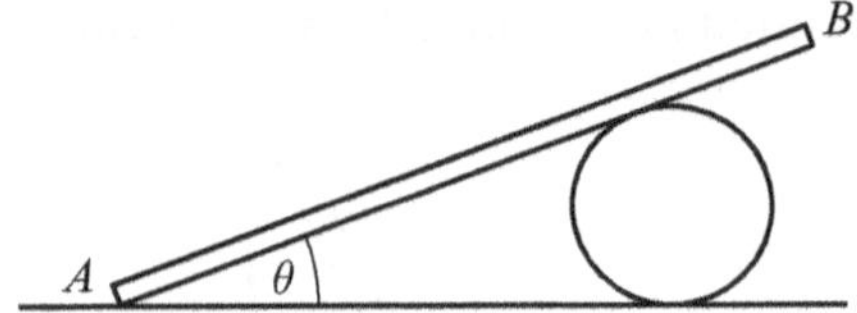

The plank rests on the cylinder at C where AC is 4 m. Given that the plank is in limiting equilibrium at A and is inclined at an angle of θ to the horizontal where $\tan\theta = \frac{1}{\sqrt{3}}$,

a find the normal reaction between the plank and the ground

b show that the coefficient of friction between the plank and the ground is $\frac{5\sqrt{3}}{17}$.

Challenge

A uniform rectangular lamina $PQRS$, of mass 10 kg rests in equilibrium with the vertex P on rough horizontal ground and vertex S, against a smooth vertical wall. $PQ = RS = 1$ m and $QR = PS = 3$ m. The lamina lies in a vertical plane perpendicular to the wall and makes an angle θ with the ground. The coefficient of friction between the lamina and the ground is 0.2. Given that the lamina is in limiting equilibrium, work out the value of θ.

Exam practice

Mathematics
International Advanced Subsidiary/ Advanced Level Mechanics 2

Time: 1 hour 30 minutes
You must have: Mathematical Formulae and Statistical Tables, Calculator
Answer ALL questions

---

1 A projectile is launched from a point A which is 10 m above horizontal ground, with a velocity of 20 m s^{-1} at an angle of 30° to the horizontal. After launch, the projectile moves freely under gravity until it strikes the ground at the point B. Find:

a the greatest height above the ground reached by the projectile **(4)**

b the speed of the projectile immediately before it lands at B **(6)**

c the angle the projectile makes with the ground immediately before it lands at B. **(4)**

2 A particle moves in a straight line such that its velocity at time t seconds is given by $v = \frac{2t^2 - 3t - 2}{\sqrt{t}}$ m s^{-1} for $t \geqslant 1$.

Work out:

a the distance the particle travels between $t = 1$ and $t = 5$ seconds **(7)**

b the acceleration of the particle at $t = 2$ s. **(5)**

3 A uniform lamina $ABCD$ is made by removing an isosceles triangle from a right-angled triangle as shown.

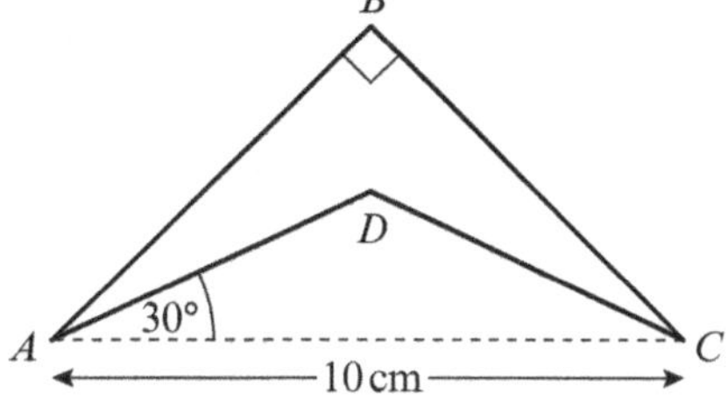

The lamina is freely suspended from A. Work out the angle AB makes with the vertical. **(12)**

4 A lorry of mass 5500 kg is travelling at a constant speed of 20 m s^{-1} along a straight road which is inclined at $\theta°$ to the horizontal. The engine is working at a rate of 110 kW. The non-gravitational resistance to motion is 1.05 kN.

a Work out the value of θ. **(5)**

The rate of working of the engine is now increased to 150 kW. Assuming the resistances to motion are unchanged,

b calculate the initial acceleration of the lorry. **(5)**

5 Two balls P and Q have masses $2m$ and m respectively. They are moving in the same direction along the same straight line on a smooth level floor. Immediately before they collide, P has speed 4 m s^{-1} and Q has speed 2 m s^{-1}. The coefficient of restitution between P and Q is 0.8. By modelling the balls as smooth spheres of equal size and the floor as a smooth horizontal plane,

a find the speed and direction of motion of P and Q immediately after the collision **(8)**

b work out the percentage of kinetic energy lost in the collision. **(6)**

6 A uniform beam PQ of length l m and mass m kg rests with the end A on rough horizontal ground where the coefficient of friction between the beam and the ground is μ. The beam rests on a smooth peg at the point R where $PR = \frac{4}{5}PQ$. The beam makes an angle of θ with the horizontal.

a Show that the normal reaction at the peg is $\frac{5}{8}mg\cos\theta$. **(5)**

b Show that if $\theta = 30°$, then $\mu = \frac{5\sqrt{3}}{17}$. **(5)**

c Given that $l = 2$ m and $\theta = 30°$, work out, in terms of m, the largest mass that can be placed at Q without the beam losing contact with the ground. **(3)**

TOTAL FOR PAPER: 75 MARKS

GLOSSARY

acceleration positive **rate of change** of **velocity**

adjacent next to

air resistance the frictional force that air exerts against a moving object

angle of projection the angle between the **horizontal** and the line on which a **particle** is projected

arbitrary not given a specific value

assumption accepted as true

at rest stationary

body an object

coalesce join together

centre of mass the point through which the **mass** of an object is concentrated

centroid (of a triangle) the position of the centre of **mass** of a triangle

coefficient of friction a measure of how rough a surface is, usually denoted μ

coefficient of restitution the ratio of speed after a collision to speed before a collision

collision when two or more objects crash into each other

component part

composite lamina a figure that consists of two or more geometric shapes

constant unchanging

coplanar in the same **plane**

coordinates a set of values that show an exact position

deceleration negative **rate of change** of **velocity**

derivative rate of change

differentiation the process of finding the **derivative**, or **rate of change**, of a function

direct collision a **collision** in which a moving object hits another object head on

displacement distance in a particular direction

elastic collision a **collision** where the speed of approach and the speed of separation of the colliding objects are related by **Newton's Law of restitution** with $0 \leqslant e \leqslant 1$

elevation the height of something in relation to a certain level

equation of motion $F = ma$

equilibrium having zero **resultant force**

exert to apply power and influence something

force a push or a pull

friction **resistance** due to how rough the surface is on which a **body** is moving

function a relation or expression involving one or more variables

gravitational relating to **gravity**

gravitational potential energy the energy that an object has as a result of its position

gravity the **force** between any object and the Earth

horizontal at right angles to the vertical

impact **collision**

impulse the change in **momentum**

impulse–momentum principle states that the change in **momentum** of an object equals the **impulse** applied to it

inclined at an angle

inelastic collision a **collision** where the two colliding objects coalesce

inextensible doesn't stretch

instantaneously for an infinitely short length of time

integration the opposite of **differentiation**

intersection the point at which two or more lines or curves cross (intersect)

interval the period of time between two actions or events

joule unit of measurement of energy

kinematics branch of mathematics that deals with motion

kinetic energy (also **K.E.**) the energy a **body** possesses as a result of its motion of a particle

lamina a two-dimensional object whose thickness can be ignored

LHS left-hand side

magnitude size

mass the measure of how much matter is in an object

maximum (plural **maxima**) largest possible value

median a line from a **vertex** of a triangle to the **midpoint** of the opposite side

midpoint the point at equal distance from both ends of something, e.g. a line

minimum (plural **minima**) lowest possible value

model an attempt to describe a system using a set of variables and a set of equations that establish relationships between the variables

moment the turning effect of **force** on a **rigid body**

momentum the **mass** of an object multiplied by its **velocity**

motion the act of moving

negative less than zero

newton unit of **force**

Newton's law of restitution defines how the speeds of particles after a **collision** depend on the nature of the particles as well as their speeds before the collision

non-uniform a **body** with **mass** that is not evenly distributed

normal reaction the **force** exerted on (i.e. applied to) an object by the surface on which it rests

numerical value a quantity expressed in numbers

opposite the reverse of

parabola a U–shaped curve

parallel lines that are the same distance apart along their whole length

particle an object with negligible dimensions (i.e. dimensions too small to be significant)

perpendicular at a 90° angle to something

pivot the point on which an object, usually a **rod**, rests

plane surface

point of application the position where an action occurs

positive more than zero

power the rate of doing work

principle of conservation of momentum states that in a **collision** between two objects, the total momentum of the two objects before the collision is equal to the total **momentum** of the two objects after the collision

project to make something move up or forwards with great **force**

projectile an object that rises, then falls under the influence of gravity

range the **horizontal** distance travelled by a **projectile**

rate of change the ratio between a change in one variable in relation to a corresponding change in another variable

rebound to bounce back after hitting something

relative to when compared to

resistance opposition

resistive force a force that acts in the opposite direction to which a particle is moving/trying to move

resolve to split (a **vector**) into its components

resultant the sum of two or more quantities

rigid body a solid **body** in which any changes in state are too small to be significant

rod a thin straight object that does not bend

rough surface a surface that has **friction**

scalar a quantity with only **magnitude**

sector part of a circle

smooth without **friction**

speed measure of how quickly a **body** moves

speed of projection measure of how quickly a **body** moves at the instant of realease

static equilibrium at rest and in a state of **equilibrium**

stationary not moving

successive occurring one after another

suspend to hang from

suvat equations equations that deal with motion

symmetry the quality of having two halves that match exactly

tension the pulling **force** passed through a string or other continuous object

thrust the **force** acting on an object that is being pushed

time of flight the time from take off until landing

tilting having one side or end higher than the other

topple to become unsteady and fall over

uniform the same throughout, in all times and parts

validity how true something is

variable changing or likely to change

vector a quantity that has both magnitude and direction

velocity rate of change of **displacement**

vertex (plural **vertices**) the point at which two or more edges meet

vertical straight up and down

watt unit of measurement of power

weight the downward **force** on an object caused by **gravity**

work energy transfer that occurs when an object is moved over a distance by an external **force**

ANSWERS

CHAPTER 1

Prior knowledge check

1 **a** 11.5 m
b 3.1 s
2 $x = v\cos\theta$, $y = v\sin\theta$
3 **a** **i** $\cos\theta = \frac{12}{13}$ **ii** $\tan\theta = \frac{5}{12}$
b **i** $\sin\theta = \frac{8}{17}$ **ii** $\cos\theta = \frac{15}{17}$

Exercise 1A

1 **a** 122.5 m **b** 100 m
2 **a** $x = 36$ m, $y = 19.6$ m **b** 41 m
3 $u = 16.6\,\text{m s}^{-1}$
4 $77.5\,\text{m s}^{-1}$
5 0.59 s
6 1.9 m
7 **a** 0.5 s **b** 0.42 **c** 3.5 m

Exercise 1B

1 **a** $u_x = 19.2\,\text{m s}^{-1}$, $u_y = 16.1\,\text{m s}^{-1}$ **b** $(19.2\mathbf{i} + 16.1\mathbf{j})\,\text{m s}^{-1}$
2 **a** $u_x = 16.9\,\text{m s}^{-1}$, $u_y = -6.2\,\text{m s}^{-1}$ **b** $(16.9\mathbf{i} - 6.2\mathbf{j})\,\text{m s}^{-1}$
3 **a** $u_x = 32.3\,\text{m s}^{-1}$, $u_y = 13.5\,\text{m s}^{-1}$ **b** $(32.3\mathbf{i} + 13.5\mathbf{j})\,\text{m s}^{-1}$
4 **a** $u_x = 26.9\,\text{m s}^{-1}$, $u_y = -7.8\,\text{m s}^{-1}$ **b** $(26.9\mathbf{i} - 7.8\mathbf{j})\,\text{m s}^{-1}$
5 $10.8\,\text{m s}^{-1}$, 56.3°
6 $6.4\,\text{m s}^{-1}$, 51.3° below the horizontal
7 **a** 33.7° **b** $k = 3$ or $k = -3$

Exercise 1C

1 3.1 (2 s.f.)
2 8.5 m (2 s.f.)
3 **a** 44 m (2 s.f.) **b** 79 m
4 **a** 2.7 s (2 s.f.) **b** 790 m (2 s.f.)
5 **a** 10 m (2 s.f.) **b** 41 m (2 s.f.)
6 **a** 3.9 s (2 s.f.) **b** 56 m (2 s.f.)
7 55° (nearest degree)
8 **a** $(36\mathbf{i} + 27.9\mathbf{j})$ m **b** $13\,\text{m s}^{-1}$ (2 s.f.)
9 **a** 22° (2 s.f.) **b** 97 m (2 s.f.)
10 **a** 16 (2 s.f.) **b** 1.6 s (2 s.f.)
11 **a** 4.4 **b** 88 **c** 50° (2 s.f.)
12 **a** 1.1 s (2 s.f.) **b** 34 m (2 s.f.)
13 $\alpha = 40.6°$ (nearest 0.1°)
$U = 44$ (2 s.f.)
14 **a** $15.6\,\text{m s}^{-1}$ **b** 2.92 s **c** $22.3\,\text{m s}^{-1}$
15 **a** $k = 7.35$
b **i** $13.6\,\text{m s}^{-1}$ **ii** 72.9°
16 **a** $10.7\,\text{m s}^{-1}$ **b** e.g. weight of the ball; air resistance

Challenge

R(→): $s = 12t$ and $s = (20\cos\alpha)t$
so $\cos\alpha = 0.6$ and $\sin\alpha = 0.8$
R(↑): $s = -4.9t^2 + 40$ and $s = (20\sin\alpha)t - 4.9t^2$
So $t = \frac{40}{20\sin\alpha} = \frac{40}{16} = 2.5$ seconds

Exercise 1D

1 R(↑): $v^2 = U^2\sin^2\alpha - 2gh$
At maximum height, $v = 0$ so $0 = U^2\sin^2\alpha - 2gh$
Rearrange to give $h = \frac{U^2\sin^2\alpha}{2g}$
2 **a** R(→): $x = 21\cos\alpha \times t$, so $t = \frac{x}{21\cos\alpha}$
R(↑): $y = 21\sin\alpha \times \frac{x}{21\cos\alpha} - \frac{1}{2}g\left(\frac{x}{21\cos\alpha}\right)^2$
$y = x\tan\alpha - \frac{x^2}{90\cos^2\alpha}$
b $\tan\alpha = 1.25$
3 **a** R(↑): $s = U\sin\alpha t - \frac{g}{2}t^2$
When particle strikes plane, $s = 0 = t(U\sin\alpha - \frac{g}{2}t)$
So $t = 0$ or $t = \frac{2U\sin\alpha}{g}$
b R(→): $s = ut = U\cos\alpha\left(\frac{2U\sin\alpha}{g}\right) = \frac{U^2\sin 2\alpha}{g}$
c Range $s = \frac{U^2\sin 2\alpha}{g}$ is greatest when $\sin 2\alpha = 1$
Occurs when $2\alpha = 90° \Rightarrow \alpha = 45°$
d 12° and 78°
4 Using $v = u + at$, at max height $t = \frac{v}{g}$
So time taken to return to the ground $= \frac{v}{g}$
Using $s = ut + \frac{1}{2}at^2$, distance travelled by one part $= 2v\left(\frac{v}{g}\right) = \frac{2v^2}{g}$
So two parts of firework are $\frac{2v^2}{g} + \frac{2v^2}{g} = \frac{4v^2}{g}$ apart.
5 **a** R(→): $x = U\cos\alpha \times t$, so $t = \frac{x}{U\cos\alpha}$
R(↑): $y = U\sin\alpha \times t - \frac{1}{2}gt^2$
Substitute for $t \Rightarrow y = U\sin\alpha\left(\frac{x}{U\cos\alpha}\right) - \frac{1}{2}g\left(\frac{x}{U\cos\alpha}\right)^2$
Use $\tan\alpha = \frac{\sin\alpha}{\cos\alpha}$ and rearrange to give
$y = x\tan\alpha - \frac{gx^2}{2U^2\cos^2\alpha}$.
b 13.7 m
6 **a** R(→): $x = U\cos\alpha \times t$, so $t = \frac{x}{U\cos\alpha}$
R(↑): $y = U\sin\alpha \times t - \frac{1}{2}gt^2$
Substitute for $t \Rightarrow y = U\sin\alpha\left(\frac{x}{U\cos\alpha}\right) - \frac{1}{2}g\left(\frac{x}{U\cos\alpha}\right)^2$
Use $\tan\alpha = \frac{\sin\alpha}{\cos\alpha}$ and $\frac{1}{\cos\alpha} = \sec\alpha$, and rearrange to give $y = x\tan\alpha - \frac{gx^2}{2U^2}\sec^2\alpha$.
Use $\sec^2\alpha = 1 + \tan^2\alpha$, and rearrange to give
$y = x\tan\alpha - \frac{gx^2}{2U^2}(1 + \tan^2\alpha)$.
b 93.8 m **c** 4.4 s
7 **a** R(→): $x = 9 = U\cos\alpha \times t$, so $t = \frac{9}{U\cos\alpha}$
R(↑): $y = U\sin\alpha \times t - \frac{1}{2}gt^2$
Substitute for $t \Rightarrow y = U\sin\alpha\left(\frac{9}{U\cos\alpha}\right) - \frac{1}{2}g\left(\frac{9}{U\cos\alpha}\right)^2$
Use $\tan\alpha = \frac{\sin\alpha}{\cos\alpha}$ and $y = 0.9$. Rearrange to give
$0.9 = 9\tan\alpha - \frac{81g}{2U^2\cos^2\alpha}$.
b $10.3\,\text{m s}^{-1}$
8 **a** R(→): $x = kt$, so $t = \frac{x}{k}$
R(↑): $y = 2kt - \frac{gt^2}{2}$
Substitute for $t \Rightarrow y = 2x - \frac{gx^2}{2k^2}$
b **i** $\frac{4k^2}{g}$ m **ii** $\frac{2k^2}{g}$ m

Challenge

For the projectile: $y = x\tan\alpha - \frac{gx^2}{2U^2\cos^2\alpha}$ so for $\alpha = 45°$
$y = x - \frac{gx^2}{U^2}$
For the slope: $y = -x$
Projectile intersects the slope when $-x = x - \frac{gx^2}{U^2} \Rightarrow x = \frac{2U^2}{g}$,
$y = -\frac{2U^2}{g}$
Distance $= \sqrt{\left(\frac{2U^2}{g}\right)^2 + \left(\frac{2U^2}{g}\right)^2} = \sqrt{8\left(\frac{U^2}{g}\right)^2} = \frac{2\sqrt{2}U^2}{g}$

Chapter review 1

1 1.9 s
2 **a** 4.1 s (2 s.f.) **b** $40\,\text{m s}^{-1}$ (2 s.f.)
c air resistance

3 a i a = gradient of line. Using the formula for the gradient of a line, $a = \frac{v-u}{t}$, which can be rearranged to give $v = u + at$
ii s = area under the graph. Using the formula for the area of a trapezium, $s = \left(\frac{u+v}{2}\right)t$
b i Substitute $t = \frac{v-u}{a}$ into $s = \left(\frac{u+v}{2}\right)t$
ii Substitute $v = u + at$ into $s = \left(\frac{u+v}{2}\right)t$
iii Substitute $u = v - at$ into $s = \left(\frac{u+v}{2}\right)t$
4 $h = 39$ (2 s.f.)
5 $u = 8$
6 a 45 m b 6.1 s
7 $h = 35$ (2 s.f.)
8 a 36 m b 30 m (2 s.f.)
9 a 140 m (2 s.f.) b 36 m s^{-1} (2 s.f.)
10 a R(↑): $s = U\sin\theta\, t - \frac{g}{2}t^2$
When particle strikes plane, $s = 0 = t(U\sin\theta - \frac{g}{2}t)$
So $t = 0$ or $t = \frac{2U\sin\theta}{g}$
R(→): $s = Ut = U\cos\theta\left(\frac{2U\sin\theta}{g}\right) = \frac{U^2\sin 2\theta}{g}$
b $\frac{U^2}{g}$
c 20.9°, 69.1° (nearest 0.1°)
11 a 2.0 s (2 s.f.) b 3.1 s (2 s.f.) c 36 m s^{-1} (2 s.f.)
12 a $\lambda = \frac{132}{g}$
b i 20.8 m s^{-1} (3 s.f.) ii $\tan^{-1}\left(\frac{17}{20}\right)$ to the ground
13 a 2 m b 5.77° or 84.2°
14 a 0.65 s b 1.5 m c 23.8 m s^{-1}
15 a Particle P: $x = 18t$, Particle Q: $x = 30\cos\alpha\, t$
When particles collide: $18t = 30\cos\alpha\, t \Rightarrow \cos\alpha = \frac{3}{5}$
b $\frac{4}{3}$ s

Challenge
62 m s^{-1}

CHAPTER 2

Prior knowledge check

1 a $6x - 5$ b $\frac{1}{\sqrt{x}} - \frac{12}{x^3}$
2 a (1.5, −4.75) b (1, 9) and (3, 5)
3 a $\frac{5x^2}{2} + 8x + 1$ b $x^3 - x^2 + 5x + 7$
4 $(2\mathbf{i} - 5\mathbf{j})$ m s^{-1}

Exercise 2A

1 a 8 m b $t = 0$ and $t = \pm 3$
2 a 4 m b 6 m
3 a 7 m s^{-1} b 9.25 m s^{-1}
c −11 m s^{-1}; body is travelling in opposite direction.
4 a 0.8 m b 4 s
c 1.6 m d $0 \leqslant t \leqslant 4$
5 a 8 m s^{-1} b $t = \frac{4}{3}$ and $t = 2$
c $t = \frac{1}{3}$ and $t = 3$ d 8 m s^{-1}
6 a 4 s b 8 m s^{-1}
7 $T = 3$: returns to starting point and $s = 0$ when $t = 0$ and $t = 3$.
8 a $t = \frac{1}{3}$ and $t = 3$ b $\frac{16}{15}$ m s^{-1}

Exercise 2B

1 a i $v = 16t^3 + \frac{1}{t^2}$ ii $a = 48t^2 - \frac{2}{t^3}$
b i $v = 2t^2 - \frac{2}{t^3}$ ii $a = 4t + \frac{6}{t^4}$
c i $v = 18t^2 + 30t - 2$ ii $a = 36t + 30$
d i $v = \frac{9t^2}{2} - 2t - \frac{5}{2t^2}$ ii $a = 9t - 2 + \frac{5}{t^3}$
2 a 46 m s^{-1} b 24 m s^{-2}
3 7 m s^{-2} in the direction of x decreasing.
4 6.75 m
5 a $k = 4$ b $a = -4$ m s^{-2}
6 1.7 cm
7 a 0.25 s b 4.54 m c $v = -1.88$ m s^{-1}
8 a The body returns to its starting position 4 s after leaving it.
b Since $t \geqslant 0$, t^3 is always positive.
Since $t \leqslant 4$, $4 - t$ is always non-negative.
c 27 m
9 a

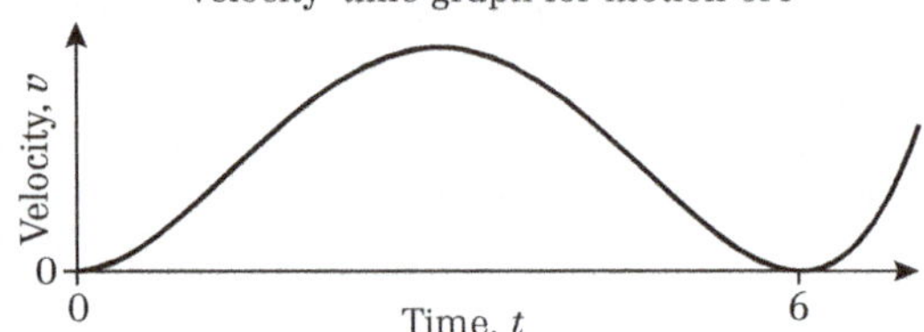

b $v = 81$ m s^{-1} when $t = 3$ s
10 a Discriminant of $2t^2 - 3t + 5$ is <0, so no solutions for $v = 0$
b 3.88 m s^{-1} (3 s.f.)
11 a

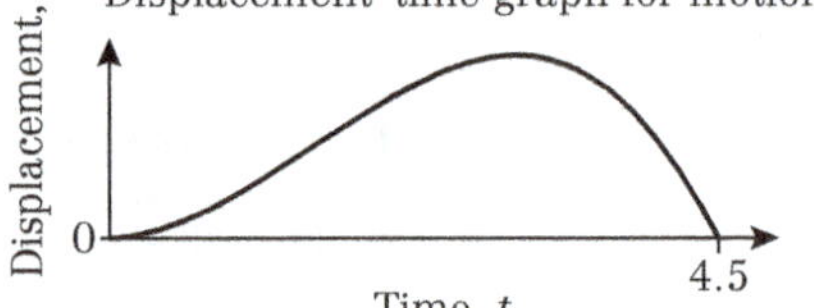

b s is a distance so cannot be negative.
c 13.5 m
d 9 m s^{-2}
12 Max distance is when $\frac{ds}{dt} = 3.6 + 3.52t - 0.06t^2 = 0$, so $t = 59.7$ (3 s.f.)
∴ Max distance = 2.23 km (3 s.f.), so the train never reaches the end of the track.
13 a $3t^{\frac{2}{3}} \geqslant 0$, $\frac{2}{e^{3t}} \geqslant 0$ b 1.18 m s^{-1}
c −0.15 m s^{-1} b −0.75 N
14 a 0.5 m s^{-1} b 0.1 m s^{-1}
15 a 12.9 m s^{-1} in the direction of s increasing
b 24 m s^{-1} in the direction of s decreasing
c 132 m
d 20.8 m and 118.5 m
16 a $k = 40$, $T = 25$ b 4 m s^{-1}
c $v = \frac{20}{\sqrt{t}}$, so for small t, the value of v is large
e.g. $t = 0.01$, $v = 200$ m s^{-1}, so not realistic for small t.
17 a $k = \frac{1}{2}$ b $t = \pi, 3\pi$
c $a = 4\cos\left(\frac{t}{2}\right)$, $4a^2 = 64\cos^2\left(\frac{t}{2}\right)$
$v = 2 + 8\sin\left(\frac{t}{2}\right)$, $(v-2)^2 = 64\sin^2\left(\frac{t}{2}\right)$
$4a^2 = 64 - (v-2)^2 \Rightarrow 64\cos^2\left(\frac{t}{2}\right) = 64 - 64\sin^2\left(\frac{t}{2}\right)$
$\Rightarrow \cos^2\left(\frac{t}{2}\right) = 1 - \sin^2\left(\frac{t}{2}\right)$
$\Rightarrow \cos^2 T + \sin^2 T = 1$
d 10 m s^{-1}, 4 m s^{-2}

Exercise 2C

1 a $s = t^3 - t$ b $s = \frac{t^4}{2} - \frac{t^3}{2}$ c $s = \frac{4}{3}t^{\frac{3}{2}} + \frac{4t^3}{3}$
2 a $v = 4t^2 - \frac{2t^3}{3}$ b $v = 6t + \frac{t^3}{9}$
3 12 m
4 a $v = 6 + 16t - t^2$ b −6
5 42.9 m (3 s.f.)
6 12.375 m
7 a $10\frac{2}{3}$ b 13 m
8 $t = \frac{3}{2}$ and $t = 5$
9 a $t = 1$ and $t = 5$ b 6 m
10 $T = 1.5$ s
11 a $v = \frac{t^2}{2} - 3t + 4$ b $t = 2$ and $t = 4$ c $\frac{2}{3}$ m
12 a 86 m b 60 m s^{-1}
13 a $v = t + \frac{\cos\pi t}{\pi} - \frac{1}{\pi}$ b $s = \frac{t^2}{2} + \frac{\sin\pi t}{\pi^2} - \frac{t}{\pi}$
14 a $v = -\frac{\cos 3\pi t}{3\pi} + \frac{2}{3\pi}$ b $\frac{1}{\pi}$
c $s = -\frac{\sin 3\pi t}{9\pi^2} + \frac{2t}{3\pi} + 1$
15 a $v = -\frac{\sin 4\pi t}{4\pi}$ b $\frac{1}{4\pi}$
c $s = \frac{\cos 4\pi t}{16\pi^2} - \frac{1}{16\pi^2}$ d $\frac{1}{8\pi^2}$ e 16
16 3.31 s
17 a 24 m s^{-1} b 54.4 m c 8.43 s d 101.2 m

Challenge

$\frac{200}{3}$ m

Exercise 2D

1 a $(3\mathbf{i} + 23\mathbf{j})$ m s^{-1} b $18\mathbf{j}$ m s^{-2}
2 $(24\mathbf{i} + 6\mathbf{j})$ mN
3 a 0.305 s b 6 m s^{-1}
c **i**-component of velocity is negative, **j**-component of velocity = 0
4 a 20 m s^{-1} b 10 m s^{-2}
b $a = 8\mathbf{i} - 6\mathbf{j}$, no dependency on t therefore constant. $|a| = 10$ m s^{-2}
5 a $6\sqrt{5}$ m s^{-1} b $t = 2$
c $(-16\mathbf{i} + 4\mathbf{j})$ m d 15.2 N (3 s.f.)
6 a $k = -0.5, -8.5$
b 10 m s^{-2} for both values of k
7 a 52 m s^{-1} b $(12\mathbf{i} + \frac{15}{2}\mathbf{j})$ m s^{-2}
8 a 4 b $(-36\mathbf{i} + 8\mathbf{j})$ m s^{-2}
9 $\mathbf{a} = 4\mathbf{i} + 2\mathbf{j}$, no t dependency so constant. $|\mathbf{a}| = 2\sqrt{5}$ m s^{-2}
10 a $\sqrt{15}$ b $(-4\sqrt{30}\mathbf{i} + 2\sqrt{15}\mathbf{j})$ m s^{-2}

Exercise 2E

1 a $16\mathbf{i}$ m s^{-1} b $128\mathbf{i} - 192\mathbf{j}$ m
2 a 13 m b $(4\mathbf{i} + 6\mathbf{j})$ m s^{-2}
3 a $\mathbf{v} = \left(t^2 - 4t + 2\pi - \frac{\pi^2}{4}\right)\mathbf{i} - 6\cos t\mathbf{j}$
b $2\pi^2 - 4\pi$
4 a $\left(\left(\frac{5t^2}{2} - 3t + 2\right)\mathbf{i} + \left(8t - \frac{t^2}{2} - 5\right)\mathbf{j}\right)$ m s^{-1}
b $t = \frac{1}{2}$ c $\frac{9\sqrt{2}}{8}$ m s^{-1}
5 a $\left((t^2 + 6)\mathbf{i} + \left(2t - \frac{t^3}{3}\right)\mathbf{j}\right)$ m b $6\mathbf{j}$ m
6 a $(-\mathbf{i} - \frac{1}{4}\mathbf{j})$ m s^{-1} b $(94\mathbf{i} - 654\mathbf{j})$ m
7 a $((2t^4 - 3t^2 - 4)\mathbf{i} + (4t^2 - 3t - 7)\mathbf{j})$ m s^{-1}
b $t = \frac{7}{4}$
8 a $\mathbf{r} = (2t^2 - 3t + 1)\mathbf{i} + (4t + 2)\mathbf{j}$ m
b i 3.4 ii $\mathbf{r} = 36\mathbf{i} + 22\mathbf{j}$

Challenge

$\mathbf{r} = \left(\frac{3\pi}{2} + 1\right)\mathbf{i} + \left(\frac{5\pi^2}{8} + 1\right)\mathbf{j}$

Exercise 2F

1 $v = \int a\,dt = at + c$
$a \times 0 + c = 0 \Rightarrow c = 0 \Rightarrow v = at$
$s = \int v\,dt = \int at\,dt = \frac{1}{2}at^2 + k$
$\frac{1}{2}a \times 0^2 + k = x \Rightarrow k = x$
so $s = \frac{1}{2}at^2 + x$
2 a $a = 5$, $v = \int 5\,dt = 5t + c$; when $t = 0$, $u = 12$ so $c = 12$, $v = 12 + 5t$
b $s = \int 12 + 5t\,dt = 12t + \frac{5t^2}{2} + d$, when $t = 0$, $s = 7$ so $d = 7$, $s = 12t + 2.5t^2 + 7$
3 $v = \frac{ds}{dt} = u + at$; $\frac{dv}{dt} = a$ so constant acceleration a
4 A $a = 4 - 6t$, not constant
B $a = 0$. no acceleration
C $a = \frac{1}{2}$, constant
D $a = -\frac{12}{t^4}$ not constant
E $v = 0$, particle stationary
5 a 4 m s^{-2}
b $p = 2$, $q = 5$, $r = 0$
6 a 680 m
b $\frac{ds}{dt} = 25 - 0.4t \Rightarrow \frac{d^2s}{dt^2} = -0.4 \therefore a$ is constant
c 420 m from A

Chapter review 2

1 a $t = 5$ b 37.5 m
2 a 30 m s^{-2} b 75 m
3 a Displacement $= 8t + t^2 - \frac{t^3}{3}$
b Max displacement when $t = 4$, $s = 26\frac{2}{3}$ m, which is less than 30 m so P does not reach B.
c $t = 6.62$ s
d $53\frac{1}{3}$ m
4 a $\frac{32}{3}$ m s^{-1} b $\frac{40}{3}$ m
5 a $(-3t^2 + 22t - 24)$ m s^{-1} b $t = \frac{4}{3}$ and $t = 6$
c $t = \frac{11}{3}$
d

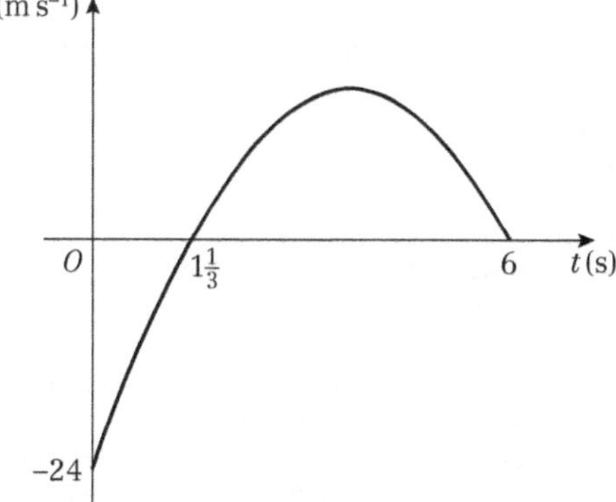

e $0 \leq t < 0.38$, $\frac{10}{3} < t < 4$
6 a $t = \frac{5}{3}$ and $t = 2$ b 13 m s^{-2} c $\frac{433}{27}$ m
7 a $v = \frac{t^4}{2} - 4t^2 + 6$ b $s = \frac{t^5}{10} - \frac{4t^3}{3} + 6t$
c $t = \sqrt{2}$ and $t = \sqrt{6}$

8 max = 8.64 m, min = 1.14 m
9 $a = 1500$, $b = 800$, $c = -16$
10 a $v = \int(20 - 6t)\,dt = 20t - 3t^2 + c$
At $t = 0$, $v = 7$ so $c = 7$ and $v = 7 + 20t - 3t^2$
b The greatest speed is $40\frac{1}{3}$ m s^{-1}
c 196 m
11 $v = \int k(7 - t^2)dt \Rightarrow v = k\left(7t - \frac{t^3}{3}\right) + c$
$t = 0, v = 0 \therefore c = 0$; $t = 3, v = 6 \therefore k = \frac{1}{2}$
$v = \frac{7}{2}t - \frac{t^3}{6}$
$s = \int v\,dt = \int\left(\frac{7}{2}t - \frac{t^3}{6}\right)dt = \frac{7t^2}{4} - \frac{t^4}{24} + c$
$t = 0, s = 0 \therefore c = 0$
$s = \frac{7t^2}{4} - \frac{t^4}{24} = \frac{1}{24}t^2(42 - t^2)$
12 a Time cannot be negative so $t \geqslant 0$
at $t = 5$ $s = 0$ so mouse has returned to its hole.
b 39.1 m
13 a Mass is not constant as fuel is used.
Gravity is not constant so weight not constant.
Thrust may not be constant.
b $v = (1.69 \times 10^{-7})t^4 - (1.33 \times 10^{-4})t^3 + 0.0525t^2 + 0.859t + 274$ m s^{-1}
c $v = 5990$ m s^{-1}
d 510 seconds (2 s.f.) after launch
14 a V_P is $(6t\mathbf{i} + 2\mathbf{j})$ m s^{-1} and V_Q is $(\mathbf{i} + 3t\mathbf{j})$ m s^{-1}
b 12.2 m s^{-1} (3 s.f.)
c $t = \frac{1}{3}$
d Equate **i**-components and solve to get $t = 1$.
Equate **j**-components and solve to get $t = \frac{1}{3}$ or 1.
So $t = 1$ and $r = (7\mathbf{i} + \frac{3}{2}\mathbf{j})$ m
15 a Differentiate: $\mathbf{v} = 6t\mathbf{i} - 8t\mathbf{j}$, $\mathbf{a} = 6\mathbf{i} - 8\mathbf{j}$ so constant
b 10 m s^{-2}
143.1° (nearest 0.1°)
16 a $-6\mathbf{i}$ m s^{-1}
b Differentiate: $\mathbf{v} = -6\sin 3t\mathbf{i} - 6\cos 3t\mathbf{j}$,
$\mathbf{a} = -18\cos 3t\mathbf{i} + 18\sin 3t\mathbf{j}$
$|\mathbf{a}| = \sqrt{18^2(\cos^2 t + \sin^2 t)} = 18$ m s^{-2} so constant
17 a Differentiate: $\mathbf{a} = 4c\mathbf{i} + (14 - 2c)t\mathbf{j}$,
$\mathbf{F} = m\mathbf{a} = \frac{1}{2}(4c\mathbf{i} + (14 - 2c)t\mathbf{j})$
b $4, \frac{234}{29} \approx 8.07$
18 a $((2t^4 - 3t^2 - 4)\mathbf{i} + (4t^2 - 3t - 7)\mathbf{j})$ m s^{-1}
b $t = \frac{7}{4}$
19 $10\sqrt{41}$ m s^{-1}
20 a $\mathbf{v} = (2t^2 + 3)\mathbf{i} + (3t + 13)\mathbf{j}$ b 3.11 s (3 s.f.)

Challenge
1 32.75 m
2 91 m s^{-1}
3 a 20
b $v = 0$ when $t = 1.64$ s (or -2.44 s) so only changes direction once
c $-2\sqrt{20}(\sqrt{20} + 1)^{\frac{1}{2}}$
4 a Differentiate: $\dot{\mathbf{r}} = (6\omega\cos\omega t)\mathbf{i} - (4\omega\sin\omega t)\mathbf{j}$
$|\dot{\mathbf{r}}|^2 = (36\omega^2\cos^2\omega t)\mathbf{i} + (16\omega^2\sin^2\omega t)\mathbf{j}$
use $\sin^2\omega t + \cos^2\omega t = 1$ and $2\cos^2\omega t = \cos 2\omega t + 1$
b $v = \sqrt{26\omega^2 + 10\omega^2\cos 2\omega t}$ max when $\cos 2\omega t = 1$ and min when $\cos 2\omega t = -1$
c 109.8° (1 d.p.)

CHAPTER 3

Prior knowledge check
1 $x = 3$, $y = 2$
2 3.6 m
3 144 cm^2

Exercise 3A
1 (3.2, 0)
2 (0, 2.5)
3 (1.1, 0)
4 $2\frac{1}{3}$ m
5 $m = 6$
6 0.7 kg
7 6.5
8 (0, −2)
9 $m_1 = 2$, $m_2 = 3$
10 $1 \times (2m + 5) = ((m - 1) \times -1) + ((5 - m) \times 1) + (2 \times m) + ((m + 1) \times 0)$
$2m + 5 = 6$
$2m = 1$
$m = 0.5$ kg

Challenge
$(2 \times PQ) + \left(3 \times \left(PQ + \frac{3}{2}PQ\right)\right) = (6 \times PG)$
$\frac{19}{2}PQ = 6PG$
$19PQ = 12PG$
$PQ : PG$ is $12 : 19$

Exercise 3B
1 (3, 2)
2 (0.5, −0.75)
3 (4.6, 4.2)
4 3i + 2.5j
5 (2.1, 0.3)
6 a 1 b 1.5
7 $p = 1$, $q = -2$
8 (1, 3)
9 $3\frac{3}{4}$ cm from AB and $4\frac{1}{3}$ cm from AD.
10 a 3 g b 3.2 cm

Challenge
0.2 kg

Exercise 3C
1 a (2, 3) b (3, 4)
c $\left(\frac{1}{3}, 1\right)$ d $\left(\frac{8a}{3}, 3a\right)$
2 Centre of mass is on the axis of symmetry at a distance $\frac{16}{3\pi}$ cm from the centre.
3 $a = 3$, $b = 3$
4 a Distance a from AB, distance $\frac{2a}{3}$ from BC.
b Distance $\frac{a}{3}$ from BC, distance $\frac{4a}{3}$ from AB.
c On the line of symmetry, $\frac{4a}{3}$ from the line AB.
d $\left(3a, \frac{2a}{3}\right)$ with A as the origin and AC as the x-axis.
5 (2, 7) and (−13, 4)
6 a B has coordinates $\left(\frac{18}{5}, \frac{41}{5}\right)$, D has coordinates $\left(\frac{12}{5}, -\frac{1}{5}\right)$
b (3, 4)
7 a (3, 5), (3, −3)

Online Worked solutions are available in SolutionBank.

b $\left(3, \frac{7}{3}\right), \left(3, -\frac{1}{3}\right)$

8 $\bar{y} = \frac{y}{3} = y - \frac{4\sqrt{3}}{3}$

$\frac{2}{3}y = \frac{4\sqrt{3}}{3}$

$y = 2\sqrt{3}$

Centre of mass is at $x = 2$

So, using Pythagoras:

$AB^2 = BC^2 = \sqrt{2^2 + (2\sqrt{3})^2}$

$AB = BC = 4$

So the triangle is equilateral.

Exercise 3D

1 a $\left(\frac{5}{2}, \frac{13}{14}\right)$ **b** $(1.7, 2.6)$ **c** $\left(\frac{113}{30}, \frac{49}{30}\right)$ **d** $\left(\frac{7}{3}, 2\right)$

e $\left(\frac{32}{9}, \frac{53}{18}\right)$ **f** $\left(2, \frac{63}{17}\right)$ **g** $\left(\frac{25}{8}, 3\right)$

2 $\frac{2}{9}a$

3 $2.89a$ (3 s.f.)

4 a 10 cm

b 12 cm horizontally to the right of A and 4.5 cm vertically above A.

5 a 4.5 cm horizontally to the right of O, and 3.5 cm vertically above O.

b 5.83 cm horizontally to right of O, and 4.32 cm vertically above O.

6 a $12\begin{pmatrix}1\\3\end{pmatrix} + 24\begin{pmatrix}5\\4\end{pmatrix} + 16\begin{pmatrix}9\\2\end{pmatrix} = 52\begin{pmatrix}\bar{x}\\\bar{y}\end{pmatrix}$

$\begin{pmatrix}\bar{x}\\\bar{y}\end{pmatrix} = \begin{pmatrix}\frac{69}{13}\\\frac{41}{13}\end{pmatrix}$

$\left(\frac{69}{13}, \frac{41}{13}\right)$

b Since $\bar{y} = \frac{41}{13}$ for original plate, holes must be symmetrically placed about the line $y = \frac{41}{13}$

c $\frac{70}{9}$

7 $x = a$

Challenge

Area of hexagon $ABCDEF = \frac{3\sqrt{3}}{2}x^2$

Using Pythagoras, find the height of the midpoint:

$x^2 = \left(\frac{x}{2}\right)^2 + h^2 \Rightarrow h = \frac{\sqrt{3}}{2}$

So the centre of mass of $ABCDEF$ is $\begin{pmatrix}\frac{x}{2}\\\frac{\sqrt{3}x}{2}\end{pmatrix}$

Area of triangle to be removed $= \frac{1}{2} \times \sqrt{3}x \times \frac{x}{2} = \frac{\sqrt{3}}{4}x^2$

So the centre of mass of DEF is $\begin{pmatrix}x + \frac{x}{6}\\\frac{\sqrt{3}x}{2}\end{pmatrix}$

$\bar{y} = \frac{\sqrt{3}x}{2}$ due to symmetry

$\frac{3\sqrt{3}}{2}x^2\begin{pmatrix}\frac{x}{2}\\\frac{\sqrt{3}x}{2}\end{pmatrix} - \frac{\sqrt{3}}{4}\begin{pmatrix}\frac{7x}{6}\\\frac{\sqrt{3}x}{2}\end{pmatrix} = \frac{5\sqrt{3}}{4}\begin{pmatrix}\bar{x}\\\frac{\sqrt{3}x}{2}\end{pmatrix}$

$\bar{x} = \frac{11x}{30}$, so $N\left(\frac{11x}{30}, \frac{\sqrt{3}x}{2}\right)$

$MN = \frac{x}{2} - \frac{11x}{30} = \frac{2}{15}x$

Exercise 3E

1 a $\left(\frac{43}{16}, \frac{21}{16}\right)$ **b** $\left(\frac{53}{18}, \frac{20}{9}\right)$

c (2.67, 2.26) (3 s.f.) **d** (3.73, 3.00) (3 s.f.)

2 Centre of mass is on line of symmetry through O, and a distance of $\frac{9(\sqrt{3} + 2)}{6 + \pi}$ from O.

3 $\frac{11}{7}a$

4 a 5.83 cm (3 s.f.)

b i 817 g (3 s.f.) **ii** 4.41 cm (3 s.f.)

5 $\frac{12}{5}$ m horizontally to the right of A and $\frac{19}{10}$ m vertically below A.

6 Centre of mass is on the line of symmetry at a distance of $\frac{3}{2\pi}$ below the line AB.

7 a 1.75 m

b $(20 \times 1.75) - (1 \times 0.5) = 19\bar{y} \Rightarrow \bar{y} = \frac{69}{38}$

$\Delta y = \frac{69}{38} - 1.75 = \frac{5}{76}$

Challenge

$\frac{1}{5}\sqrt{5}$ cm

Exercise 3F

1 a 20.4° (3 s.f.) **b** 24.4° (3 s.f.) **c** 56.8° (3 s.f.)

2 63.0° (3 s.f.)

3 80.5° (3 s.f.)

4 a $\frac{26}{7}$ cm

b $\frac{18}{7}$ cm

c $\alpha = 22.2°$ (3 s.f.)

5 67.1° (3 s.f.)

6 65.3° (3 s.f.)

7 a 1.5 cm **b** 1.5 cm **c** 45°

8 $\theta = \tan^{-1}\left(\frac{5}{16}\right)$

Challenge

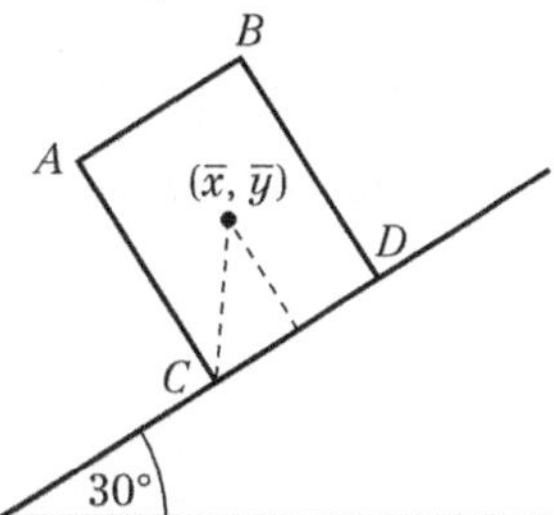

Take C as (0, 0)

$M\begin{pmatrix}2\\3\end{pmatrix} + kM\begin{pmatrix}0\\6\end{pmatrix} = M(1 + k)\begin{pmatrix}\bar{x}\\\bar{y}\end{pmatrix}$

where $(\bar{x}, \bar{y})$ is the centre of mass of the lamina and mass

$\begin{pmatrix}(1 + k)\bar{x}\\(1 + k)\bar{y}\end{pmatrix} = \begin{pmatrix}2\\3 + 6k\end{pmatrix}$

So $\bar{x} = \frac{2}{1 + k}$ and $\bar{y} = \frac{3 + 6k}{1 + k}$

$\tan 30° = \frac{\bar{x}}{\bar{y}}$

so $\bar{y} = \sqrt{3} \times \bar{x}$

$\frac{3 + 6k}{1 + k} = \sqrt{3}\left(\frac{2}{1 + k}\right)$

$3 + 6k = 2\sqrt{3}$

$6k = 2\sqrt{3} - 3$

$k = \frac{\sqrt{3}}{3} - \frac{1}{2}$

$k = \frac{1}{\sqrt{3}} - \frac{1}{2}$ as required

Exercise 3G

1 21.8° (3 s.f.)
2 104° (3 s.f.)
3 a $x = 8$ cm b 0.2
4 30.6° (3 s.f.)
5 28.5° (3 s.f.)
6 19.4°
7 a $T_1 = \frac{5}{8}W$, $T_2 = \frac{3}{8}W$ b 78.7° (3 s.f.)
8 a $T_1 = \frac{7}{18}W$, $T_2 = \frac{11}{9}W$ b 25.5° (3 s.f.)

Challenge
$\theta = 26.6°$ (1 d.p.)

Exercise 3H

1 $\frac{7}{3}$ cm horizontally to the right of AD and 2 cm vertically below AB.
2 2.4 cm vertically below AB and 5.6 cm to the right of (original) AD.
3 20.1° (3 s.f.)
4 $\left(\frac{35}{11}, \frac{12}{11}\right)$
5 (2.89, 1.31) (3 s.f.)
6 50.2° (3 s.f.)
7 a 0.07 m
  b C, as the centre of mass of the composite lamina is between O and C.
  c At B: $17.2g$; At C: $2.8g$
8 a 61.2° (3 s.f.) b 27.6° (3 s.f.)

Challenge
a Let width = 2, then height = $\frac{1}{\tan 22.5°} = \frac{1}{\sqrt{2} - 1} = \sqrt{2} + 1$
b On central axis, a distance $0.54x$ from the bottom edge of the paper.

Chapter review 3

1 a 0.413 m (3 s.f.)
  b 12° (nearest degree)
  c 0.275 (3 s.f.)
2 $\theta = 36.9°$ (3 s.f.)
3 $\left(-\frac{1}{7}, \frac{3}{2}\right)$
4 a $1.7a$ b $1.1a$
5 a 39.0° (3 s.f.) b 7.8° (3 s.f.)
6 a 2.5 cm
  b Take F as (0, 0) from part **a**.
  $\begin{pmatrix}\bar{x}\\ \bar{y}\end{pmatrix} = \begin{pmatrix}1\\ 2.5\end{pmatrix}$
  The lamina will be on the point of toppling when the centre of mass lies directly above F.
  so $\tan\theta = \left(\frac{1}{2.5}\right)$
  and $\theta = \tan^{-1}\left(\frac{2}{5}\right)$ as required
7 Take F as (0, 0) the mass is attached at E (2, 0).
The centre of mass of the mass and lamina is given by
$M\begin{pmatrix}1\\ 2.5\end{pmatrix} + M\begin{pmatrix}2\\ 0\end{pmatrix} = 2M\begin{pmatrix}\bar{x}\\ \bar{y}\end{pmatrix}$
$\begin{pmatrix}\bar{x}\\ \bar{y}\end{pmatrix} = \frac{1}{2}\begin{pmatrix}3\\ 2.5\end{pmatrix}$
$\begin{pmatrix}\bar{x}\\ \bar{y}\end{pmatrix} = \begin{pmatrix}1.5\\ 1.25\end{pmatrix}$
The lamina will be on the point of toppling when the centre of mass lies directly above F.
so $\tan\theta = \left(\frac{1.5}{1.25}\right)$
and $\theta = \tan^{-1}\left(\frac{6}{5}\right)$ as required
8 $AB = 0.25M$, $BC = \frac{2}{3}M$, $CD = 0.5M$, $AD = M$
$0.25M\begin{pmatrix}3\\ 3\end{pmatrix} + \frac{2}{3}M\begin{pmatrix}5\\ 5\end{pmatrix} + 0.5M\begin{pmatrix}8\\ 3\end{pmatrix} + M\begin{pmatrix}6\\ 1\end{pmatrix} = \frac{29}{12}M\begin{pmatrix}\bar{x}\\ \bar{y}\end{pmatrix}$
$\begin{pmatrix}\bar{x}\\ \bar{y}\end{pmatrix} = \begin{pmatrix}\frac{169}{29}\\ \frac{79}{29}\end{pmatrix}$
$\tan\theta = \frac{\left(\frac{169}{29} - 3\right)}{\left(5 - \frac{79}{29}\right)} = \frac{\left(\frac{82}{29}\right)}{\left(\frac{66}{29}\right)} = \frac{41}{33}$
$\theta = \arctan\left(\frac{41}{33}\right)$
9 a $T_1 = 1.2Mg$ and $T_2 = 0.8Mg$ b 45°
10 a $T_1 = Mg$ and $T_2 = 2Mg$ b 47.3° (3 s.f.)

Challenge
Lamina does not topple as critical value is 24.4° (3 s.f.)

Review exercise 1

1 $\theta = \tan^{-1}\left(\frac{1}{7}\right)$
2 3
3 a Ball will momentarily be at rest 25.6 m above A.
  $0^2 = u^2 + 2 \times 9.8 \times 25.6$, $u = 22.4$
  b 4.64 (3 s.f.)
  c 6380 (3 s.f.)
  d

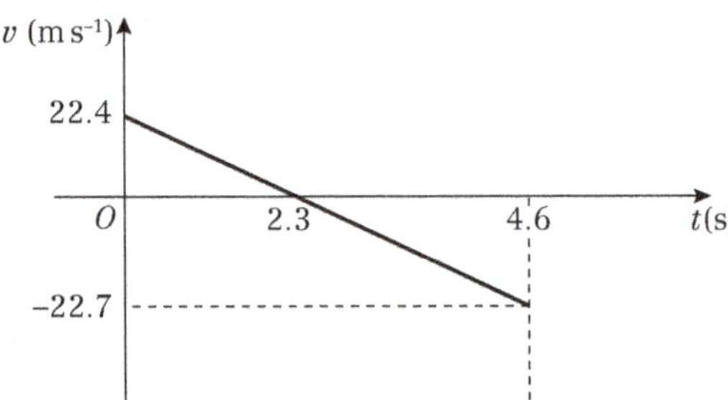

  e Consider air resistance due to motion under gravity.
4 a 0.404 s (3 s.f.) b 0.808 m (3 s.f.)
5 a 19.8 ms^{-1} (3 s.f.)
  b The ball as a projectile has negligible size and is subject to negligible air resistance.
  Free fall acceleration remains constant during flight of ball.
6 a 2.7 s (2 s.f.) b 790 m (2 s.f.)
7 a $u = 3.6$ (2 s.f.) b $k = 86$ (2 s.f.)
  c 43° (2 s.f.)
8 a $u_y = u\sin\alpha$, $s_y = 0$, $a = -g$
  Using $s = ut + \frac{1}{2}at^2$: $0 = ut\sin\alpha - \frac{1}{2}gt^2$
  $\frac{1}{2}gt = u\sin\alpha$ so $t = \frac{2u\sin\alpha}{g}$
  b $u_x = u\cos\alpha$, $s_x = R$, $a = 0$, $t = \frac{2u\sin\alpha}{g}$
  Using $s = ut + \frac{1}{2}at^2$:
  $R = ut\cos\alpha = \frac{2u^2\sin\alpha\cos\alpha}{g} = \frac{u^2\sin 2\alpha}{g}$
  c $R = \frac{u^2\sin 2\alpha}{g}$ so $\frac{dR}{d\alpha} = \frac{2u^2\cos 2\alpha}{g}$
  At a maximum $\frac{dR}{d\alpha} = 0$ so $\cos 2\alpha = 0$ and $\alpha = 45°$

d 12° and 78° (nearest degree)
9 6 s
10 a $4.5\,\text{m s}^{-1}$ **b** 4.5 s
11 a $t = 1$ and $t = \frac{5}{3}$ **b** $16\,\text{m s}^{-2}$ **c** 4 m
12 a $6 - 3t^{\frac{1}{2}}$ **b** $3t^2 - \frac{4}{5}t^{\frac{5}{2}}$
13 a $(t^2 + 2)\mathbf{i} + t\mathbf{j}$ **b** $27.5\,\text{m s}^{-1}$ (3 s.f.)
c $4.12\,\text{m s}^{-2}$ at an angle of 14° to the positive unit **i** vector
14 a $(24\mathbf{i} + 12\mathbf{j})\,\text{m s}^{-1}$ **b** $\frac{\mathrm{d}^2r}{\mathrm{d}t^2} = 8\mathbf{i} + 4\mathbf{j}$
15 24 m
16 a $\mathbf{v} = \left(t^2 - \frac{3}{4}t^4 + 3\right)\mathbf{i} - (4t^2 + 4t - 1)\mathbf{j}$
b 0.207 s (3 s.f.)
17 a $\mathbf{v} = (-2t^2 + 8)\mathbf{i} - 2t\mathbf{j}$ **b** $t = 2\,\text{s}$
18 a Taking moments about the x-axis:
$2(8 + \lambda) = 3 \times 4 + 5 \times 0 + 4\lambda$ so $\lambda = 2$
b $k - 1.1$
19 $x = y = \frac{1}{2}$
20 $(3\mathbf{i} + 2.5\mathbf{j})\,\text{m}$
21 a The total mass is $2M + M + kM = (3 + k)M$
$\text{M}(Oy)$: $(3 + k)M \times 3 = 2M \times 6 + M \times 0 + kM \times 2$
$\Rightarrow 9 + 3k = 12 + 2k \Rightarrow k = 3$
b $c = -\frac{1}{3}$
22 a 10.7 cm (3 s.f.) **b** 25° (nearest degree)
23 a The area of rectangle $ABDE$ is $6a \times 8a = 48a^2$
The area of $\triangle BCD$ is $\frac{1}{2} \times 6a \times 4a = 12a^2$
The area of lamina $ABCDE$ is $48a^2 + 12a^2 = 60a^2$

Shape	Lamina	Rectangle	Triangle
Mass ratios	$60a^2$	$48a^2$	$12a^2$
Displacement from X	GX	$4a$	$-\frac{4}{3}a$

$\text{M}(X)$: $60a^2 \times GX = 48a^2 \times 4a + 12a^2 \times \left(-\frac{4}{3}a\right)$
$= 192a^3 - 16a^3 = 176a^3$
$GX = \frac{176a^3}{60a^2} = \frac{44}{15}a$
b $\frac{11}{15}$
24 a $\frac{6l}{5}$ **b** l **c** 51° (nearest degree)
25 a $\frac{19}{15}a$ **b** $\frac{7}{45}M$
26 a 6 cm **b** 22.6° (1 d.p.)
27 a 6.86 cm (2 d.p.) **b** 32.1° (1 d.p.)
28 a Let the distances of the centre of mass of L, say G, from AD and AB be $\bar{x}$ and $\bar{y}$ respectively.
The mass of L is $3m + 4m + m + 2m = 10m$

Shape	L	$ABCD$	A	B	C
Mass	$10m$	$3m$	$4m$	m	$2m$
Distances (x)	$\bar{x}$	$2.5a$	0	$5a$	$5a$
Distances (y)	$\bar{y}$	a	0	0	$2a$

b $0.7a$ **c** 20° (nearest degree)
d $\text{M}(O)$: $P \times 2a = 10mg \times (2.5a - \bar{x}) = 10mg \times 0.25a$
$\Rightarrow P = \frac{2.5mga}{2a} = \frac{5}{4}mg$
e $\frac{5\sqrt{65}}{4}mg$
29 a i $\frac{5}{2}a$ **ii** $\frac{4}{3}a$ **b** 15° (nearest degree)
30 a i $\frac{5l}{12}$ **ii** $\frac{l}{3}$ **b** 15° (nearest degree)
31 a 3.28 cm (3 s.f.) **b** 0.211 (3 s.f.)

Challenge
1 a 43.8 cm (3 s.f.)
b $T_1 = 0.452\,\text{N}$ (3 s.f.), $T_2 = 0.548\,\text{N}$ (3 s.f.)
c $k = 7.45$ (3 s.f.)
2 Max distance: 12.3 m, $t = 323.13$
3 $\text{R}(\rightarrow)$: $d\cos\theta = u\sin\theta \times t \Rightarrow t = \frac{d\cos\theta}{u\sin\theta}$
$\text{R}(\uparrow)$: $-d\sin\theta = u\cos\theta \times t - \frac{g}{2}t^2$
Substitute for t.
Rearrange to show that $d = \frac{2u^2}{g}\tan\theta\sec\theta$

CHAPTER 4

Prior knowledge check

1 a 107 N (3 s.f.) **b** $3.22\,\text{m s}^{-2}$ (3 s.f.)
c 40.2 m (3 s.f.)
2 0.58 (2 d.p.)

Exercise 4A

1 2.52 J **2** 8.5 N **3** 24.0 J (3 s.f.)
4 588 J **5** 330 J **6** 73.5 J
7 38.3 m (3 s.f.)
8 a 228 J (3 s.f.)
b Assumption that there is no frictional force between sled and ice, reasonable assumption as coefficient of friction with ice will be very low.
9 0.255 (3 s.f.)
10 64.7 J (3 s.f.)
11 23 400 J (3 s.f.)
12 281 J (3 s.f.)
13 2.48 J (3 s.f.)
14 a 21.7 N (3 s.f.) **b** 326 J (3 s.f.) **c** 452 J (3 s.f.)
15 0.559 (3 s.f.)
16 112 J (3 s.f.)
17 a 35.3 J (3 s.f.) **b** 16.5 J (3 s.f.)
c $7.19\,\text{m s}^{-1}$ (3 s.f.)

Exercise 4B

1 a 33.8 J (3 s.f.) **b** 6 J **c** 500 J
d 200 J **e** 160 000 J
Order: **e**, **c**, **d**, **a**, **b**
2 a 44.1 J, gain **b** 8085 J, gain
c 22 050 J, loss **d** 34 104 J, loss
3 76.8 J **4** 168 750 J **5** 8
6 $4.53\,\text{m s}^{-1}$ (3 s.f.)
7 a 728 J (3 s.f.)
b No air resistance. Valid for low speeds but not large speeds.
8 a 11.8 J (3 s.f.) **b** 4.85 J (3 s.f.)
9 384 000 J (3 s.f.)
10 a 65 625 J **b** 1 837 500 J
11 a 20.0 m (3 s.f.)

Challenge
1 a K.E. = $48.0t^2$, P.E. = $-48.0t^2$, measured from the top of the cliff.
b K.E. + P.E. = $4.9t^2 - 4.9t^2 = 0$

Exercise 4C

1 a 27.4 J (3 s.f.) **b** $11.7\,\text{m s}^{-1}$ (3 s.f.)
2 a 36 J **b** 36 J **c** 7.35 m
3 a 56.3 J (3 s.f.) **b** 56.3 J (3 s.f.) **c** 5.63 m (3 s.f.)
4 a 9.6 J **b** 9.6 J **c** 0.350
5 a 54 J **b** 54 J **c** 4.59 m
6 $9.90\,\text{m s}^{-1}$ (3 s.f.)
7 20.4 m (3 s.f.)
8 10.6 (3 s.f.)
9 a 56.2 J (3 s.f.) **b** 56.2 J (3 s.f.)
c $4.74\,\text{m s}^{-1}$ (3 s.f.)

10 0.408 (3 s.f.)
11 8.27 (3 s.f.)
12 7.94 m s^{-1} (3 s.f.)
13 2.33 m (3 s.f.)
14 a 284 N
b Resistive forces could depend on speed.
15 128 N (3 s.f.)
16 11.8 m (3 s.f.)

Challenge
1130 m s^{-1} (3 s.f.)

Exercise 4D

1 18 kW
2 15 000 W (or 15 kW)
3 278 N
4 25 m s^{-1}
5 a 20 000 W (or 20 kW)
b Typically resistance increases with velocity.
6 550 N
7 a 1.10 m s^{-2} b 0.294 m s^{-2} c 25.7 m s^{-1}
8 11 400 W (or 11.4 kW)
9 $R = 300$ N
10 10 m s^{-1}
11 a 175 N (3 s.f.) b 0.854 m s^{-2} (3 s.f.)
12 a 0.868 m s^{-2} (3 s.f.) b 13.2 m s^{-1} (3 s.f.)
13 a 18 000 W (or 18 kW) b 6.80 m s^{-1}
14 192 W
15 a 6.11 m s^{-2} b 0.342 m s^{-2}
16 37.9
17 a 6.11 m s^{-1} (3 s.f.) b 0.342 m s^{-2} (3 s.f.)

Chapter review 4

1 20.2 N (3 s.f.)
2 a 2940 J b 98 J s^{-1} (or 98 W)
3 a 20 J b 0.163
4 a 4.48 m s^{-2} (3 s.f.) b 1.51 m (3 s.f.)
5 a 0.708 m s^{-2} (3 s.f.) b 0.521 m s^{-2} (3 s.f.)
6 a 11.4 kW (3 s.f.) b 21.3 (3 s.f.)
7 a $\frac{9mgs}{5}$ b $\frac{14gs}{25}$
8 a 2.95 m s^{-1} (3 s.f.) b 61.2 J (3 s.f.)
9 0.2 m s^{-2}
10 32 600 000 J (or 32 600 kJ) (3 s.f.)
11 a 16.2 J b 16.2 J c 4.05 N
12 a 250 J b 0.638 (3 s.f.)
13 a 480 N b 25.4 m s^{-1} (3 s.f.)
14 0.15 m s^{-1}
15 a 7.42 N (3 s.f.) b 435 J (3 s.f.)
c 14.0 m s^{-1} (3 s.f.)
16 a 33.3 m s^{-1} (3 s.f.)
b 0.222 m s^{-2} (3 s.f.)
17 a 12 kW b 24 kW c 10.8 (3 s.f.)

Challenge
a 588 000 sinθ W
b When $\theta = 0°$, no force to act against so no power required. When $\theta = 90°$, maximum power is needed.

CHAPTER 5

Prior knowledge check

1 a 1.25 m s^{-1} direction reversed.
b 0.9 N s
2 25 m s^{-1}
3 31.9 m (3 s.f.)

Exercise 5A

1 $(44\mathbf{i} - 24\mathbf{j})$ m s^{-1} 2 $(8\mathbf{i} + 8\mathbf{j})$ m s^{-1}
3 $(\mathbf{i} - 2\mathbf{j})$ m s^{-1} 4 $(3\mathbf{i} - 4\mathbf{j})$ m s^{-1}
5 $(18\mathbf{i} - 24\mathbf{j})$ N s, $(7\mathbf{i} - 7\mathbf{j})$ m s^{-1}
6 $(10\mathbf{i} - 5\mathbf{j})$ N s, $(25\mathbf{i} + 2\mathbf{j})$ m s^{-1}
7 $(-12\mathbf{i} - 12\mathbf{j})$ N s 8 $(-6\mathbf{i} + 4.5\mathbf{j})$ N s
9 $|\mathbf{Q}| = 30$ $\alpha = 37°$ (nearest degree)
10 $|\mathbf{Q}| = \sqrt{5} = 2.24$ (3 s.f.) $\alpha = 27°$ (nearest degree)
11 $6\sqrt{10}$ or 19.0 N s (3 s.f.)
12 $(-5\mathbf{i} + 30\mathbf{j})$ m s^{-1}
13 $\mathbf{v} = (14\mathbf{i} + 20\mathbf{j})$ m s^{-1}
14 $\mathbf{v} = \mathbf{i} + 2\mathbf{j}$ 18° (nearest degree)
15 $6\mathbf{i}$ m s^{-1}
16 $\frac{3}{7}\sqrt{2}$

Challenge
$\mathbf{I} = m(c - a)\mathbf{i} + m(d - b)\mathbf{j}$; $\tan 45 = \frac{d - b}{c - a} = 1$; $b + c = a + d$

Exercise 5B

1 a $\frac{2}{3}$ b $\frac{1}{2}$ c $\frac{1}{3}$
2 a $v_1 = 0, v_2 = 3$
b $v_1 = 2.5, v_2 = 3$
c $v_1 = 4, v_2 = 6$
d $v_1 = -4, v_2 = 4$
e $v_1 = -5, v_2 = -2$
3 a 3.5 ms^{-1} b 1
4 5 m s^{-1} and 3 m s^{-1} both in the direction that B was moving before the impact.
18 N s
5 $\frac{u}{2}$ direction reversed, $\frac{1}{4}$
6 $\frac{u}{3}(5 - 4e)$, $\frac{u}{3}(5 + 2e)$
7 Using conservation of linear momentum gives
$2mu - 3mu = -mv_1 + mv_2 \Rightarrow -v_1 + v_2 = -u$ (1)
Newton's law of restitution gives $e = \frac{v_2 + v_1}{5u}$ (2)
Eliminating v_1 from (1) and (2) gives
$v_2 = \frac{1}{2}u(5e - 1)$
Since v_2 is in the positive direction $v_2 > 0$, so $\frac{1}{2}u(5e - 1) > 0$, and $e > \frac{1}{5}$
8 a $u\left(1 - \frac{3k}{10}\right)$
b Newton's law of restitution gives
$eu = \frac{3}{10}u - u\left(1 - \frac{3k}{10}\right)$, so $e = \frac{1}{10}(3k - 7)$
Since $0 \leqslant e \leqslant 1$,
$0 \leqslant \frac{1}{10}(3k - 7) \leqslant 1 \Rightarrow \frac{7}{3} \leqslant k \leqslant \frac{17}{3}$
9 a $u(5 - 3k)$
b Newton's law of restitution gives
$eu = ku - u(5 - 3k)$, so $e = 4k - 5$
Since $0 \leqslant e \leqslant 1$, $0 \leqslant 4k - 5 \leqslant 1$
$\Rightarrow \frac{5}{4} \leqslant k \leqslant \frac{3}{2}$
10 a Using conservation of linear momentum gives
$4mu + 6mu = mv_1 + 3mv_2 \Rightarrow v_1 + 3v_2 = 10u$ (1)
Newton's law of restitution gives
$e = \frac{v_2 - v_1}{2u}$ (2)
Eliminating v_1 from (1) and (2) gives
$v_2 = \frac{u}{2}(5 + e)$
b $\frac{u}{2}(5 - 3e)$
c After the collision P has speed $\frac{u}{2}(5 + 3e)$ which is always positive, therefore P continues to move in the same direction.

d $e = \frac{1}{3}$

Challenge
Using conservation of linear momentum gives
$6m - mu = 3mv + 2mv \Rightarrow 6 - u = 5v$
so $v = \frac{1}{5}(6 - u)$ (1)
Newton's law of restitution gives
$\frac{1}{4} = \frac{v}{2 + u}$ so $v = \frac{1}{4}(2 + u)$ (2)
Eliminating v from (1) and (2) gives $u = \frac{14}{9}$

Exercise 5C

1 a $\frac{2}{5}$ **b** $\frac{1}{2}$
2 a $3.5\,\text{m s}^{-1}$ **b** $3\,\text{m s}^{-1}$
3 a $8\,\text{m s}^{-1}$ **b** $8\,\text{m s}^{-1}$
4 $e = 0.75$
5 0.77 (2 s.f.)
6 a 0.1875 **b** Particle would rebound higher.
7 $\frac{1}{2}$
8 2.94 s
9 $\left(\sqrt{gh} - \frac{1}{2}g\right)$m

Challenge
Use $v^2 = u^2 + 2as$ with $u = 0\,\text{m s}^{-1}$, $a = g\,\text{m s}^{-2}$ and $s = h\,\text{m}$
$v = \sqrt{2gh}$
Newton's law of restitution gives
speed of separation from floor $= e\sqrt{2gh}\,\text{m s}^{-1}$
Use $v^2 = u^2 + 2as$ with $u = e\sqrt{2gh}\,\text{m s}^{-1}$, $a = -g\,\text{m s}^{-2}$ and $v = 0\,\text{m s}^{-1} \Rightarrow s = he^2$

Exercise 5D

1 a $6\,\text{m s}^{-1}$ **b** 1.5 J
2 A has speed $\frac{7u}{3}$ with direction of travel reversed and B has speed $\frac{u}{3}$ towards A.
Loss of K.E. is $\frac{5mu^2}{3}$
3 60 J
4 0.225 J
5 a $2\,\text{m s}^{-1}$ **b** 12 060 J (12.06 kJ)
6 a $\frac{5}{3}\,\text{m s}^{-1}$ **b** $1606\frac{2}{3}$ J
7 a $N = 3$ **b** $\frac{3}{8}$
8 a $0.3\,\text{m s}^{-1}$ **b** $\frac{1}{5}$ **c** 3600 J
9 $v - u$ and $v + \frac{1}{2}u$
10 a Conservation of linear momentum:
$8 + 3 = 2u + 3v$
$\Rightarrow 11 = 2u + 3v \Rightarrow \frac{11 - 3v}{2} = u$
Kinetic energy:
$16 + 1.5 - 3 = u^2 + 1.5v^2$
Substitute for u:
$\Rightarrow \left(\frac{11 - 3v}{2}\right)^2 + 1.5v^2 = 14.5$
$\Rightarrow 3.75v^2 - \frac{33}{2}v + \frac{121}{4} = 14.5$
$\Rightarrow 5v^2 - 22v + 21 = 0$
b A moves with speed $1\,\text{m s}^{-1}$ and B moves with speed $3\,\text{m s}^{-1}$ in the same direction as before the impact. A cannot travel faster than B as it cannot pass through it.
11 a $2\,\text{m s}^{-1}$ **b** 35 J
12 Common speed before string becomes taut $= \frac{mu}{M + m}$
Kinetic energy before collision $= \frac{1}{2}mu^2$
Kinetic energy after collision $= \frac{1}{2}(m + M)\left(\frac{mu}{m + M}\right)^2$
$= \frac{m^2u^2}{2(m + M)}$
Loss of kinetic energy $= \frac{1}{2}mu^2 - \frac{m^2u^2}{2(m + M)} = \frac{mMu^2}{2(m + M)}$
13 a $7.5\,\text{m s}^{-1}$ **b** 375 J
14 a 0.32 s, $2.5\,\text{m s}^{-1}$ **b** 0.375 J

Challenge
2.5 J

Exercise 5E

1 a $u = 3, v = 5, x = 4, y = 4.5$
b $u = 2, v = 4, x = 3.5, y = 4$
2 $-1.5\,\text{m s}^{-1}$, $0.5\,\text{m s}^{-1}$, $5\,\text{m s}^{-1}$,
3 a $\frac{1}{2}u(1 - e)$, $\frac{1}{4}u(1 + e)(1 - e)$ and $\frac{1}{4}u(1 + e)^2$
b A will catch up with B provided that $2 > 1 + e$. Since $e < 1$ this condition holds and A will catch up with B, resulting in a further collision.
4 a $u(1 + 3e) > 3u \Rightarrow e > \frac{2}{3}$
b The direction of A is reversed by the collision.
5 $\frac{5u}{8}$ and $\frac{7u}{12}$
6 a $-3u$ and $5u$ **b** $\frac{17u}{4}$ and $\frac{43u}{12}$
7 a i 19.6 cm **ii** 9.604 cm
b A series of bounces with a constant ratio of heights.
c 1.17 m (3 s.f.)
d Model predicts an infinite number of bounces, this is unrealistic.
8 a e^2H **b** e^4H **c** $\frac{H(1 + e^2)}{(1 - e^2)}$
9 $\frac{d}{2}\left(\frac{1}{2} + \frac{1}{e_1} + \frac{1}{e_1e_2}\right)$

Challenge
After the particles collide: $v(P) = 0.25\,\text{m s}^{-1}$, $v(Q) = 1.25\,\text{m s}^{-1}$
t = time in seconds after the particles collide.
Q collides with W_2 after travelling 2 m, i.e. after $t = 1.6$ s.
P will take $t = 8$ s to reach W_1.
After Q collides with W_2, velocity $= 0.64\,\text{m s}^{-1}$.
For particle P, distance from W_2 at time $t = 2 + 0.25t$
For particle Q, distance from W_2 at time $t = 0.64(t - 1.6)$
P and Q will collide when $2 + 0.25t = 0.64(t - 1.6) \Rightarrow t = 8.4$ s
Time to collide > time for P to reach W_1.
P will hit W_1 before colliding with Q for a second time.

Chapter review 5

1 26 Ns, 23°(nearest degree)
2 $213\mathbf{j}\,\text{m s}^{-1}$
3 $(7\mathbf{i} + 56\mathbf{j})\,\text{m s}^{-1}$
4 a $61\,\text{m s}^{-1}$
b $(68\mathbf{i} + 23\mathbf{j})\,\text{m s}^{-1}$
5 Using conservation of linear momentum gives
$mu_1 - mu_2 = mv$
$u_1 - u_2 = v$ (1)
Newton's law of restitution gives
$\frac{1}{3} = \frac{v}{u_1 + u_2}$ so $u_1 + u_2 = 3v$ (2)
Solving (1) and (2) gives
$u_1 = 2v$ and $u_2 = v$
So the ratio of speeds is 2:1
6 4
7 a $V = \frac{mv}{M}$

b Kinetic energy $= \frac{mv^2}{2} + \frac{MV^2}{2}$

Substitute for V: $\frac{mv^2}{2} + \frac{m^2v^2}{2M} = \frac{m(m+M)v^2}{2M}$

c The assumption that the boat is initially at rest is unlikely.

8 $2.5\,\text{ms}^{-1}$, $42.5\,\text{J}$

9 a Using conservation of linear momentum gives
$3mu = 3mv_1 + mv_2$
$3v_1 + v_2 = 3u$ (1)
Newton's law of restitution gives
$v_2 - v_1 = eu$ (2)
Eliminating v_2 from (1) and (2) gives
$v_1 = \frac{u(3-e)}{4}$

b $v_2 = \frac{3u}{4}(e+1)$

Kinetic energy before collision $= \frac{3}{2}mu^2$
Kinetic energy after collision
$= \frac{3mu^2}{32}(3-e)^2 + \frac{9mu^2}{32}(e+1)^2$
Loss of kinetic energy
$= \frac{3}{2}mu^2 - \left(\frac{3mu^2}{32}(3-e)^2 + \frac{9mu^2}{32}(e+1)^2\right)$
$= \frac{3}{2}mu^2 - \frac{mu^2}{32}[3(3-e)^2 + 9(e+1)^2]$
$= \frac{3}{2}mu^2 - \frac{mu^2}{32}[3(9 - 6e + e^2) + 9(e^2 + 2e + 1)]$
$= \frac{3}{2}mu^2 - \frac{mu^2}{32}(12e^2 + 36)$
$= \frac{3}{2}mu^2 - \frac{3mu^2}{8}(e^2 + 3)$
$= \frac{12}{8}mu^2 - \frac{3mu^2}{8}(e^2 + 3)$
$= \frac{12}{8}mu^2 - \frac{3}{8}mu^2e^2 + \frac{9}{8}mu^2$
$= \frac{3}{8}mu^2 - \frac{3}{8}mu^2e^2$
$= \frac{3mu^2(1-e^2)}{8}$

c $\frac{3mu(1+e)}{4}\,\text{Ns}$

10 a $6\,\text{ms}^{-1}$ and $1\,\text{ms}^{-1}$ in the direction of the 100 g mass prior to the impact.
b Loss of K.E. = 2.45 J

11 3 s after the 10 kg sphere has stopped moving.

12 First collision: A with B
$4mV = 4mv_1 + 3mv_2$
$\frac{v_2 - v_1}{V} = \frac{3}{4} \Rightarrow v_2 = \frac{3V}{4} + v_1$
$4V = 4v_1 + \frac{9V}{4} + 3v_1$
$4V = 7v_1 + \frac{9V}{4}$
$v_1 = \frac{1}{4}V, v_2 = V$
$v_2 > v_1$ so second collision is B with C
$3mV = 3mv_3 + 3mv_4$
$\frac{v_4 - v_3}{V} = \frac{3}{4} \Rightarrow v_4 = \frac{3V}{4} + v_3$
$3V = 3v_3 + \frac{9V}{4} + 3v_3$
$3V = 6v_3 + \frac{9V}{4}$
$v_3 = \frac{1}{8}V,\ v_4 = \frac{7}{8}V$
$v_4 > v_3$ but $v_1 > v_3$ so third collision is A with B
$mV + \frac{3}{8}mV = 4mv_5 + 3mv_6$
$\frac{v_6 - v_5}{\frac{1}{4}V - \frac{1}{8}V} = \frac{3}{4} \Rightarrow v_6 = \frac{3}{32}V + v_5$
$\frac{11}{8}V = 4v_5 + \frac{9}{32}V + 3v_5$
$\frac{11}{8}V = 7v_5 + \frac{9}{32}V$
$v_5 = \frac{5}{32}V,\ v_6 = \frac{1}{4}V$
As $\frac{5}{32}V < \frac{1}{4}V < \frac{7}{8}V$ there are no further collisions.

13 a 9072 J
b *Either* heat *or* sound.

14 a $\frac{u}{7}(4 - 3e)$ and $\frac{4}{7}u(1+e)$
b Impulse = change in momentum
$2mu = \frac{12mu}{7}(1+e)$
$1 + e = \frac{7}{6}$
$e = \frac{1}{6}$

15 a Using conservation of linear momentum gives
$mkV + \lambda mV = \lambda mv$ so $V = \frac{\lambda v}{\lambda + k}$ (1)
Newton's law of restitution gives
$e = \frac{v}{V(k-1)}$ (2)
Eliminating V from (1) and (2) gives
$e = \frac{\lambda + k}{\lambda(k-1)}$
b $\frac{\lambda + k}{\lambda(k-1)} < 1$
$\lambda + k < \lambda(k-1)$
$\lambda + k < \lambda k - \lambda$
$2\lambda - \lambda k < -k$
$\lambda k - 2\lambda > k$
$\lambda(k-2) > k$
$\lambda > \frac{k}{k-2}$ and since $\lambda > 0$, $k - 2 > 0$, $k > 2$

16 a Both balls change directions, the first moves up with speed $0.7\,\text{ms}^{-1}$ and the second moves down with speed $3.5\,\text{ms}^{-1}$.
b 91% (2 s.f.)

17 a 0.5 m **b** $\frac{2}{\sqrt{g}}$ s or 0.64 s (2 s.f.)
c $\frac{1}{4}\sqrt{g}\,\text{ms}^{-1}$ or $0.78\,\text{ms}^{-1}$ (2 s.f.)

18 To find time taken for ball to make first contact with floor use
$s = ut + \frac{1}{2}at^2$ with $u = 0\,\text{ms}^{-1}$, $s = h\,\text{m}$ and $a = g\,\text{ms}^{-2}$
$t = \sqrt{\frac{2h}{g}}$
To find speed of ball on first contact with floor use
$v^2 = u^2 + 2as$ with $u = 0\,\text{ms}^{-1}$, $s = h\,\text{m}$ and $a = g\,\text{ms}^{-2}$
$v = \sqrt{2gh}$
Newton's law of restitution gives
speed of separation from floor $= e\sqrt{2gh}$
To find time between first contact and maximum height reached, use $v = u + at$ with $u = e\sqrt{2gh}\,\text{ms}^{-1}$, $v = 0\,\text{ms}^{-1}$ and $a = -g\,\text{ms}^{-2}$
$t = e\sqrt{\frac{2h}{g}}$
So the time between the first and second bounce is
$2e\sqrt{\frac{2h}{g}}$

By similar reasoning the time between the second and third bounce is $2e^2\sqrt{\frac{2h}{g}}$ and the time between the third and fourth bounce is $2e^3\sqrt{\frac{2h}{g}}$

So the total time until the ball comes to a standstill is given by

$$t = \sqrt{\frac{2h}{g}} + 2e\sqrt{\frac{2h}{g}} + 2e^2\sqrt{\frac{2h}{g}} + 2e^3\sqrt{\frac{2h}{g}} + \dots$$

$$= \sqrt{\frac{2h}{g}} + 2\sqrt{\frac{2h}{g}}(e + e^2 + e^3 + \dots) = \sqrt{\frac{2h}{g}} + 2\sqrt{\frac{2h}{g}}\left(\frac{e}{1-e}\right)$$

$$= \sqrt{\frac{2h}{g}}\left(\frac{1-e}{1-e}\right) + 2\sqrt{\frac{2h}{g}}\left(\frac{e}{1-e}\right) = \frac{1+e}{1-e}\sqrt{\frac{2h}{g}}$$

19 $\frac{25}{32}$

20 $V = \sqrt{\frac{2ME}{m(M+m)}}$ m s^{-1}

21 a Use $v^2 = u^2 + 2as$ with $u = 0$ m s^{-1}, $a = g$ m s^{-2} and $s = H$ m
$v = \sqrt{2gH}$
Newton's law of restitution gives
speed of separation from floor = $e\sqrt{2gH}$ m s^{-1}
Use $v^2 = u^2 + 2as$ with $u = e\sqrt{2gH}$ m s^{-1}, $a = -g$ m s^{-2} and $v = 0$ m s^{-1} and $s = h$ m
$2gh = 2e^2gH$
$e = \sqrt{\frac{h}{H}}$

b $\frac{h^2}{H}$

c Infinite bounces with a constant ratio of heights.

22 a B has speed $\frac{5}{12}\sqrt{2g}$ m s^{-1} in the direction it was originally moving.
C has speed $\frac{7}{6}\sqrt{2g}$ m s^{-1} in the direction B was originally moving.

b $\frac{7g}{24}$ J

c The amount of kinetic energy lost would increase.

23 6:19

Challenge

1 P changes direction after impact
$km(u - v) = m(v + u)$ so $k = \frac{u+v}{u-v}$

a k must be positive so $u > v$

b If $k = \frac{u+v}{u-v}$ then Q changes direction after impact.
If $k = \frac{u-v}{u+v}$ then P changes direction after impact.

2 Using conservation of momentum, with common final velocity v:
$m_3u = (m_1 + m_2 + m_3)v$
So $V = \frac{m_3u}{(m_1 + m_2 + m_3)}$
Kinetic energy when all three strings are taut
$= \frac{1}{2}(m_1 + m_2 + m_3)\left(\frac{m_3u}{m_1 + m_2 + m_3}\right)^2 = \frac{m_3^2u^2}{2(m_1 + m_2 + m_3)}$

CHAPTER 6

Prior knowledge check

1 a 8 Nm anticlockwise   b 500 Nm clockwise

2 $R_A = \frac{1}{3}mg$, $R_B = \frac{2}{3}mg$

3 2.4 cm

Exercise 6A

1 34.6 N, 50 N, 17.3 N, 0.35

2 a 22.8 N   b 98 N, 22.8 N   c 0.233

d The weight of a uniform ladder passes through its midpoint.

3 a 41.6°   b 24.0°

c No friction at the wall.

4 a $5\frac{1}{3}$ m

b i The ladder may not be uniform.
ii There would be friction between the ladder and the vertical wall.

5 a 1.99 N   b 0.526

c There is no friction between the rail and the pole.

6 R(↑): $R = mg$, R(→): $N = \mu mg$
Let the length of the ladder be $2l$
Taking moments about A,
$mgl\cos\theta = 2\mu mgl\sin\theta$
Cancelling and rearranging gives
$2\mu\tan\theta = 1$

7 104 N, 64.5 N, 0.620

8 R(↑): $R + \mu_2N = mg$, R(→): $N = \mu_1R$
Taking moments about the base of the ladder
$2N\tan\theta + 2\mu_2N = mg$
$2\tan\theta + 2\mu_2 = \frac{1}{\mu_1} + \mu_2$
$2\tan\theta = \frac{1}{\mu_1} - \mu_2$
$\tan\theta = \frac{1 - \mu_1\mu_2}{2\mu_1}$

9 a $\frac{W\sqrt{3}}{6}$

b $F = P = \frac{W\sqrt{3}}{6}$, $R = W$, $F \leqslant \mu R$, $\mu \geqslant \frac{\sqrt{3}}{6}$

c $\frac{W}{4}$

10 a 6.50 N, 10.8 N   b $\frac{1}{\sqrt{3}}$

11 a $\frac{19}{2\sqrt{3}}W$ b $\left(\frac{19}{2\sqrt{3}} - 2\right)W \leqslant P \leqslant \left(\frac{19}{2\sqrt{3}} + 2\right)W$

c It allows us to assume that the weight of the ladder acts through its midpoint.

d i The magnitude of the reaction at Y will get smaller.
ii The range of values for P will get smaller.

12 a Moments about A:
$F \times \frac{3}{5} \times 4a = mg2a\cos 30° + kmg4a\cos 30°$
$\Rightarrow F \times \frac{12a}{5} = 2mga \times \frac{\sqrt{3}}{2}(1 + 2k)$
$\Rightarrow F = \frac{5\sqrt{3}}{12}mg(1 + 2k)$

b i $H = \frac{5\sqrt{3}}{24}mg(1 + 2k)$

ii $V = mg(1 + k) - \frac{5}{8}mg(1 + 2k) = \left\{\frac{mg}{8}(3 - 2k)\right\}$

c $k = \frac{9 - \sqrt{3}}{18 + 2\sqrt{3}}$

Chapter review 6

1 R(→): $F = N$
Taking moments about A:
$Fa\sin\theta + \frac{5}{2}mga\cos\theta = 5aN\sin\theta$
$\frac{5}{2}mg\cos\theta = 4F\sin\theta$
$\frac{5}{8}mg = F\tan\theta$

2 $\frac{7}{24}$

3 a $\frac{8W}{9}$

b R(↑): $R + \mu N \geqslant W$, R(→): $N = \frac{W}{3}$

$\frac{\mu W}{3} \geqslant W - \frac{8W}{9}$

c The ladder has negligible thickness/the ladder does not bend.

4 a Taking moments about point where ladder touches the ground:
R(↑): $R = W$, R(→): $N = 0.3R$
$1.5W = 1.2W$. This cannot be true so the ladder cannot rest in this position.

b R(→): $F = N$
Taking moments about point where ladder touches the ground $1.5W = 4N$, $F = N = \frac{3W}{8}$

c $\frac{W}{4g}$

5 18 N (2 s.f.)

Challenge

a 94.3 N (3 s.f.)

b 80.6 N (3 s.f.), 54.2° (3 s.f.) to the horizontal

Review exercise 2

1 a $\mathbf{v} = 10\mathbf{i} + 20\mathbf{j}$ m s^{-1} b 63.4° (3 s.f.)
c 40 J

2 a 7.5 N b $\mathbf{v} = 39\mathbf{i} - 42\mathbf{j}$ m s^{-1}

3 a 5.83 N s (3 s.f.) b 31°
c 35 J

4 a $v = (2t + 2)\mathbf{i} + (3t^2 - 4)\mathbf{j}$ b 13 m s^{-1}

5 a 610 N (3 s.f.) b 458 kJ (3 s.f.)
c 801 kJ (3 s.f.)

6 a 1568 J
b It will be less than 1568 J.

7 Work done = $mgh \Rightarrow 19600 = 1000 \times 9.8 \times 25 \sin\theta$

$\Rightarrow \sin\theta = \frac{19\,600}{1000 \times 9.8 \times 25} = \frac{2}{25} \Rightarrow \theta = \arcsin\left(\frac{2}{25}\right)$

8 a 175 J b 196 kJ

9 a 8.4 m s^{-1} (3 s.f.) b $\mu = 0.42$ (2 s.f.)

10 a 41 J (2 s.f.) b $\mu = 0.67$ (2 s.f.)

11 a 22.4 J b 6.4 m s^{-1} (2 s.f.)
c 4.3 m s^{-1} (2 s.f.)

12 a $\frac{7}{5}mgh$ b $\frac{3}{5}gh$

13 a $50 = F \times 25 \Rightarrow F = 2000$
For the car and trailer combined:
R(→): $F - 750 - R = 0$
$R = F - 750 = 2000 - 750 = 1250$

b 1.4 m s^{-2} c 850 N d 335 kJ

e The resistance could be modelled as varying with speed.

14 a 0.8 b $\frac{245}{3}$
c Resistance usually varies with speed.

15 a Energy lost = 7832 J; Work done = 10 000 J
Total work done = 2168 J = 2200 J (2 s.f.)
b 200 W

16 220

17 a 0.7 m s^{-2} b 44.4 kW

18 a 35 m s^{-1} b 15 m s^{-1} (2 s.f.)

19 a Let F N be the magnitude of the driving force produced by the engine of the car.
12 kW = $F \times 15 \Rightarrow F = 800$
R(→): $\mathbf{F} = m\mathbf{a}$
$F - R = 1000 \times 0.2 \Rightarrow R = 600$
b 20 (2 s.f.)

20 a 14.3 kW (3 s.f.) b 6.6 kW (2 s.f.)

21 a 0.15 m s^{-2} b 35.1 m s^{-1} (3 s.f.)

22 a K.E. gained = P.E. lost
$\frac{1}{2}\cancel{m}v^2 - \frac{1}{2}\cancel{m}u^2 = \cancel{m}gh$
$\frac{1}{2} \times 24.5^2 - \frac{1}{2}u^2 = 9.8 \times 15$
$u^2 = 24.5^2 - 2 \times 9.8 \times 15 = 306.25$
$u = \sqrt{306.25} = 17.5$ (3 s.f.)
b 55° (nearest degree)
c 60 m

23 a 52 (2 s.f.) b 3 s
c 48 m d 24 m s^{-1} (2 s.f.)

24 2.46 m (3 s.f.)

25 $\mu = \frac{1}{2\tan\alpha}$

26 44.4 N (3 s.f.)

27 $e = \frac{3}{5}$

28 a Using conservation of momentum:
$mu + km\lambda u = kmv$
$u(1 + k\lambda) = kv$ (1)
Newton's law of restitution:
$v = e(u - \lambda u) = eu(1 - \lambda)$ (2)
Eliminate v from (1) and (2)
$u(1 + k\lambda) = keu(1 - \lambda)$
$e = \frac{1 + k\lambda}{k(1 - \lambda)}$

b $e \leqslant 1$
$\Rightarrow 1 + k\lambda \leqslant k - k\lambda$
$\frac{1}{1 - 2\lambda} \leqslant k$
but $0 < \lambda < \frac{1}{2} \Rightarrow 0 < 1 - 2\lambda < 1$ and $k > 1$

29 a Using conservation of momentum:
$mu = mv_S + 2mv_T$
$u = v_S + 2v_T$ (1)
Newton's law of restitution:
$eu = v_T - v_S$ (2)
Eliminating v_S from (1) and (2):
$u + eu = 3v_T$
$v_T = \frac{1}{3}u(1 + e)$

b i Speed of S is $\frac{1}{3}u(2e - 1)$
ii Direction of motion reversed.

30 a Using conservation of momentum:
$3m \times 2u - 2mu = 3mu_P + 2mu_Q$
$\therefore 4u = 3u_P + 2u_Q$ (1)
Newton's law of restitution:
$e(2u + u) = u_Q - u_P$
$3eu = u_Q - u_P$ (2)
Eliminate u_P from (1) and (2)
$4u = 3(u_Q - 3eu) + 2u_Q$
$4u = 5u_Q - 9eu$
$u_Q = \frac{1}{5}u(9e + 4)$

b $\frac{2}{3} < e \leqslant 1$

c $e = \frac{7}{9}$

31 14.4 m

32 a $e = \frac{2}{3}$

b If the plane was rough and the conditions described above are otherwise identical, then the coefficient of restitution must be greater. If it were the same then the speed of the particle at each point on the return journey would be $< \frac{2}{3}$ of the minimum speed of the particle on the outward journey.

33 a $\frac{\sqrt{70}}{10}$

b Time t_1 until first bounce
$50 = \frac{9.8}{2}t_1^2 \Rightarrow t_1 = \frac{10\sqrt{5}}{7}$
Time t_2 to reach maximum height after first bounce
$35 = \frac{9.8}{2}t_2^2 \Rightarrow t_2 = \frac{5\sqrt{14}}{7}$

Total time from first to second bounce

$T = t_1 + 2t_2$

$= \frac{10\sqrt{5}}{7} + 2 \times \frac{5\sqrt{14}}{7} = \frac{10}{7}(\sqrt{5} + \sqrt{14})$ s

34 a A has speed $\frac{u}{2}$

B has speed $\frac{3u}{2}$

b $\frac{45}{2}mu^2$ J

35 a $e = \frac{1}{4}$

b $\frac{9}{4}mu^2$ J

36 a Using conservation of momentum:

$mu = mv_A + 3mv_B$

$u = v_A + 3v_B$ (1)

Newton's law of restitution:

$eu = v_B - v_A$ (2)

Eliminating v_A from (1) and (2) gives $u + eu = 4v_B$

$v_B = \frac{1}{4}(1 + e)u$

b $v_A = \frac{1}{4}(1 - 3e)u$

c $e = \frac{1}{3}$

d $v_A = \frac{u}{4}(1 - 3e) = \frac{u}{4}\left(1 - 3 \times \frac{1}{3}\right) = 0$

$\therefore A$ is at rest

37 41.5 kJ

38 $\frac{1}{12}$ J

39 a $e = \frac{2}{3}$

b From part **a**, $v_A = -\frac{2}{3}u$

For spheres B and C:

Conservation of momentum:

$4mu = 2mv_B + 5mv_C$

Newton's law of restitution:

$\frac{v_C - v_B}{2u} = \frac{3}{5}$

$v_C - v_B = \frac{6}{5}u$

$v_C = v_B + \frac{6}{5}u$

Substitute v_C in the conservation of momentum equation:

$4u = 2v_B + 6u + 5v_B$

$\Rightarrow v_B = -\frac{2}{7}u$

$\Rightarrow v_C = \frac{32}{35}u$

$v_A < v_B < v_C$ $\therefore$ there are no further collisions.

40 a Using conservation of momentum:

$2mu = 2mv_P + mv_Q$

$2u = 2v_P + v_Q$ (1)

Newton's law of restitution:

$\frac{1}{3}u = v_Q - v_P$ (2)

$(1) + 2 \times (2): \frac{8u}{3} = v_Q + 2v_Q$

$3v_Q = \frac{8u}{3}$

$v_Q = \frac{8u}{9}$

Using (2)

$v_P = v_Q - \frac{1}{3}u$

$v_P = \frac{8u}{9} - \frac{1}{3}u$

$v_P = 5\frac{u}{9}$

b $e = \frac{25}{32}$

c Q is now moving towards the wall; after it rebounds off the wall it will return to collide with P once more.

41 a Using conservation of momentum:

$2m \times 5u = 2mv_P + 3mv_Q$

$10u = 2v_P + 3v_Q$ (1)

Newton's law of restitution:

$e \times 5u = v_Q - v_P$ (2)

(1) + 2 × (2)

$10u + 10eu = 3v_Q + 2v_Q$

$10u + 10eu = 5v_Q$

$v_Q = 2u + 2eu = 2(1 + e)u$

b From (2)

$v_P = v_Q - 5eu = 2(1 + e)u - 5eu$

$v_P = 2 \times 1.4u - 5 \times 0.4u = 0.8u$

$v_P > 0$ $\therefore P$ moves towards wall and will collide with Q after Q rebounds from the wall.

c $e = 0.8$

$v_P = 2 \times 1.8u - 5 \times 0.8u = -0.4u$

Q hits the wall

$v_Q = 3.6uf$

For a second collision

$3.6uf > 0.4u$

$f > \frac{0.4}{3.6} = \frac{1}{9}$

Range of values for f is

$\frac{1}{9} < f \leqslant 1$

42 a Using conservation of momentum:

$2m \times 2u + 3m \times u = 2mv_A + 3mv_B$

$7u = 2v_A + 3v_B$ (1)

Newton's law of restitution:

$e(2u - u) = v_B - v_A$

$eu = v_B - v_A$ (2)

(1) + 2 × (2)

$7u + 2eu = 3v_B + 2v_B$

$v_B = \frac{1}{5}u(7 + 2e)$

b $v_A = \frac{1}{5}u(7 - 3e)$

c $\frac{1}{5}u(7 - 3e) = \frac{11u}{10}$

$14u - 6eu = 11u \Rightarrow 6eu = 3u$

$e = \frac{1}{2}$

d $\frac{5d}{16}$

e After B hits the barrier

$v_B = \frac{11}{16} \times \frac{8u}{5} = \frac{11u}{10}$

Equal speeds, opposite directions.

$\therefore A$ and B will collide at midpoint of the distance from A to the barrier at the instant B hits the barrier, i.e. they collide at distance $\frac{5d}{32}$ from the barrier.

43 a $\frac{82}{9}$ m

b This model predicts an infinite number of bounces which is not realistic.

44 First collision: A with B

Using conservation of momentum:

$4m = 2mv_B + mv_A$

$4 = 2v_B + v_A$ (1)

45 a 5g

b

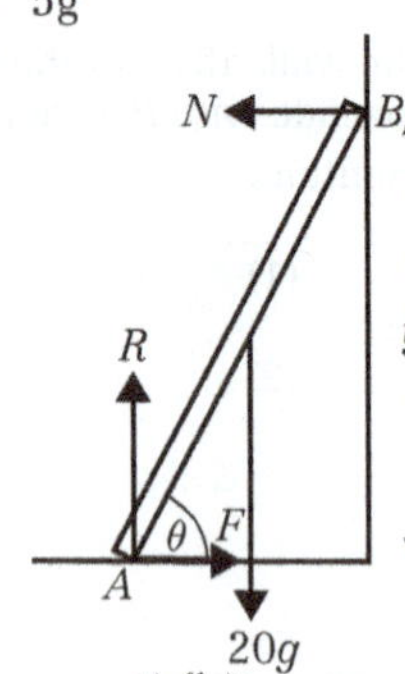

Res (→) $N = 0.25F$
Res (↑) $R = 20g$
Res (→) $N = 0.25F = 5g$
Taking moments about A:
$20gx = 5gy$
$20g \times 5\cos\theta = 5g \times 10\sin\theta$
$\frac{\sin\theta}{\cos\theta} = \frac{100g}{50g}$
$\tan\theta = 2$
$\theta = \tan^{-1}(2)$ as required

46

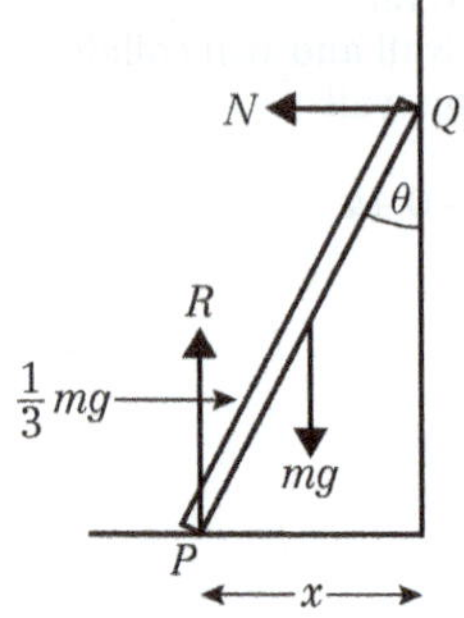

Res (↑) $R = mg$
Res (→) $N = \frac{1}{3}mg$
Taking moments about Q:
$Rx = \frac{1}{2}mgx + \frac{1}{3}mg \times \frac{3}{4}y$
$mgx = \frac{1}{2}mgx + \frac{1}{4}mgy$
$x = \frac{1}{2}y$
$l\sin\theta = \frac{1}{2}l\cos\theta$
$\frac{\sin\theta}{\cos\theta} = \frac{1}{2}$
$\tan\theta = \frac{1}{2}$
$\theta = \tan^{-1}\left(\frac{1}{2}\right)$ as required

47 $\theta = \tan^{-1}\left(\frac{11}{24}\right)$

48 a $\frac{25}{4}g$N **b** $\frac{15}{2\sqrt{2}}g$N **c** $\frac{15}{4}g$N

49

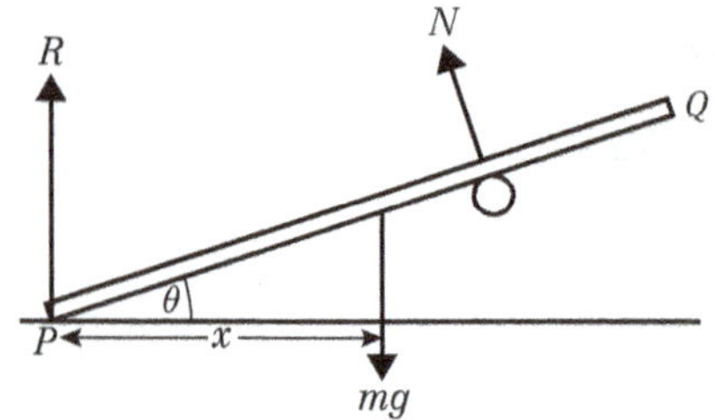

Res (↑) $R + N\cos\theta = mg$
Taking moments about P:
$mgx = 0.75lN$
and
$x = \frac{1}{2}l\cos\theta$
so
$\frac{1}{2}mgl\cos\theta = \frac{3}{4}lN$
$N = \frac{2}{3}mg\cos\theta$ as required

50 a $\frac{85}{32}g$N

b From part **a**
$\mu R = N\sin\theta$
the reaction at A is $R = \frac{85}{32}g$
and reaction at C is $N = \frac{25\sqrt{3}}{16}g$
Substituting gives
$\mu \times \frac{85}{32}g = \frac{25\sqrt{3}}{16}g \times \frac{1}{2}$
$\mu = \frac{25\sqrt{3}}{85}$
$\mu = \frac{5\sqrt{3}}{17}$ as required

Challenge

$\tan^{-1}\left(\frac{15}{11}\right)$

Exam Practice

1 a $\frac{250}{49}$ m **b** $2\sqrt{149}$ m s^{-1} **c** 44.8° (3 s.f.)

2 a 49.2 m (3 s.f.) **b** 6.01 m s^{-2}

3 17.3°

4 a 4.74° **b** $\frac{4}{11}$ m s^{-2}

5 a P has speed 2.8 m s^{-1} in the same direction as its original motion.
P has speed 4.4 m s^{-1} in the same direction as its original motion.

b 2.67 % (3 s.f.)

6 a

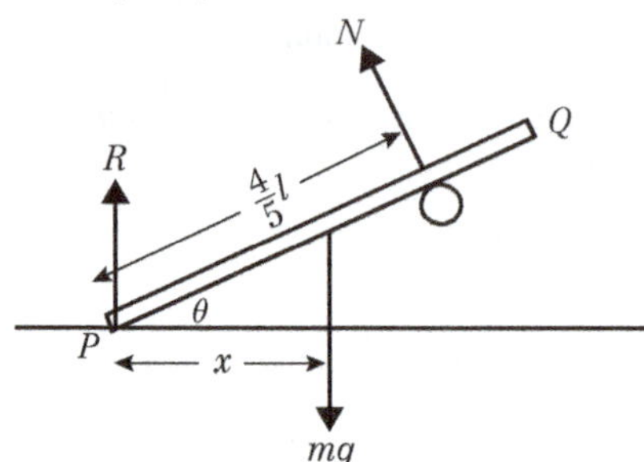

Taking moments about P
$\frac{4}{5}lN = mgx$
substituting $x = \frac{1}{2}l\cos\theta$
gives $\frac{4}{5}lN = \frac{1}{2}mgl\cos\theta$
so $N = \frac{5}{8}mg\cos\theta$ as required

b Res ↑ $R + N\cos\theta = mg$
so $R = mg - N\cos\theta$
Res → $\mu R = N\sin\theta$
so $R = \frac{N\sin\theta}{\mu}$
$mg - N\cos\theta = \frac{N\sin\theta}{\mu}$
so $\mu = \frac{N\sin\theta}{mg - N\cos\theta}$
since $\theta = 30°$

$$\mu = \frac{N \times \frac{1}{2}}{mg - N \times \sqrt{\frac{3}{2}}} = \frac{N}{2mg - \sqrt{3}N}$$

and

$$N = \frac{5}{8}mg\cos\theta = \frac{5}{8}mg \times \frac{\sqrt{3}}{2} = \frac{5\sqrt{3}}{16}mg$$

substituting for N gives

$$\mu = \frac{\frac{5\sqrt{3}}{16mg}}{2mg} - \frac{\sqrt{3} \times 5\sqrt{3}}{16mg} = \frac{\frac{5\sqrt{3}}{16}}{\frac{32-15}{16}} = \frac{5\sqrt{3}}{17} \text{ as required}$$

c $1.5m$

INDEX

A

B

C

D

E

F

G

H

I

K

L